Integrated Approach to Sedimentary Basin Formation

Integrated Approach to Sedimentary Basin Formation

Edited by **Cortez Ford**

R CALLISTO REFERENCE

New York

Published by Callisto Reference,
106 Park Avenue, Suite 200,
New York, NY 10016, USA
www.callistoreference.com

Integrated Approach to Sedimentary Basin Formation
Edited by Cortez Ford

© 2015 Callisto Reference

International Standard Book Number: 978-1-63239-429-3 (Hardback)

Contents

Preface

An integrated approach towards the formation of sedimentary basin has been presented in this profound book. The purpose of this book is to provide an overview on the techniques of sedimentary basin formation on active plate margins, which display enormous diversity representing intricate tectonic processes. An interdisciplinary approach regarding basin-forming mechanism has been based on geology, geophysics and sedimentology. It is devoted to the genetic assessment of sedimentary basins in distorted zones in fore-arc and intra-arc regions. It also deals with basin formation in peripheral regions of Eurasia and intra-arc / foreland basins under the influence of the fluctuation of stress regimes. Geophysical methods to basin analysis have also been presented right from microscopic to regional scales. The extensive content of this book provides the readers with contemporary accomplishments of basin researches on active margins by Earth scientists.

After months of intensive research and writing, this book is the end result of all who devoted their time and efforts in the initiation and progress of this book. It will surely be a source of reference in enhancing the required knowledge of the new developments in the area. During the course of developing this book, certain measures such as accuracy, authenticity and research focused analytical studies were given preference in order to produce a comprehensive book in the area of study.

This book would not have been possible without the efforts of the authors and the publisher. I extend my sincere thanks to them. Secondly, I express my gratitude to my family and well-wishers. And most importantly, I thank my students for constantly expressing their willingness and curiosity in enhancing their knowledge in the field, which encourages me to take up further research projects for the advancement of the area.

Editor

Genetic Analyses of Sedimentary Basins in Wrench Deformation Zones

Strike-Slip Basin – Its Configuration and Sedimentary Facies

Atsushi Noda

Additional information is available at the end of the chapter

1. Introduction

Plate convergent margins are areas of concentrated lithospheric stress. They include areas of compression accompanied by thrusting, mountain building, and related foreland/forearc basin development, and also areas of strike-slip movement associated with transpressional uplift, transtensional subsidence, or pure strike-slip displacement. The degree of shortening and uplift or extension and subsidence depends on the modes of convergence between oceanic plates, island arcs, and continental crusts. Strike-slip faulting is one of the most important mechanisms of sedimentary basin formation at plate convergent margins, where localized extension can cause topographic depressions. Sedimentary strata deposited in these basins record the history of lithospheric response to the convergence.

Sedimentary successions of archetypal examples of strike-slip basins, such as the Ridge Basin in California, have been characterized in terms of the dominance of axial sediment supply and continuous depocenter migration in a direction opposite to that of the sediment supply [1]. Their basin lengths are typically about three times longer than the basin widths [2]. The Izumi Group of the Cretaceous turbidite successions in southwestern Japan [e.g., 3 and references therein] has the same characteristics in the sedimentary succession as the Ridge Basin, and is therefore considered to contain strata that were deposited in a strike-slip basin. However, the basin geometry is quite different from that of the Ridge Basin, whose shape is more elongated with a length of more than 300 km and a width of less than 20 km, and whose southern margin has been truncated by post-depositional strike-slip fault displacement. Strike-slip basins thus present a wide diversity in terms of their geometry, evolution, and filling processes.

Since many local examples were collected after the 1980s [4], advanced research techniques including subsurface exploration including seismic survey and borehole drilling [e.g., 5], analog experiment [e.g., 6], and numerical simulation [e.g., 7] revealed that basin formation and filling processes were not simple but variable. In this paper, I try to review some of representative strike-slip basins along convergent margins, especially focusing on basin formation and filling processes, as the first step for comprehensive understandings of the tectono-sedimentary evolution in strike-slip basins.

2. Strike-slip faults

Much research has been published regarding the classification and terminology of strike-slip faults [e.g., 8–12]. Here, I use three tectonic settings as a basis for classifying strike-slip faults along convergent plate margins: subduction, continental collision, and plate-boundary transform zones.

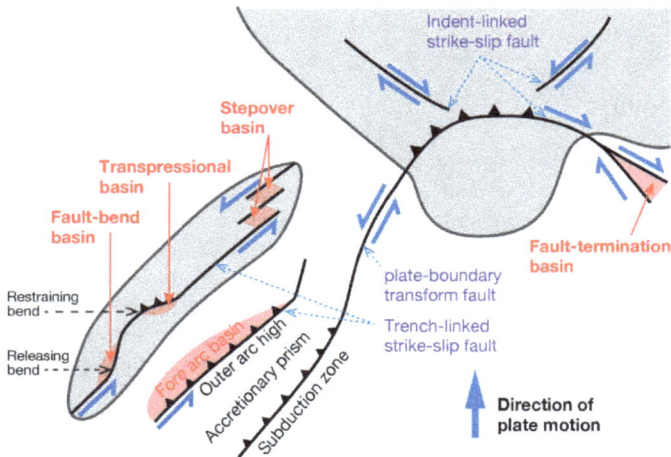

Figure 1. Tectonic settings of strike-slip faults and related strike-slip basins

2.1. Subduction zone

Subduction zones, where oceanic plates obliquely subduct underneath continental or island arc crusts, are sometimes accompanied by strike-slip faults separating elongate forearc slivers from continental margins or island arcs. This type of strike-slip fault is referred to as a trench-linked [9] or trench-parallel [13] strike-slip fault (Figure 1).

Trench-linked strike-slip faults lie parallel to the trench in the accommodating part of the trench-parallel component of oblique convergence of subducting plates [14–16]. The basic principle is considered to be that a trench-linked strike-slip fault is able to concentrate shear in a much more efficient way than distributing shear across the much larger and more gently dipping interface of the subducting plate. Therefore, the low dip angles and high friction coefficients of subduction zones favor slip on vertical strike-slip faults rather than on gently dipping slabs [9].

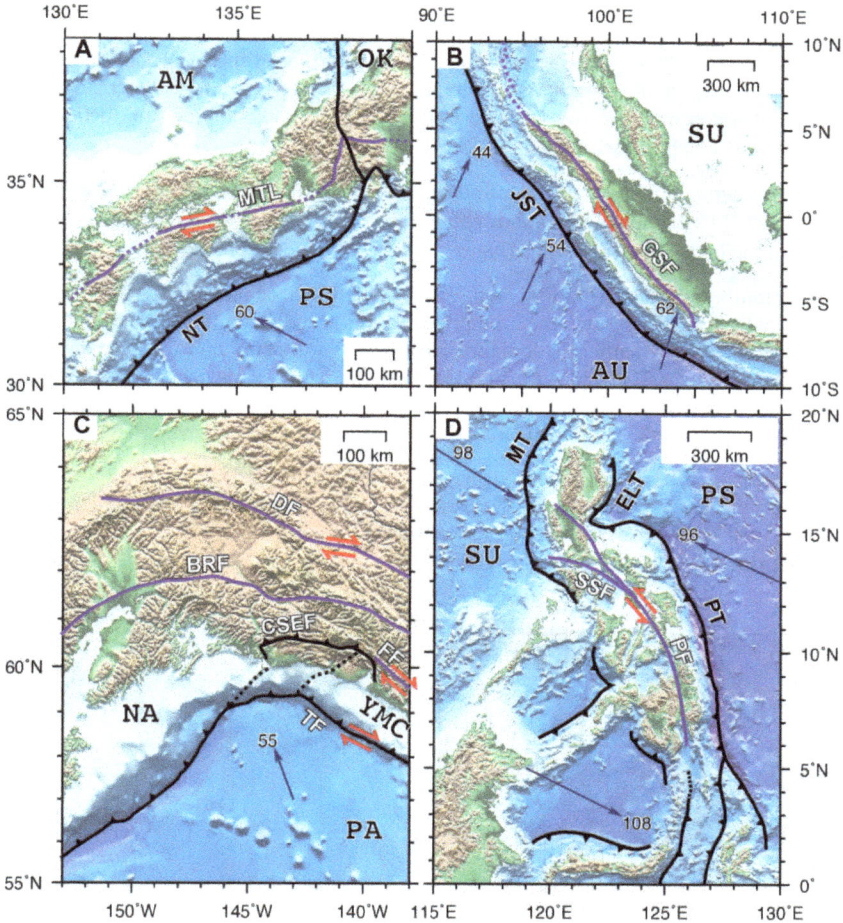

Figure 2. Modern examples of trench-linked strike-slip faults. (A) The Median Tectonic Line (MTL) active fault system in southwestern Japan, related to oblique subduction of the Philippine Sea Plate (PS) along the Nankai Trough (NT). (B) The Great Sumatra Fault system (GSF) along the Java–Sumatra Trench (JST). (C) Strike-slip faults in Alaska. Fault names: DF, Denali; BRF, Boarder Ranges; CSEF, Chugach St. Elias; FF, Fairweather; TF, Transition. (D) The Philippine Fault system (PF). Abbreviations: SSF, Sibuyan Sea Fault; MT, Manila Trench; PT, Philippine Trench; ELT, East Luzon Trough. Plate names: AM, Amur; OK, Okhotsk; PS, Philippine Sea; AU, Australian; SU, Sundaland; NA, North American; PA, Pacific; YMC, Yukutat microcontinent. Black and purple lines are subduction zones and trench-linked strike-slip faults, respectively. All maps were drawn using SRTM and GEBCO with plate boundary data [30]. Blue arrows indicate the direction and velocity of relative plate motion (mm yr^{-1}) based on [31].

These strike-slip faults sometimes separate a narrow sliver plate from the remainder of the over-riding forearc plate. Modern examples include the Median Tectonic Line in Japan, the Sumatra Fault in Indonesia, the Denali Fault in Alaska and the Philippine Fault in Philippine (Figure 2). An ancient example can be observed in the Bering Sea [17]. The strike-slip faults in these settings are typically long (hundreds of kilometers) but occasionally segmented [9].

The Median Tectonic Line (MTL) active fault system is the longest and most active arc-parallel, right-lateral strike-slip fault system in Japan [e.g.18, 19]. The MTL extends over a distance of 500 km and accommodates the trench-parallel component of oblique subduction of the Philippine Sea Plate (Figure 2A).

The Great Sumatra Fault is a right-lateral strike-slip fault more than 1900 km in length (Figure 2B). It is related to the northward subduction of the Australian Plate beneath the Sundaland Plate along the Java–Sumatra Trench (Figure 2B) [20–22].

Strike-slip faulting in Alaska has involved several widely spaced major faults, with an overall seaward migration of activity: the Denali Fault was initiated in the Late Cretaceous or Paleocene and the Fairweather Fault in the Pleistocene (Figure 2C) [23, 24]. These right-lateral strike-slip faults are associated with the Alaskan subduction zone where the Pacific Plate subducts northwestward. Although current strike-slip movement takes place predominantly on the Fairweather Fault, the Denali Fault also shows some Holocene movement.

The Philippine Fault is a left-lateral strike-slip fault sandwiched between the Manila and Philippine trenches [25–27]. It completely traverses the Philippine archipelago and extends for more than 1000 km (Figure 2D). Several strike-slip basins are developed along releasing bends or overstepped faults in this fault zone [28, 29].

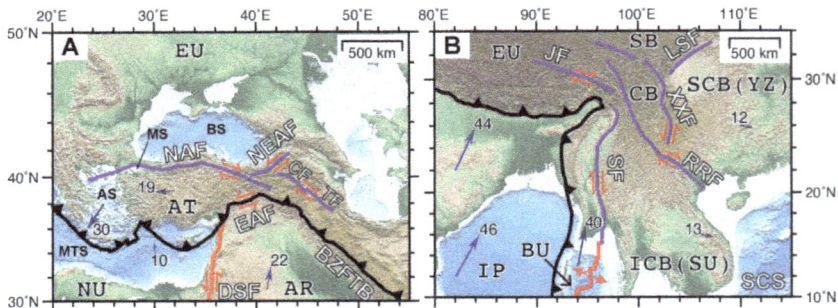

Figure 3. Indent-linked strike-slip fault systems. (A) North Anatolian Fault (NAF) caused by the collision of the Arabian Plate (AR) with the Eurasian Plate (EU). Abbreviations: EAF, East Anatolian Fault; NEAF, Northeast Anatolian Fault; CF, Chalderan Fault; TF, Tabriz Fault; DSF, Dead Sea Fault; BZFTB, Bitlis–Zagros Fold and Thrust Belt; BS, Black Sea; MTS, Mediterranean Sea; MS, Marmara Sea; AS, Aegean Sea. (B) The Red River Fault (RRF) zone, caused by the collision of the Indian Plate (IP) with the Eurasian Plate. Abbreviations: SCB, South China Block; SB, Songpan Block; CB, Chuandian Block; ICB, Indochina Block; SF, Sagaing Fault; JF, Jiali Fault; XXF, Xianshuihe–Xiaojiang Fault; LSF, Longmen Shan Fault. Faults and blocks are based on [33]. Plate names: EU, Eurasian; AR, Arabian; NU, Nubia (Africa); AT, Anatolian; SU, Sundaland; BU, Burma; YZ, Yangtze. Maps were drawn using SRTM and GEBCO with plate boundary data [30]. Black, red, and purple lines are plate convergent margins, plate-boundary transform faults, and indent-linked strike-slip faults, respectively. Blue arrows indicate the direction and velocity of plate motion (mm yr⁻¹) relative to the Eurasian Plate based on [31]

2.2. Continental collision zones

Continental collision can cause crustal shortening and thickening by thrusting and escape or by extruding crustal blocks along conjugate strike-slip faults within the plate. These types of collision-related strike-slip faults between continental blocks are classified as indent-linked strike-slip faults (Figure 1) [9].

Modern examples include several strike-slip faults in Turkey where the Arabian Plate is converging with the Eurasian Plate, and in southern China where the Indian Plate is colliding with the Eurasian Plate (Figure 3). In the latter example, the collision originally formed the left-lateral Red River Fault associated with the southeastward extrusion of the Indochina Block [32]. After the propagation of the indent, the South China Block was extruded along the pre-existing Red River Fault as a block boundary with right-lateral movement.

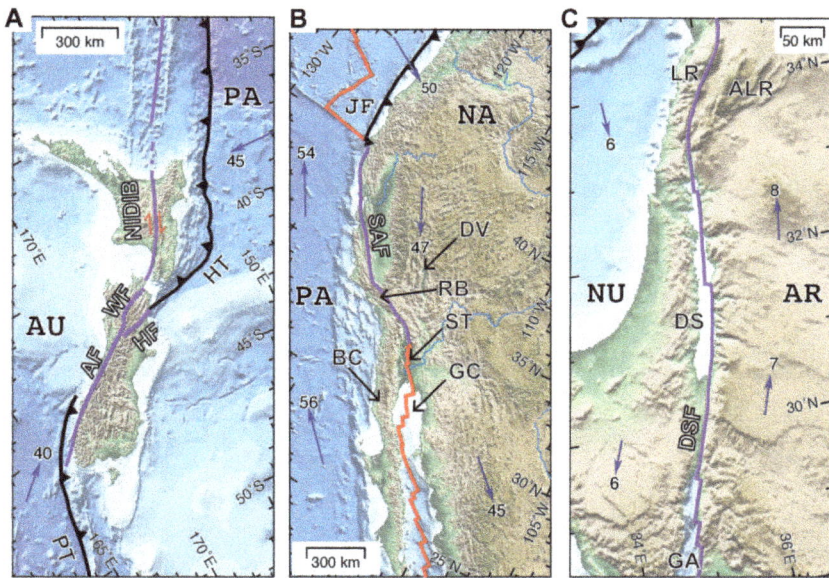

Figure 4. Plate-boundary transform fault systems. (A) Alpine Fault (AF) in New Zealand. Abbreviations: HF, Hope Fault; WF, Wairau Fault; NIDFB, North Island Dextral Fault Belt; HT, Hikurangi Trough; PT, Puysegur Trench. Faults are from [34] and [35]. (B) San Andreas Fault systems (SAF) in North America. Abbreviations: DV, Death Valley; RB, Ridge Basin; ST, Salton Trough; GC, Gulf of California; BC, Baja California Peninsula. (C) Dead Sea Fault systems. Abbreviations: DS, Dead Sea; GA, Gulf of Aqaba; LR, Lebanon Range; ALR, Anti-Lebanon Range. Plate names: PA, Pacific; AU, Australian; NA, North American; JF, Juan de Fuca; AR, Arabian; NU, Nubian (African). All maps were drawn by using SRTM and GEBCO with plate boundary data [30]. Black, red, and purple lines are subduction zones, oceanic spreading ridges, and plate-boundary transform faults, respectively. Blue arrows indicate the direction and velocity of relative plate motion (mm yr⁻¹) based on [31]

2.3. Plate-boundary transform zones

Plate-boundary transform faults develop between two plates rotating around the poles that define the relative motion between them [9]. The San Andreas Fault, Dead Sea Fault, Alpine Fault (New Zealand), and the northern and southern margins of the Caribbean Plate are modern examples of this type (Figure 4).

3. Strike-slip basins

Strike-slip faults can accommodate localized compression or extension at continental margins, in island arcs, and also within continents. Sedimentary basins commonly develop where the fault kinematics are divergent with respect to the plate vector along strike-slip faults. Since the 1980s, various classifications of strike-slip basins have been formulated [4, 11, 36–40]. Common characteristics of strike-slip basins [4, 39] include: (1) elongated geometry, (2) asymmetry of both sediment thickness and facies pattern, (3) dominance of axial infilling, (4) coarser-grained marginal facies along the active master fault, (5) finer-grained main facies, (6) depocenter migration opposite to the direction of axial sediment transport, (7) very thick strata relative to the burial depth, (8) high sedimentation rate, (9) abrupt lateral and vertical facies changes and unconformities, (10) compositional changes that reflect horizontal movement of the provenance, (11) abundant syn-sedimentary slumping and deformation, and (12) rapid subsidence in the initial stage of basin formation.

There are many strike-slip basins along plate convergent margins (Figure 5 and Table 1). Here I classify strike-slip basins into four types, discussed in turn below.

	Indent-linked strike-slip faults	Trench-linked strike-slip faults	Plate boundary transform faults
Fault-bend basin	Vienna Basin[1] Marmara Sea[2]	Izumi Group*[3] St. George Basin*[4] Suwa Lake[5]	Ridge Basin*[6] Death Valley[7]
Stepover basin	Thai Basin[8]	Matsuyama Plain[9] Salan Grande Basin[10]	Dead Sea Basin[11] Cayman Trough[12] Cariaco Basin[13] Salton Trough[14]
Fault-termination basin	Yinggehai Basin[15] Malay Basin[16]	Beppu Bay[17]	Gulf of California[18]
Transpressional basin		Aceh Basin[19] Tokushima Plain[20]	

Table 1. Modern and ancient (*) examples of strike-slip basins according to the types of strike-slip faults. Numbers correspond to those in Figure 5

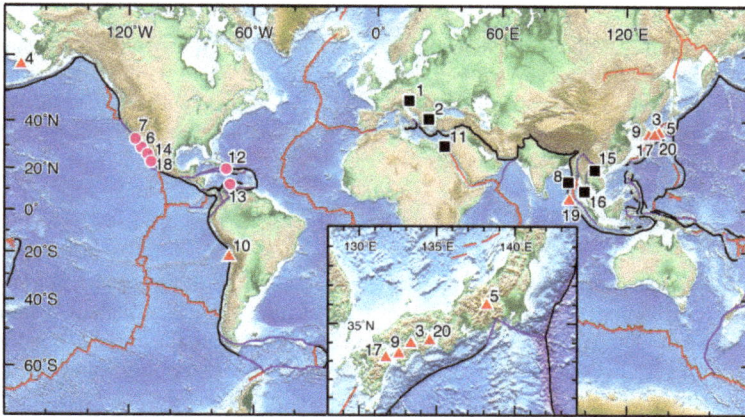

Figure 5. Strike-slip basins at plate convergent margins. Red triangles, trench-linked; black squares, indent-linked; purple circles, plate-boundary transform faults. Numbers correspond to those in Table 1

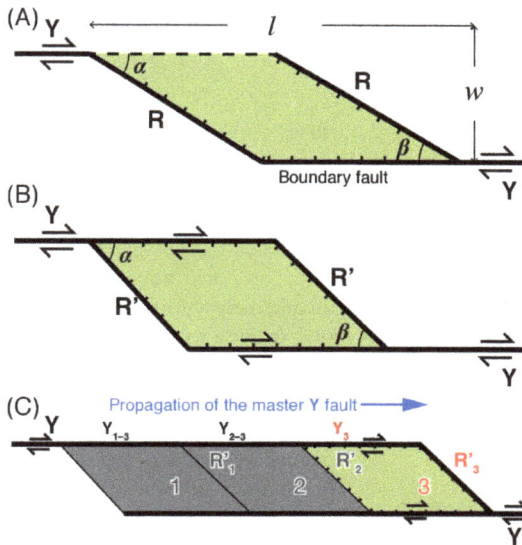

Figure 6. Geometrical models of (A) a spindle-shaped fault-bend basin and (B) a rhomb-shaped stepover strike-slip basin. Colored areas indicate subsiding basins. (C) Multistage evolution of stepover basins. As a result of step-wise propagation of one of the master faults, a new basin (3) is created, but the pre-existing basins 1 and 2 become inactive, resulting in a long (high l/w ratio) strike-slip basin with progressive depocenter migration. Diagrams are modified from [42]

3.1. Fault-bend basins

Fault-bend basins result from vertical displacement of normal faults in front of releasing bends corresponding to gentle transverse **R** (synthetic Riedel) faults connected to stepped master **Y** (principal displacement) faults (Figures 1 and 6A). The basin geometry is generally spindle-shaped or lazy-Z-shaped in plan view [38]. This type is considered to represent an early stage of the evolution of a pull-apart basin [12].

3.2. Stepover basins

As the master faults continue to propagate, they overlap and pull the crustal blocks farther apart, with lengthening geometries that gradually change from lazy-Z-shaped fault-bend basins to rhomboid-shaped stepover basins (Figure 6B). The basins subside by extension along strike-slip fault systems where the sense of *en échelon* segment stepping coincides with the sense of the slip (i.e., right-stepping faults have dextral displacement). The term 'pull-apart basin' was originally introduced to explain a depression in the Death Valley whose sides were pulled apart along releasing bends or oversteps of faults [41]. According to the pull-apart mechanism, two sides of the basin are bounded by faults with primarily horizontal displacement, and the other two sides are bounded by listric or transverse faults.

Stepover basins generally maintain their length/width ratio [2], as expressed by the following relationship between the length (l) and width (w) of a pull-apart basin based on the dimensions of natural pull-apart basins (Figure 6):

$$\log l = c_1 \log w + \log c \tag{1}$$

The best fitting constants have been found to be $c_1 = 1.0$ and $c_2 = 3.2$, which yield $l/w \approx 3.2$ with a 95% confidence interval about the ratio of $2.4 < l/w < 4.3$.

In sandbox experiments [42], a spindle-shaped basin appears in the first stage of basin evolution and is bounded by master **Y** faults and their synthetic Riedel (**R**) faults. Subsequently, antithetic Riedel (**R′**) faults replace **R** faults, leading to a rhomb-shaped basin. The l/w ratio depends on the angles α and β (Figure 6):

$$l/w = 1/\tan\alpha + 1/\tan\beta \tag{2}$$

The mean angle between **R** and **Y** faults in the experiments is $a = \beta = 30°$; that is, $l/w=3.5$. This value is consistent with those of natural basins.

As overlapped offsets of the master strike-slip faults propagate, basins elongate and finally become long pull-apart basins. The Dead Sea Basin, with a length of 132 km and a width of 18 km (l/w=7.2), is considered to have been formed by the coalescence of three successive and adjacent sedimentary basins whose depocenters migrated northward with time [43]. Although each sub-basin has a l/w ratio typical of a pull-apart basin (2.4, 3.3, and 2.6 from south to north), propagation of the master fault, accompanied by the creation of a new stepover basin, has resulted in the basins high l/w ratio (Figure 6C) [42].

Very high l/w ratios can also result from continuous transtension leading to extreme thinning of the crust and rupture. This process induces magmatic activity, high heat flow, and then the generation of new oceanic crust that may be younger than the overlying sedimentary succession (e.g., the Cayman Trough or the Gulf of California).

3.3. Fault-termination basins

Fault-termination basins are developed in transtensional stress domains at the ends of strike-slip faults where normal or oblique slip faults diffuse or splay off to terminate the deformation field [44]. If a part of a crustal block undergoes translation within the block, it results in shortening/uplift at one end and extension/subsidence at the other (Figure 7). Basins formed by such subsidence are referred to as fault-termination basins or transtensional fault-termination basins [44].

Modern examples include the Yinggehai Basin (Song Hong Basin) along the Red River Fault zone [45], the Malay and Pattani basins in the Gulf of Thailand [46], several segmented basins in the Gulf of California [44, 47], the northern Aegean Sea [48, 49], and Beppu Bay along the Median Tectonic Line (Figure 5 and Table 1) [50].

3.4. Transpressional basins

Transpressional basins tend to develop along oblique convergent margins whose subsidence results from flexural loading of the hanging-wall crust, similar to foreland basins adjacent to uplifted blocks [52–54]. Such basins are usually long, narrow structural depressions that lie parallel to the master faults.

The Sumatra forearc basins are modern examples of this type. Uplift of outer arc highs bounded by trench-linked strike-slip faults may cause flexural subsidence on the forearc side and generate elongate wedge-shaped sedimentary basins.

4. Examples of strike-slip basins

The wide variability of strike-slip faults makes it difficult to develop a simple model of the formation of strike-slip basins and their sedimentary facies. Although the geometries of such basins depend on the amount of fault displacement, the angle and distance between overstepped faults, and the depth of detachment of the faults, the basins are generally elongate, narrow, and deep. Several representative examples of the strike-slip basins described in this section show a range of basin evolutionary paths and filling processes.

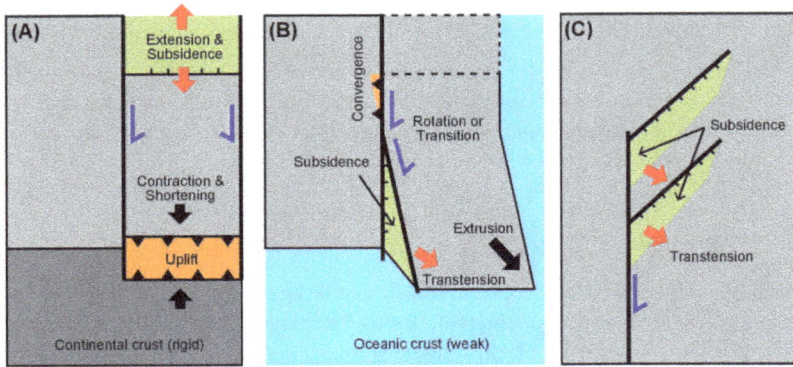

Figure 7. Typical geometry of termination areas of strike-slip faults. (A) If the block collides with a rigid continental crust, it shortens and is uplifted, accompanied by thrusts. At the opposite side to the uplift, upper crust is mechanically pulled away, leading to subsidence. (B) If the block extrudes with a rotational component into a weak oceanic crust (a transtensional setting), a sedimentary basin forms at the end of the strike-slip fault. Examples include the Yinggehai Basin, which is related to the extrusion of the Indochina Block, and the Gulf of California, which is related to the transtensional movement of the Baja California Peninsula. (C) The strike-slip fault diffuses its displacement through splayed extensional normal faults at its end [44]. An example is the Cerdanya clastic basin formed by late Miocene normal faulting at the termination of the La Tet strike-slip fault, Spain [51].

4.1. Fault-bend basin: The Ridge Basin

4.1.1. Geology

The Ridge Basin, which is one of the best-studied examples of a strike-slip basin [37, 55], is situated along the San Andreas Fault, a right-lateral plate-boundary transform fault between the Pacific and the North American plates, and along the San Gabriel Fault, a major strand of the San Andreas Fault (Figure 4B).

The San Gabriel Fault is a listric, ESE-dipping, oblique-slip fault rather than a subvertical, strike-slip fault [56]. The Ridge Basin is a type of fault-bend basin developed in front of a releasing bend on the San Gabriel Fault, along which the upper crust stretched and subsided to form a space in which sediments could be accommodated. The bottom of the basin is bounded by the subhorizontal San Gabriel Fault at a depth of ~4 km.

The basin originated in the late Miocene as a narrow depression within the broad San Andreas transform belt in southern California. The basin has a length of 45 km and a width of 15 km; the length/width ratio of 3 is a typical value for pull-apart basins [2]. The strata are exposed as a northwest-dipping homoclinal sequence that becomes younger to the northwest. The exposed sediment thickness reaches ~14 km, somewhat larger than the basin depth (~4 km) [56].

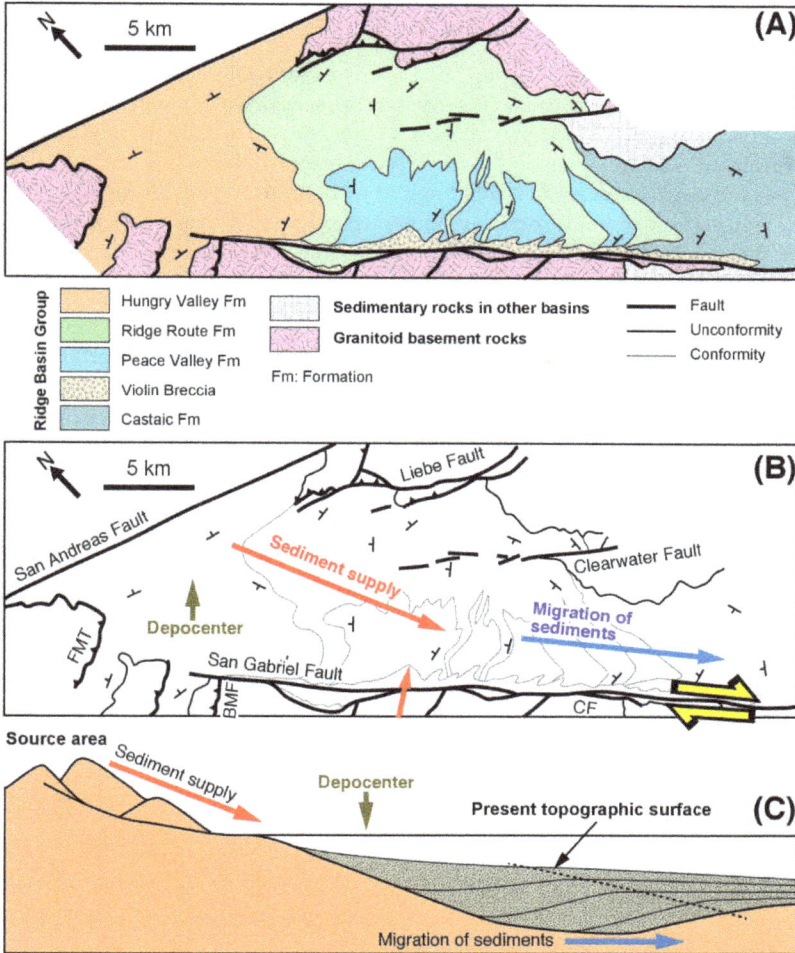

Figure 8. (A) Simplified geological map showing formations in the Ridge Basin [58]. (B) Conceptual basin-filling process for the Ridge Basin [8]. Abbreviations: FMT, Frazier Mountain Thrusts; BMF, Bear Mountain Fault; CF, Canton Fault. (C) Cross-sectional profiles showing continuous axial sediment supply and migration of sediments with relatively fixed depocenters [61]

4.1.2. Basin filling processes

The strata within the Ridge Basin are assigned to the Ridge Basin Group, which includes five formations: Castaic, Peace Valley, Violin Breccia, Ridge Route, and Hungry Valley (Figure 8A). Sedimentation began in the late Miocene (ca. 11 Ma) with deposition of the marine Castaic

Formation. The younger Castaic Formation is interfingered with the older Violin Breccia, which consists of conglomerates adjacent to the San Gabriel Fault scarp [57].

The main part of the Ridge Basin Group consists of the Peace Valley and Ridge Route formations. The Peace Valley Formation consists mainly of sandstone and mudstone of lacustrine, fluvial, deltaic, and alluvial facies, accompanied by minor carbonaceous deposits. The Ridge Route Formation, which crops out in the northeastern part of the Basin, is composed of alluvial sandstone and conglomerate, and is interfingered with the Peace Valley Formation. Both the Peace Valley and Ridge Route formations are interfingered southwestward into the Violin Breccia. The uppermost unit of the Ridge Basin, the Hungry Valley Formation, conformably overlies the Peace Valley and Ridge Route formations. The deposition of this formation, including alluvial conglomerate, sandstone, and mudstone, ended at ca. 4 Ma.

The Ridge Basin Group presents a 14-km-thick stratigraphic section of gently (20–25°) northwest-dipping beds; it shows the dominance of axial sediment supply and migration of the deposits by dextral movement of the San Gabriel Fault (Figure 8B). The releasing bend may have a paired restraining bend on the northwestern side of the fault. Within the restraining bend, highlands were formed, which in turn provided sediment to be transported into the basin. Most of the sediment filling the basin was carried by rivers draining source areas located to the northeast. The sediments forming the Ridge Basin Group were deposited at a rate of about 2 m kyr^{-1}.

The right-lateral displacement of the San Gabriel Fault carried the basin, together with the sediments, southeastward, resulting in a northwestward migration of the depocenter and successively younger beds onlapping onto the basin floor (Figure 8C) [58, 59]. Nearly constant values of vitrinite reflectance data (Ro = 0.5) throughout the group [60] support the continuous removal of sedimentary strata deposited in a relatively fixed depocenter and transported to the southeast along the San Gabriel Fault. More than 45 km of lateral displacement is estimated, based on the distribution of the Violin Breccia. This displacement, and basin migration, ended in the early Pliocene.

4.2. Stepover basin: The Dead Sea Basin

4.2.1. Geology

The Dead Sea Fault system is located along a plate-boundary transform zone that separates the Arabian Plate from the African Plate (Figures 4C and 9) [12]. Movement along the Dead Sea Fault commenced in the Miocene in response to the opening of the Red Sea. The very low rate of relative plate motion between Arabia and Africa (6–8 mm yr^{-1}) has yielded only 30 km of displacement during the past 5 Myr, and about 105 km of total offset during the past 18 Myr.

The Dead Sea Fault system includes both transpressional and transtensional domains (Figure 9). Several strike-slip basins are developed along the steps of segmented faults in the transtensional domain, while the Lebanon and Anti-Lebanon ranges have been uplifted in the transpressional domain related to the restraining bend. The Dead Sea Basin is the largest strike-slip basin in the system, and is partly overlain by a deep hypersaline lake located at Earth's lowest continental elevation (418 m below sea level at the lake surface) [43, 62–64].

The Dead Sea Basin is 132 km long and 7–18 km wide, yielding a high length/width ratio (> 7). The basin is segmented into sequential sub-basins by deep transverse normal faults rather than by listric faults. The length of the basin is greater than the total offset length (~105 km) of the fault system, which is atypical of pull-apart basins [65].

The basin has a cross-sectional asymmetry, with a steep eastern slope and a gentle western slope. Seismic refraction and gravity data indicate that the southern Dead Sea Basin is unusually deep, containing about 14 km of sedimentary fill [66]. Geophysical data suggest that the deep basin is probably bordered on all sides by vertical faults that cut deep into the basement [67]. The thick sediment accumulation yields a large negative Bouguer gravity anomaly (lower than –100 mGal) [64]. Given the depth of the basin, ductile deformation in the lower crust would be expected; however, the present-day heat flow in the Dead Sea Basin is low (~40 mW m^{-2}) [68], suggesting that the lower crust may still be cool and brittle, and that the Moho is not elevated beneath the basin. These inferences are consistent with seismic activity at depths of 20–32 km.

The Dead Sea Basin has traditionally been considered a classic example of a stepover basin [2], but other interpretations have been proposed, including propagating basins [67], stretching basins [64], and sequential basins [63]. The sequential basin model, in which several active sub-basins are delimited by boundary master faults and transverse faults, and simultaneously

Figure 9. The Dead Sea Basin developed in a transtensional domain of the Dead Sea Fault system. On the northern side of the basin, the Lebanon and Anti-Lebanon ranges were uplifted in a transpressional domain. The locations of faults are taken from [12]. Abbreviations: AmF, Amaziyahu Fault; ArF, Arava Fault; WIF, Western Intrabasinal Fault; EBF, Eastern Boundary Fault; WBF, Western Boundary Fault; MS, Mount Sedom; LD, Lisan Peninsula; LR, Lebanon Range; ALR, Anti-Lebanon Range; JR, Jordan River; SG, Sea of Galilee [69]; HV, Hula Valley [70]. Plate names: AR, Arabian; NU, Nubian (African).

become larger and deeper as the master faults propagate, could explain why the Dead Sea Basin is longer than the total amount of slip along the Dead Sea Fault.

4.2.2. Basin-filling processes

The depositional environments of the Dead Sea Basin are affected by the arid climatic conditions, with the area having an average annual rainfall of 50–75 mm. The modern sediments are transported to the basin mainly from the north by the Jordan River and from other directions by marginal tributaries. The mean annual discharges from the north, east, west, and south are 1100, 203, 4–5, and 4 mm, respectively [71].

In the middle to late Miocene, fluvial clastics of the Hazeva Formation were deposited in the southern sector of the basin (Figure 9). The formation consists of fluvial sandstones and conglomerates, including pre-Cretaceous components, transported from distant sources south and southeast of the Dead Sea Basin [43, 64]. During the Pliocene, the evaporitic Sedom Formation accumulated in estuarine–lagoonal environments in the Dead Sea basin; the formation consists mostly of lacustrine salts, gypsum, and carbonates interbedded with some clastics, and is found in the central sector of the basin. The 2–3 km thickness of this evaporitic formation may have formed in < 1 Myr; therefore, the sedimentation rate was probably higher than 2 m kyr^{-1}.

In the Pleistocene and Holocene, fluvial and lacustrine deposits, alternating with evaporites and locally sourced clastics, accumulated in lakes that post-date the formation of the Sedom lagoon. The Amora, Lisan, and younger formations consist of laminated evaporitic (gypsum) and aragonite sediments that continue to accumulate in the modern Dead Sea in the northern sector of the basin. The average sedimentation rate in this stage reached 1–1.5 m kyr^{-1} [64]. On the whole, the depocenters have migrated northward since the Miocene.

The margins of the Dead Sea are dominated by alluvial fans. The modern basin margin environments consist of (1) talus slopes, (2) incised and confined stream channels, and (3) coarse-grained and relatively high-gradient alluvial fans. In contrast, sediments in the offshore environment are composed of thick sequences of evaporitic salt intercalated with thin beds of laminated aragonite and detrital silt [72].

4.3. Fault-termination basin: The Yinggehai Basin

4.3.1. Geology

The Yinggehai Basin (Song Hong Basin) is an example of a fault-termination basin, and is situated at the southeastern end of the Red River Fault zone (RRFZ) [73, 74]. The RRFZ, extending for some 1000 km, separates the South China Block to the north from the Indochina Block to the south (Figure 10), and is considered to be related to the continental collision between the Indian and Eurasian plates [e.g., 75]. The formation of the Yinggehai Basin was controlled by the successive clockwise extrusions of the Indochina Block and the South China Block.

The RRFZ was a sinistral strike-slip fault in the first stage of its evolution (34–17 Ma), associated with ductile deformation [76] and the creation of an unconformity in the Gulf of Tonkin (the offshore part of the Hanoi depression [77]). After a quiescent stage from 17 to 5 Ma, due to a slowdown in the clockwise rotation of the Indochina Block [78], the movement along the RRFZ became dextral [45, 79, 80]. The right-lateral shearing is indicated by geomorphic fault traces and large river offsets [81–83], as well as GPS observations [80]. This change from a sinistral to a dextral sense of movement supports the basic tenets of the two-phase extrusion model; namely, the early, collision-driven escape of Indochina towards the SE, and the subsequent change to accommodate the present-day escape of Tibet and South China [78, 84, 85].

The Yinggehai Basin, situated in the offshore extension of the RRFZ, is 500 km long and 50–60 km wide ($l/w \approx 10$); it is oriented SE–NW and is located offshore between Hainan Island to the east and the Indosinian Peninsula to the west [73]. The basin formed originally as a sinistral strike-slip basin [77, 86], but developed into a dextral strike-slip basin after the change in the sense of fault displacement of the RRFZ [87, 88]. The basin subsided by simple shear on low-angle, detached normal faults of the upper crust and by pure shear of the lower crust [45, 87].

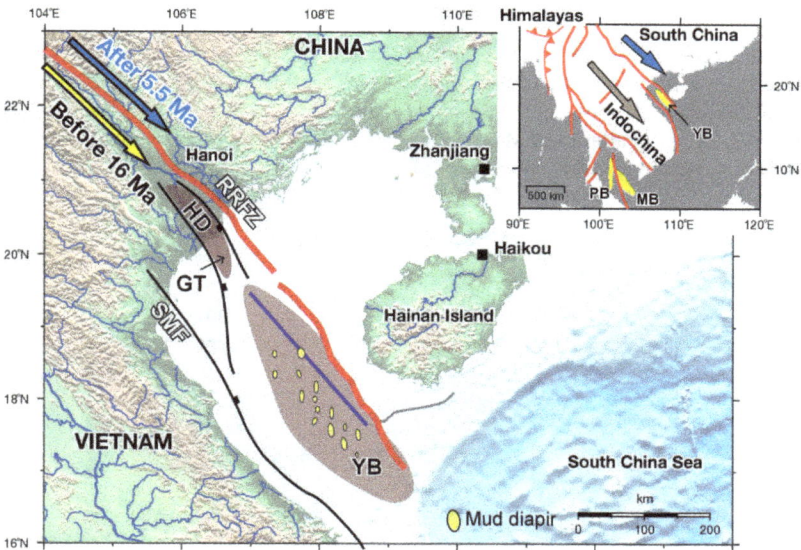

Figure 10. Geological setting of the Red River Fault zone (RRFZ) and the Yinggehai Basin. The RRFZ was originally a left-lateral strike-slip fault caused by the southeastern extrusion of the Indochina Block. The sense of displacement changed to right-lateral in response to the southeastward extrusion of the South China Block. Abbreviations: SMF, Song Ma Fault; HD, Hanoi Depression; GT, Gulf of Tonkin; YB, Yinggehai Basin; MB, Malay Basin; PB, Pattani Basin. The thick blue line marks the cross-sectional profile displayed in Figure 11. Modified from [80, 89, 90].

4.3.2. Basin-filling processes

The Yinggehai Basin is filled with 10–17-km-thick clastic deposits [45]. Basin sedimentation began in the late Eocene [45, 88]. During the Oligocene (> 21 Ma), clockwise rotation of the Indochina Block induced sinistral slip along the RRFZ and the basin rapidly subsided and started to fill. The depocenter was situated in the southern part of the basin during this syn-rift stage (Figure 11) [45, 91]. During the quiescent post-rift stage of the RRFZ, the rate of basin subsidence accordingly decreased and the depocenter gradually migrated northwestward [45]. The reactivation of the RRFZ with dextral movement triggered rapid subsidence [80] and enhanced the input of sediment [45]. The depocenter migrated from the center to the south-eastern end of the basin (Figure 11).

The infill of the Yinggehai Basin varies from alluvial, fluvial, and lacustrine deposits (before 21 Ma, the syn-rift stage) to marine sediments (after 21 Ma, the post-rift stage) [80]. Almost all the sediments in the basin are considered to have been derived from the Himalayas through the Red River drainage network [92]. Large mountain belts with high rates of sediment yield and along-fault transport networks were able to effectively supply huge volumes of detritus into the basin. Thick sediments and high sedimentation rates resulted in an over-pressured condition leading to mud diapirism [73], and also to depressed surface heat flow (~80 mW m^{-2}) [93]. The orientations of mud diapirs in the basin (Figure 13) indicate E–W extension related to right-lateral motion of the northeastern bounding fault [89, 90].

The Pattani and Malay basins in the Gulf of Thailand (inset map in Figure 10) are also considered to be fault-termination basins related to the continental collision of the Indian Plate [94, 95, 96]. Sediment supply into the basins is dominated by rivers flowing along the strike-slip faults [97, 98]. The Pattani and Malay basins contain thicknesses of sediment of more than 8 and 14 km, respectively. The subsidence of the basins was controlled at first by tectonic depression related to strike-slip deformation and then by thermal subsidence due to high surface heat flow (100–110 mW m^{-2}) [46, 99–101].

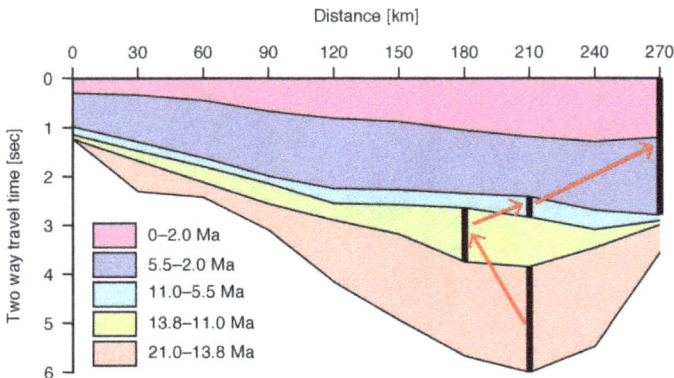

Figure 11. Sediment thickness in the Yinggehai Basin along the profile line shown in Figure 10. Red arrows suggest depocenter migration based on the thickest parts of the sediments deposited in each period (shown by solid vertical lines). Modified from [45].

4.4. Transpressional basin: The Aceh Basin

4.4.1. Geology

Sumatra is a classic example of slip partitioning due to an obliquely subducting plate [14]. The Indian and Australian plates are subducting beneath the Sundaland Plate in southeastern Asia along the Java–Sumatra Trench (Figure 2), where oblique subduction is accompanied by trench-parallel forearc translation [21]. The subduction thrust and the trench-parallel strike-slip fault (Great Sumatra Fault) isolate a wedge of forearc in the form of a sliver plate (the Burma Plate). The Great Sumatra Fault extends along the entire length of Sumatra Island (>1900 km) [22] and finally joins the West Andaman Fault (WAF), which constitutes a series of transform faults and spreading centers in the Andaman Sea [102]. In the forearc sliver, outer-arc uplift related to development of the accretionary prisms occurs on the trenchward side, and forearc basins (the Aceh and Simeulue basins) are developed on the landward side.

The Aceh Basin (Figures 12 and 13) is a wedge-shaped forearc basin with a long (> 200 km) and narrow (< 50 km) geometry (*l*/*w* > 4) bounded by the West Andaman Fault, a trench-linked strike-slip fault obliquely crossing the northward extension of the Great Sumatra Fault [103–

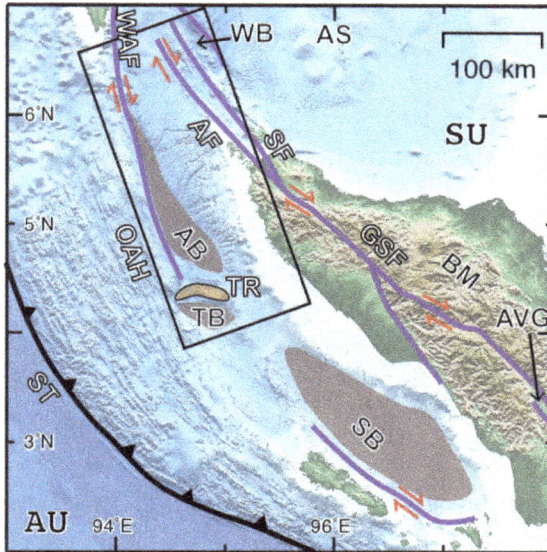

Figure 12. Physiographic map of the Sumatra region. Purple lines are strike-slip faults. Abbreviations: GSF, Great Sumatra Fault; AF, Aceh Fault; SF, Simeulue Fault; WAF, West Andaman Fault; AB, Aceh Basin; TR, Tuba Ridge; TB, Tuba Basin; SM, Simeulue Basin; WB, Weh Basin; BM, Barisan Mountain Range; AVG, Alas Valley Graben; OAH, outer-arc high; ST, Sumatra Trench; AS, Andaman Sea. The Weh Basin and Alas Valley graben are considered to be stepover pull-apart basins [22, 102]. The enclosed area is shown in Figure 13. Bathymetry is based on using SRTM and GEBCO with the data recently collected by [108–110].

106]. The bounding fault is a right-lateral transpressional fault with accompanying topographic highs of *en échelon* anticlines on the western margin and compressional ridges of Tuba Ridge on the southern margin (Figure 13). The transpressional uplift generating the outer-arc high by a thickening crustal block may be resulting in subsidence opposite to the high. This type of sedimentation is similar to that of foreland basins, where depressions are caused by the overburden pressure of the thrusted crust. Rifting and basin formation started in Sumatra during the Paleogene [107].

4.4.2. Basin-filling processes

The deposits in the basin are thickest along the boundary fault between the basin and the outer-arc high, and gradually thin with increasing distance from the faults (Figure 14B and C). Regarding the recent deposits, represented by seismic units 3 and 4 (Figure 14), unit 3 sediments in the southern part are thicker than those in the northern part, but unit 4 sediments are thicker in the northern part. Therefore, the main depocenter is considered to have migrated from the south (unit 3) to the north (unit 4). This interpretation is supported by seismic profiles of [106], who noted that the southern part of the Aceh Basin is raised above the northern part.

Most of the sediments are considered to have been supplied from Sumatra Island through small submarine channels (Figure 13). However, little is known about axial sediment redistribution within the basin.

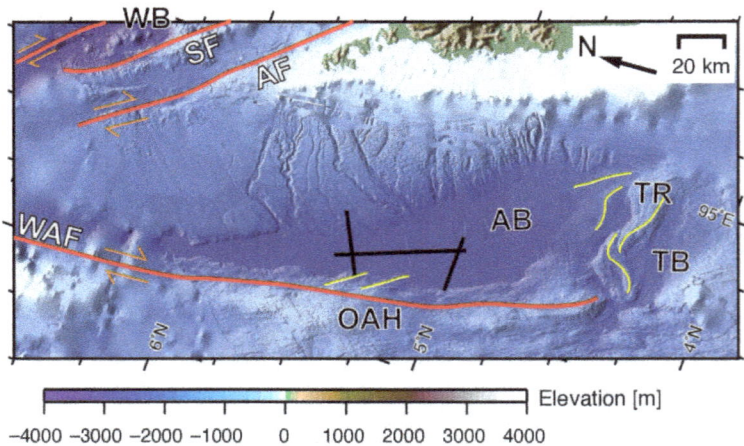

Figure 13. Detailed bathymetry around the Aceh Basin. Red and yellow lines are strike-slip faults and axes of anticlines [105], respectively. Thick solid lines in the Aceh Basin mark cross-sectional profiles shown in Figure 14. Abbreviations are the same as for Figure 12. Bathymetry is based on using SRTM and GEBCO with the data recently collected by [108–110]

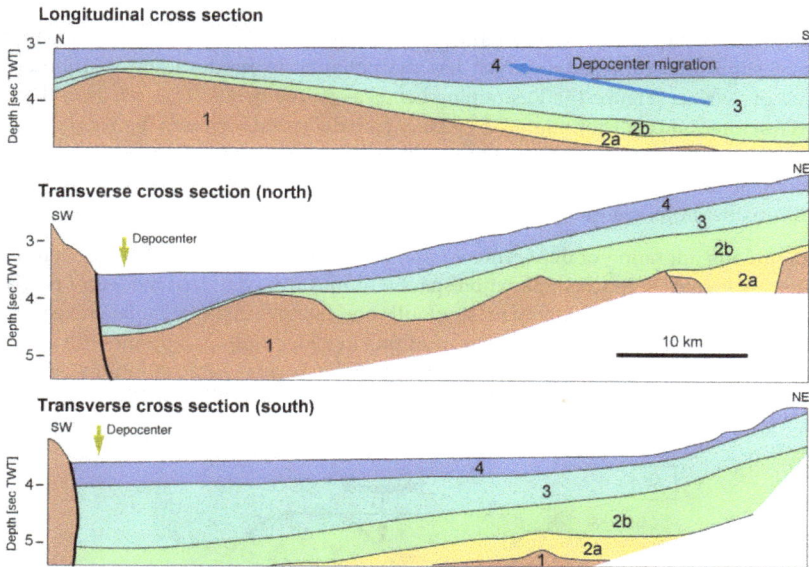

Figure 14. Cross-sectional profiles of the Aceh Basin. The uppermost seismic unit 4 was deposited predominantly in the northern part of the basin, suggesting a northward migration of the depocenter. Modified from [103]

5. Summary

This preliminary review has introduced some of the representative strike-slip basins at convergent margins from the viewpoints of basin formation and filling processes. Because strike-slip basins present a wide range of formational processes and sedimentary facies, it is difficult to establish a simple model of their evolution. To understand both modern and ancient strike-slip basins, the following factors need to be considered:

- Tectonic setting: plate boundary between continental plate or island arc microplate and subducting oceanic plates, collision between continents, within-plate

- Local stress field: compression, transpression, pure strike-slip, transtension, extension

- Fault configuration: existence of releasing or restraining bend, directions and dips of boundary and transverse faults, offset length of overstepped master faults

- Basin geometry: length, width, depth

- Thermodynamic condition: heat flux, gravity, volcanic front, mantle upwelling, ocean floor spreading

• Climate: sediment yield, modes of sediment transport, chemistry of deposits

Compressional uplift along master faults is probably needed for the dominance of axial sediment supply into the basin, which has the potential to produce and distribute huge volumes of detritus (Figure 15). Restraining bends as paired bends [12] such as in the Dead Sea Basin setting, and collision related to continental indentation such as in the Yinggehai Basin setting, are possible cases where axial sediment supply is enhanced. Conversely, a marginal high along the master fault is required for the formation of transpressional basins such as the Aceh Basin; therefore, marginal sediment supply may tend to dominate in such basins.

The continuous migration of depocenters requires that the progressive displacement of the master faults creates new accommodation space (Figure 15). In standard models of pull-apart basins, which are bounded by steep master faults and listric transverse faults, increasing the offset leads to a widening of the fault zone, resulting in wider pull-apart basins with a l/w ratio of about 3 for each basin [2, 42]. Therefore, for large l/w basins with continuous depocenter

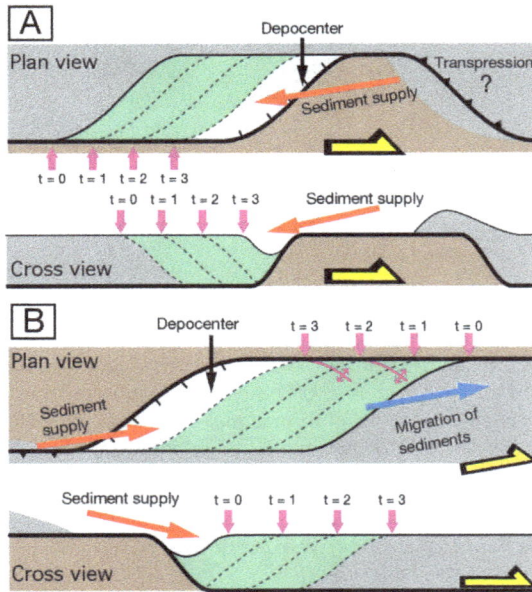

Figure 15. Conceptual models for depocenter migration and axial sediment supply in fault-bend basins. (A) Progressive right-lateral migration of paired bends on the foot-wall generates compressional uplift and extensional depression on the hanging-wall. Sediments are always supplied from the same direction along the long-axis of the basin. (B) Depocenter fixes along the releasing bend result from the right-lateral migration of sediments deposited on the foot-wall. A transpressional component would be required to generate the sediment source, and *en échelon* folds may form along the master faults. Both models generate deposits with axial sediments whose thicknesses are greater than the burial depths.

migration, such as the Dead Sea Basin and the Izumi Group, other formational mechanisms should be considered instead of the pull-apart basin model. Step-wise propagation or sequential progradation of the master fault could produce elongated stepover basins. A high l/w ratio could also potentially be produced by a transpressional basin along a trench-linked strike-slip fault. However, there is a need to establish physical models for basin formation in such settings.

Acknowledgements

Kai Berglar and his working groups kindly provided the bathymetric data of the Sumatra region. Financial support for this research was provided by the National Institute of Advanced Industrial Science and Technology (AIST). Constructive comments from Yasuto Itoh were insightful for improving the manuscript.

Author details

Atsushi Noda*

Address all correspondence to: a.noda@aist.go.jp

Geological Survey of Japan, National Institute of Advanced Industrial Science and Technology, Tsukuba, Ibaraki, Japan

References

[1] Nilsen TH, Mclaughlin RJ. Comparison of tectonic framework and depositional patterns of the Hornelen strike-slip basin of Norway and the Ridge and Little Sulphur Creek strike-slip basins of California. In: Biddle KT, Christie-Blick N (eds.) Strike-Slip Deformation, Basin Formation, and Sedimentation. Special Publication, no. 37. Tulsa, Oklahoma: SEPM; 1985. p. 79–103.

[2] Aydin A, Nur A. Evolution of pull-apart basins and their scale independence. Tectonics 1982;1(1) 91–105. doi:10.1029/TC001i001p00091.

[3] Noda A, Toshimitsu S. Backward stacking of submarine channel-fan successions controlled by strike-slip faulting: The Izumi Group (Cretaceous), southwest Japan. Lithosphere 2009;1(1) 41–59. doi:10.1130/L19.1.

[4] Nilsen TH, Sylvester AG. Strike-slip basins. In: Busby CJ, Ingersoll RV (eds.) Tectonics of Sedimentary Basins. Oxford: Blackwell Science; 1995. p. 425–457.

[5] Brothers DS, Driscoll NW, Kent GM, Harding AJ, Babcock JM, Baskin RL. Tectonic evolution of the Salton Sea inferred from seismic reflection data. Nature Geoscience 2009;2(8) 581–584. doi:10.1038/ngeo590.

[6] Dooley TP, Schreurs G. Analogue modelling of intraplate strike-slip tectonics: A review and new experimental results. Tectonophysics 2012;574–575 1–71. doi:10.1016/j.tecto.2012.05.030.

[7] Petrunin AG, Sobolev SV. Three-dimensional numerical models of the evolution of pull-apart basins. Physics of the Earth and Planetary Interiors 2008;171(1–4) 387–399. doi:10.1016/j.pepi.2008.08.017.

[8] Reading HG. Characteristics and recognition of strike-slip fault systems. In: Ballance PF, Reading HG (eds.) Sedimentation in Oblique-Slip Mobile Zones. International Association of Sedimentologists Special Publication, no. 4. Oxford: Blackwell Science; 1980. p. 7–26. doi:10.1002/9781444303735.ch2

[9] Woodcock NH, Daly MC. The role of strike-slip fault systems at plate boundaries and discussion. Philosophical Transactions of the Royal Society of London. Series A 1986;317(1539) 13–29. doi:10.1098/rsta.1986.0021.

[10] Christie-Blick N, Biddle KT. Deformation and basin formation along strike-slip faults. In: Biddle KT, Christie-Blick N (eds.) Strike-Slip Deformation, Basin Formation, and Sedimentation. Special Publication, no. 37. Tulsa, Oklahoma: SEPM; 1985. p. 1–34.

[11] Sylvester AG. Strike-slip faults. Geological Society of America Bulletin 1988;100(11) 1666–1703. doi:10.1130/0016-7606(1988)100<1666:SSF>2.3.CO;2.

[12] Mann P. Global catalogue, classification and tectonic origins of restraining-and releasing bends on active and ancient strike-slip fault systems. In: Cunningham WD, Mann P (eds.) Tectonics of strike-slip restraining and releasing bends. Special Publications, vol. 290, London: Geological Society; 2007. p. 13–142. doi:10.1144/SP290.2.

[13] Yeats RS, Sieh KE, Allen CR. Geology of Earthquakes. New York: Oxford University Press 1997, 576 pp.

[14] Fitch TJ. Plate convergence, transcurrent faults, and internal deformation adjacent to southeast Asia and the western Pacific. Journal of Geophysical Research 1972;77 4432–4460.

[15] Jarrard RD. Terrane motion by strike-slip faulting of forearc slivers. Geology 1986;14(9) 780–783. doi:10.1130/0091-7613(1986)14<780:TMBSFO>2.0.CO;2.

[16] McCaffrey R. Oblique plate convergence, slip vectors, and forearc deformation. Journal of Geophysical Research 1992;97(B6) 8905–8915.

[17] Carter K, Lerche I. Basin evolution, thermal history and hydrocarbon potential of the St. George Basin, Bering Sea, Alaska: A comparative study using one-and two-di-

mensional models. Marine and Petroleum Geology 1991;8(4) 392–409. doi: 10.1016/0264-8172(91)90062-6.

[18] Tsutsumi H, Okada A. Segmentation and Holocene surface faulting on the Median Tectonic Line, Southwest Japan. Journal of Geophysical Research 1996;101(B3) 5855–5871.

[19] Ikeda M, Toda S, Kobayashi S, Ohno Y, Nishizaka N, Ohno I. Tectonic model and fault segmentation of the Median Tectonic Line active fault system on Shikoku, Japan. Tectonics 2009;28(TC5006). doi:10.1029/2008TC002349.

[20] Karig DE, Lawrence MB, Moore GF, Curray JR. Structural frame work of the fore-arc basin, NW Sumatra. Journal of the Geological Society, London 1980;137(1) 77–91. doi: 10.1144/gsjgs.137.1.0077.

[21] McCaffrey R. The tectonic framework of the Sumatran subduction zone. Annual Review of Earth and Planetary Sciences 2009;37(1) 345–366. doi:10.1146/annurev.earth. 031208.100212.

[22] Sieh K, Natawidjaja D. Neotectonics of the Sumatran fault, Indonesia. Journal of Geophysical Research 2000;105(B12) 28,295–28,326.

[23] Bruhn RL, Pavlis TL, Plafker G, Serpa L. Deformation during terrane accretion in the Saint Elias orogen, Alaska. Geological Society of America Bulletin 2004;116(7–8) 771–787. doi:10.1130/B25182.1.

[24] Berger AL, Spotila JA, Chapman JB, Pavlis TL, Enkelmann E, Ruppert NA, Buscher JT. Architecture, kinematics, and exhumation of a convergent orogenic wedge: A thermochronological investigation of tectonic–climatic interactions within the central St. Elias orogen, Alaska. Earth and Planetary Science Letters 2008;270(1–2) 13–24. doi: 10.1016/j.epsl.2008.02.034.

[25] Allen CR. Circum-Pacific faulting in the Philippines-Taiwan region. Journal of Geophysical Research 1962;67(4795). doi:10.1029/JZ067i012p04795.

[26] Bachman SB, Lewis SD, Schweller WJ. Evolution of a forearc basin, Luzon Central Valley, Philippines. American Association of Petroleum Geologists Bulletin 1983;67(7) 1143–1162. doi:10.1306/03B5B718-16D1-11D7-8645000102C1865D.

[27] Ryan HF, Coleman PJ. Composite transform-convergent plate boundaries: description and discussion. Marine and Petroleum Geology 1992;9(1) 89–97. doi: 10.1016/0264-8172(92)90006-Z.

[28] Förster H, Oles D, Knittel U, Defant MJ, Torres RC. The Macolod Corridor: A rift crossing the Philippine island arc. Tectonophysics 1990;183(1–4) 265–271. doi: 10.1016/0040-1951(90)90420-D.

[29] Ringenbach JC, Pinet N, Stéphan JF, Delteil J. Structural variety and tectonic evolution of strike-slip basins related to the Philippine Fault System, northern Luzon, Philippines. Tectonics 1993;12(1) 187–203. doi:10.1029/92TC01968.

[30] Coffin MF, Gahagan LM, Lawver p LA No 174. Present-day plate boundary digital data compilation. Technical Report 174, University of Texas Institute for Geophysics 1998.

[31] DeMets C, Gordon RG, Argus DF. Geologically current plate motions. Geophysical Journal International 2010;181(1) 1–80. doi:10.1111/j.1365-246X.2009.04491.x.

[32] Molnar P, Tapponnier P. Cenozoic tectonics of Asia: Effects of a continental collision: Features of recent continental tectonics in Asia can be interpreted as results of the India-Eurasia collision. Science 1975;189(4201) 419–426. doi:10.1126/science.189.4201.419.

[33] Zhang PZ. A review on active tectonics and deep crustal processes of the Western Sichuan region, eastern margin of the Tibetan Plateau. Tectonophysics 2013;584 7–22. doi:10.1016/j.tecto.2012.02.021.

[34] Barnes PM, Sutherland R, Delteil J. Strike-slip structure and sedimentary basins of the southern Alpine Fault, Fiordland, New Zealand. Geological Society of America Bulletin 2005;117 411–435.

[35] Wallace LM, Barnes P, Beavan J, Van Dissen R, Litchfield N, Mountjoy J, Langridge R, Lamarche G, Pondard N. The kinematics of a transition from subduction to strike-slip: An example from the central New Zealand plate boundary. Journal of Geophysical Research 2012;117(B02405). doi:10.1029/2011JB008640.

[36] Ballance PF, Reading HG (eds.) Sedimentation in Oblique-Slip Mobile Zones. International Association of Sedimentologists Special Publication, no. 4. Oxford: Blackwell Science; 1980. 265 pp. doi:10.1002/9781444303735

[37] Crowell JC, Link MH (eds.) Geologic History of Ridge Basin, Southern California. Los Angeles, California: SEPM, Pacific Section; 1982. 304 pp.

[38] Mann P, Hempton MR, Bradley DC, Burke K. Development of pull-apart basins. Journal of Geology 1983;91(5) 529–554.

[39] Hempton MR. The evolution of thought concerning sedimentation in pull-apart basins. In: Boardman SJ (ed.) In Revolution in the Earth Sciences: Advances in the Past Half-Century. Dubuque, Iowa: Kendall-Hunt Publishers; 1983. p. 167–180.

[40] Biddle KT, Christie-Blick N (eds.) Strike-Slip Deformation, Basin Formation, and Sedimentation. Special Publication, no. 37. Tulsa, Oklahoma: SEPM; 1985. 386 pp.

[41] Burchfiel BC, Stewart JH. "Pull-apart" origin of the central segment of Death Valley, California. Geological Society of America Bulletin 1966;77(4) 439–442. doi: 10.1130/0016-7606(1966)77[439:POOTCS]2.0.CO;2.

[42] Basile C, Brun JP. Transtensional faulting patterns ranging from pull-apart basins to transform continental margins: an experimental investigation. Journal of Structural Geology 1999;21(1) 23–37. doi:10.1016/S0191-8141(98)00094-7.

[43] Zak I, Freund R. Asymmetry and basin migration in the dead sea rift. Tectonophysics 1981;80(1–4) 27–38. doi:10.1016/0040-1951(81)90140-2.

[44] Umhoefer PJ, Schwennicke T, Del Margo MT, Ruiz-Geraldo G, Ingle JC, McIntosh W. Transtensional fault-termination basins: An important basin type illustrated by the Pliocene San Jose Island basin and related basins in the southern Gulf of California, Mexico. Basin Research 2007;19(2) 297–322. doi:10.1111/j.1365-2117.2007.00323.x.

[45] Clift PD, Sun Z. The sedimentary and tectonic evolution of the Yinggehai–Song Hong basin and the southern Hainan margin, South China Sea: Implications for Tibetan uplift and monsoon intensification. Journal of Geophysical Research 2006;111(B06405). doi:10.1029/2005JB004048.

[46] Morley CK, Westaway R. Subsidence in the super-deep Pattani and Malay basins of Southeast Asia: a coupled model incorporating lower-crustal flow in response to post-rift sediment loading. Basin Research 2006;18(1) 51–84. doi:10.1111/j.1365-2117.2006.00285.x.

[47] Dorsey J Rebecca, Umhoefer J Paul. Influence of sediment input and plate-motion obliquity on basin development along an active oblique-divergent plate boundary: Gulf of California and Salton Trough. In: Busby C, Azor A (eds.) Tectonics of Sedimentary Basins: Recent Advances. Chap. 10. Chichester: Wiley-Blackwell; 2012. p. 209–225.

[48] Mann P. Model for the formation of large, transtensional basins in zones of tectonic escape. Geology 1997;25(3) 211–214. doi:10.1130/0091-7613(1997)025<0211: MFTFOL>2.3.CO;2.

[49] Koukouvelas IK, Aydin A. Fault structure and related basins of the North Aegean Sea and its surroundings. Tectonics 2002;21(1046). doi:10.1029/2001TC901037.

[50] Itoh Y, Takemura K, Kamata H. History of basin formation and tectonic evolution at the termination of a large transcurrent fault system: deformation mode of central Kyushu, Japana. Tectonophysics 1998;284(1–2) 135–150. doi:10.1016/S0040-1951(97)00167-4.

[51] Cabrera L, Roca E, Santanach P. Basin formation at the end of a strike-slip fault: the Cerdanya Basin (eastern Pyrenees). Journal of the Geological Society, London 1988;145(2) 261–268. doi:10.1144/gsjgs.145.2.0261.

[52] Beaumont C. Foreland basins. Geophysical Journal of the Royal Astronomical Society 1981;65(2) 291–329. doi:10.1111/j.1365-246X.1981.tb02715.x.

[53] DeCelles PG, Giles KA. Foreland basin systems. Basin Research 1996;8(2) 105–123. doi:10.1046/j.1365-2117.1996.01491.x.

[54] Sinclair HD, Naylor M. Foreland basin subsidence driven by topographic growth versus plate subduction. Geological Society of America Bulletin 2012;124(3–4) 368–379. doi:10.1130/B30383.1.

[55] Crowell JC (ed.) Evolution of Ridge Basin, Southern California: An Interplay of Sedimentation and Tectonics. Special Paper 367. Boulder, Colorado: Geological Society of America; 2003. 247 pp.

[56] May SR, Ehman KD, Gray GG, Crowell JC. A new angle on the tectonic evolution of the Ridge Basin, a "strike-slip" basin in Southern California. Geological Society of America Bulletin 1993;105(10) 1357–1372. doi: 10.1130/0016-7606(1993)105<1357:ANAOTT>2.3.CO;2.

[57] Link MH. Depositional systems and sedimentary facies of the Miocene-Pliocene Ridge Basin Group, Ridge Basin, southern California. In: Crowell JC (ed.) Evolution of Ridge Basin, Southern California: An Interplay of Sedimentation and Tectonics. Special Paper 367. Boulder, Colorado: Geological Society of America; 2003. p. 17–87. doi:10.1130/0-8137-2367-1.17.

[58] Crowell JC. Introduction to geology of Ridge Basin, Southern California. In: Crowell JC (ed.) Evolution of Ridge Basin, Southern California: An Interplay of Sedimentation and Tectonics. Special Paper 367. Boulder, Colorado: Geological Society of America; 2003. p. 1–15. doi:10.1130/0-8137-2367-1.1.

[59] Crowell JC. Tectonics of Ridge Basin region, southern California. In: Crowell JC (ed.) Evolution of Ridge Basin, Southern California: An Interplay of Sedimentation and Tectonics. Special Paper 367. Boulder, Colorado: Geological Society of America; 2003. p. 157–203. doi:10.1130/0-8137-2367-1.157.

[60] Link MH, Smith PR. Organic geochemistry of Ridge Basin, southern California. In: Crowell JC, Link MH (eds.) Geologic History of Ridge Basin, Southern California. Los Angeles, California: SEPM, Pacific Section; 1982. p. 191–197.

[61] Crowell JC. The tectonics of Ridge Basin, southern California. In: Crowell JC, Link MH (eds.) Geologic History of Ridge Basin, Southern California. Los Angeles, California: SEPM, Pacific Section; 1982. p. 25–42.

[62] Weber M, Abu-Ayyash K, Abueladas A, Agnon A, Alasonati-Tasárová Z, Al-Zubi H, Babeyko A, Bartov Y, Bauer K, Becken M, Bedrosian PA, Ben-Avraham Z, Bock G, Bohnhoff M, Bribach J, Dulski P, Ebbing J, El-Kelani R, Föster A, Förster HJ, Frieslander U, Garfunkel Z, Goetze HJ, Haak V, Haberland C, Hassouneh M, Helwig S, Hofstetter A, Hoffmann-Rothe A, Jäckel KH, Janssen C, Jaser D, Kesten D, Khatib M, Kind R, Koch O, Koulakov I, Laske G, Maercklin N, Masarweh R, Masri A, Matar A, Mechie J, Meqbel N, Plessen B, Möller P, Mohsen A, Oberhänli R, Oreshin S, Petrunin A, Qabbani I, Rabba I, Ritter O, Romer RL, Rümpker G, Rybakov M, Ryberg T, Saul J, Scherbaum F, Schmidt S, Schulze A, Sobolev SV, Stiller M, Stromeyer D, Tarawneh K, Trela C, Weckmann U, Wetzel U, Wylegalla K. Anatomy of the Dead Sea

Transform from lithospheric to microscopic scale. Reviews of Geophysics 2009;47(RG2002). doi:10.1029/2008RG000264.

[63] Lazar M, Ben-Avraham Z, Schattner U. Formation of sequential basins along a strike–slip fault–Geophysical observations from the Dead Sea basin. Tectonophysics 2006;421(1–2) 53–69. doi:10.1016/j.tecto.2006.04.007.

[64] Garfunkel Z, Ben-Avraham Z. The structure of the Dead Sea basin. Tectonophysics 1996;266(1–4) 155–176. doi:10.1016/S0040-1951(96)00188-6.

[65] Ben-Avraham Z, Lyakhovsky V, Schubert G. Drop-down formation of deep basins along the Dead Sea and other strike-slip fault systems. Geophysical Journal International 2010;181(1) 185–197. doi:10.1111/j.1365-246X.2010.04525.x.

[66] Ben-Avraham Z, Schubert G. Deep "drop down" basin in the southern Dead Sea. Earth and Planetary Science Letters 2006;251(3–4) 254–263. doi:10.1016/j.epsl.2006.09.008.

[67] ten Brink US, Ben-Avraham Z. The anatomy of a pull-apart basin: Seismic reflection observations of the Dead Sea Basin. Tectonics 1989;8(2) 333–350.

[68] Ben-Avraham Z, Hänel R, Villinger H. Heat flow through the Dead Sea rift. Marine Geology 1978;28(3–4) 253–269. doi:10.1016/0025-3227(78)90021-X.

[69] Hurwitz S, Garfunkel Z, Ben-Gai Y, Reznikov M, Rotstein Y, Gvirtzman H. The tectonic framework of a complex pull-apart basin: seismic reflection observations in the Sea of Galilee, Dead Sea transform. Tectonophysics 2002;359(3–4) 289–306. doi: 10.1016/S0040-1951(02)00516-4.

[70] Heimann A, Zilberman E, Amit R, Frieslander U. Northward migration of the southern diagonal fault of the Hula pull-apart basin, Dead Sea Transform, northern Israel. Tectonophysics 2009;476(3–4) 496–511. doi:10.1016/j.tecto.2009.07.024.

[71] Greenbaum N, Ben-Zvi A, Haviv I, Enzel Y. The hydrology and paleohydrology of the Dead Sea tributaries. In: Enzel Y, Agnon A, Stein M (eds.) New Frontiers in Dead Sea Paleoenvironmental Research. Special Paper 401. Boulder, Colorado: Geological Society of America; 2006. p. 63–93. doi:10.1130/2006.2401(05).

[72] Bartov Y, Bookman R, Enzel Y. Current depositional environments at the Dead Sea margins as indicators of past lake levels. In: Enzel Y, Agnon A, Stein M (eds.) New Frontiers in Dead Sea Paleoenvironmental Research. Special Paper 401. Boulder, Colorado: Geological Society of America; 2006. p. 127–140. doi:10.1130/2006.2401(08).

[73] Zhang Q, Zhang Q. A distinctive hydrocarbon basin—Yinggehai Basin, South China Sea. Journal of Southeast Asian Earth Sciences 1991;6(2) 69–74. doi: 10.1016/0743-9547(91)90097-H.

[74] Nielsen LH, Mathiesen A, Bidstrup T, VejbÃęk OV, Dien PT, Tiem PV. Modelling of hydrocarbon generation in the Cenozoic Song Hong Basin, Vietnam: a highly pro-

spective basin. Journal of Asian Earth Sciences 1999;17(1–2) 269–294. doi:10.1016/S0743-9547(98)00063-4.

[75] Tapponnier P, Peltzer G, Armijo R. On the mechanics of the collision between India and Asia. In: Coward MP, Alison C (eds.) Collision Tectonics. Special Publications, vol. 19. London: Geological Society; 1986. p. 113–157. doi:10.1144/GSL.SP. 1986.019.01.07.

[76] Tapponnier P, Lacassin R, Leloup P, Scharer U, Dalai Z, Haiwei W, Xiaohan L, Shaocheng J, Lianshang Z, Jiayou Z. The Ailao Shan/Red River metamorphic belt: Tertiary left-lateral shear between. Nature 1990;343(6257) 431–437. doi:10.1038/343431a0.

[77] Rangin C, Klein M, Roques D, Le Pichon X, Trong LV. The Red River fault system in the Tonkin Gulf, Vietnam. Tectonophysics 1995;243(3–4) 209–222. doi: 10.1016/0040-1951(94)00207-P.

[78] Leloup PH, Lacassin R, Tapponnier P, Schärer U, Zhong D, Liu X, Zhang L, Ji S, Trinh PT. The Ailao Shan-Red River shear zone (Yunnan, China), Tertiary transform boundary of Indochina. Tectonophysics 1995;251(1–4) 3–84. doi: 10.1016/0040-1951(95)00070-4.

[79] Leloup PH, Arnaud N, Lacassin R, Kienast JR, Harrison TM, Trong TTP, Replumaz A, Tapponnier P. New constraints on the structure, thermochronology, and timing of the Ailao Shan-Red River shear zone, SE Asia. Journal of Geophysical Research 2001;106(6683). doi:10.1029/2000JB900322.

[80] Zhu M, Graham S, Mchargue T. The Red River Fault zone in the Yinggehai Basin, South China Sea. Tectonophysics 2009;476(3–4) 397–417. doi:10.1016/j.tecto. 2009.06.015.

[81] Allen CR, Gillespie AR, Yuan H, Sieh KE, Buchun Z, Chengnan Z. Red River and associated faults, Yunnan Province, China: Quaternary geology, slip rates, and seismic hazard. Geological Society of America Bulletin 1984;95(6) 686–700. doi: 10.1130/0016-7606(1984)95<686:RRAAFY>2.0.CO;2.

[82] Replumaz A, Lacassin R, Tapponnier P, Leloup PH. Large river offsets and Plio-Quaternary dextral slip rate on the Red River fault (Yunnan, China). Journal of Geophysical Research 2001;106(819). doi:10.1029/2000JB900135.

[83] Schoenbohm LM, Burchfiel BC, Liangzhong C, Jiyun Y. Miocene to present activity along the Red River fault, China, in the context of continental extrusion, upper-crustal rotation, and lower-crustal flow. Geological Society of America Bulletin 2006;118(5–6) 672–688. doi:10.1130/B25816.1.

[84] Tapponnier P, Peltzer G, Le Dain AY, Armijo R, Cobbold P. Propagating extrusion tectonics in Asia: New insights from simple experiments with plasticine. Geology 1982;10(12) 611–616. doi:10.1130/0091-7613(1982)10<611:PETIAN>2.0.CO;2.

[85] Lee TY, Lawver LA. Cenozoic plate reconstruction of Southeast Asia. Tectonophysics 1995;251(1–4) 85–138. doi:10.1016/0040-1951(95)00023-2.

[86] Guo L, Zhong Z, Wang L, Shi Y, Li H, Liu S. Regional tectonic evolution around Yinggehai basin of South China Sea. Geological Journal of China University 2001;7(1) 1–12.

[87] Li S, Lin C, Zhang Q, Yang S, Wu P. Episodic rifting of continental marginal basins and tectonic events since 10 Ma in the South China Sea. Chinese Science Bulletin 1999;44(1) 10–23. doi:10.1007/BF03182877.

[88] Sun Z, Zhou D, Zhong Z, Zeng Z, Wu S. Experimental evidence for the dynamics of the formation of the Yinggehai basin, NW South China Sea. Tectonophysics 2003;372(1–2) 41–58. doi:10.1016/S0040-1951(03)00230-0.

[89] Luo X, Dong W, Yang J, Yang W. Overpressuring mechanisms in the Yinggehai Basin, South China Sea. American Association of Petroleum Geologists Bulletin 2003;87(4) 629–645. doi:10.1306/10170201045.

[90] Lei C, Ren J, Clift PD, Wang Z, Li X, Tong C. The structure and formation of diapirs in the Yinggehai–Song Hong Basin, South China Sea. Marine and Petroleum Geology 2011;28(5) 980–991. doi:10.1016/j.marpetgeo.2011.01.001.

[91] Zhong Z, Wang L, Xia B, Dong W, Sun Z, Shi Y. The dynamics of Yinggehai Basin Formation and its tectonic significance. Acta Geologica Sinica 2004;78(3) 302–309. (in Chinese with English abstract).

[92] Yan Y, Carter A, Palk C, Brichau S, Hu X. Understanding sedimentation in the Song Hong–Yinggehai Basin, South China Sea. Geochemistry, Geophysics, Geosystems 2011;12(Q06014). doi:10.1029/2011GC003533.

[93] He L, Xiong L, Wang J. Heat flow and thermal modeling of the Yinggehai Basin, South China Sea. Tectonophysics 2002;351(3) 245–253. doi:10.1016/S0040-1951(02)00160-9.

[94] Ngah K, Madon M, Tjia HD. Role of pre-Tertiary fractures in formation and development of the Malay and Penyu basins. In: Hall R, Blundell D (eds.) Tectonic Evolution of Southeast Asia. Special Publications, vol. 106. London: Geological Society; 1996. p. 281–289. doi:10.1144/GSL.SP.1996.106.01.18.

[95] Morley CK. A tectonic model for the Tertiary evolution of strike–slip faults and rift basins in SE Asia. Tectonophysics 2002;347(4) 189–215. doi:10.1016/S0040-1951(02)00061-6.

[96] Searle MP, Morley CK. Tectonic and thermal evolution of Thailand in the regional context of SE Asia. In: Ridd MF, Barber AJ, Crow MJ (eds.) The Geology of Thailand, Chap. 20. London: Geological Society; 2011. p. 539–571.

[97] Fyhn MBW, Boldreel LO, Nielsen LH. Escape tectonism in the Gulf of Thailand: Paleogene left-lateral pull-apart rifting in the Vietnamese part of the Malay Basin. Tectonophysics 2010;483(3–4) 365–376. doi:10.1016/j.tecto.2009.11.004.

[98] Miall AD. Architecture and Sequence Stratigraphy of Pleistocene Fluvial Systems in the Malay Basin, Based on Seismic Time-Slice Analysis. American Association of Petroleum Geologists Bulletin 2002;86(7) 1201–1216. doi:10.1306/61EEDC56-173E-11D7-8645000102C1865D.

[99] Madon MB, Watts. Gravity anomalies, subsidence history and the tectonic evolution of the Malay and Penyu Basins (offshore Peninsular Malaysia). Basin Research 1998;10(4) 375–392. doi:10.1046/j.1365-2117.1998.00074.x.

[100] Petersen HI, Mathiesen A, Fyhn MBW, Dau NT, Bojesen-Koefoed JA, Nielsen LH, Nytoft HP. Modeling of petroleum generation in the Vietnamese part of the Malay Basin using measured kinetics. American Association of Petroleum Geologists Bulletin 2011;95(4) 509–536. doi:10.1306/09271009171.

[101] Watcharanantakul R, Morley CK. Syn-rift and post-rift modelling of the Pattani Basin, Thailand: evidence for a ramp-flat detachment. Marine and Petroleum Geology 2000;17(8) 937–958. doi:10.1016/S0264-8172(00)00034-9.

[102] Ghosal D, Singh SC, Chauhan APS, Hananto ND. New insights on the offshore extension of the Great Sumatran fault, NW Sumatra, from marine geophysical studies. Geochemistry, Geophysics, Geosystems 2012;13. doi:10.1029/2012GC004122.

[103] Izart A, Mustafa Kemal B, Malod JA. Seismic stratigraphy and subsidence evolution of the Northwest Sumatra fore-arc basin. Marine Geology 1994;122(1–2) 109–124. doi:10.1016/0025-3227(94)90207-0.

[104] Berglar K, Gaedicke C, Lutz R, Franke D, Djajadihardja YS. Neogene subsidence and stratigraphy of the Simeulue forearc basin, Northwest Sumatra. Marine Geology 2008;253(1–2) 1–13. doi:10.1016/j.margeo.2008.04.006.

[105] Berglar K, Gaedicke C, Franke D, Ladage S, Klingelhoefer F, Djajadihardja YS. Structural evolution and strike-slip tectonics off north-western Sumatra. Tectonophysics 2010;480(1–4) 119–132. doi:10.1016/j.tecto.2009.10.003.

[106] Seeber L, Mueller C, Fujiwara T, Arai K, Soh W, Djajadihardja YS, Cormier MH. Accretion, mass wasting, and partitioned strain over the 26 Dec 2004 Mw9.2 rupture offshore Aceh, northern Sumatra. Earth and Planetary Science Letters 2007;263(1–2) 16–31. doi:10.1016/j.epsl.2007.07.057.

[107] Curray JR. Tectonics and history of the Andaman Sea region. Journal of Asian Earth Sciences 2005;25(1) 187–232. doi:10.1016/j.jseaes.2004.09.001.

[108] Henstock TJ, Mcneill LC, Tappin DR. Seafloor morphology of the Sumatran subduction zone: Surface rupture during megathrust earthquakes? Geology 2006;34(6) 485–488. doi:10.1130/22426.1.

[109] Graindorge D, Klingelhoefer F, Sibuet JC, Mcneill L, Henstock TJ, Dean S, Gutscher MA, Dessa JX, Permana H, Singh SC, Leau H, White N, Carton H, Malod JA, Rangin C, Aryawan KG, Chaubey AK, Chauhan A, Galih DR, Greenroyd CJ, Laesanpura A, Prihantono J, Royle G, Shankar U. Impact of lower plate structure on upper plate deformation at the NW Sumatran convergent margin from seafloor morphology. Earth and Planetary Science Letters 2008;275(3–4) 201–210. doi:10.1016/j.epsl.2008.04.053.

[110] Dean SM, Mcneill LC, Henstock TJ, Bull JM, Gulick SPS, Austin JA, Bangs NLB, Djajadihardja YS, Permana H. Contrasting Décollement and Prism Properties over the Sumatra 2004–2005 Earthquake Rupture Boundary. Science 2010;329(5988) 207–210. doi:10.1126/science.1189373.

Chapter 2

Variation in Forearc Basin Configuration and Basin-filling Depositional Systems as a Function of Trench Slope Break Development and Strike-Slip Movement: Examples from the Cenozoic Ishikari–Sanriku-Oki and Tokai-Oki–Kumano-Nada Forearc Basins, Japan

Osamu Takano, Yasuto Itoh and
Shigekazu Kusumoto

Additional information is available at the end of the chapter

1. Introduction

This chapter aims to elucidate variation of forearc basins in terms of basin configurations and basin-filling depositional systems with some examinations of their controlling factors, using actual examples from the Cenozoic forearc basins along the Northeast and Southwest Japan Arcs. Forearc basin is a sedimentary basin formed in the arc-trench gap between a volcanic arc and plate subduction zone (Figure 1) [1]. Although there are some notable past forearc basin studies (e.g., [1, 2]), the detailed characteristics of forearc basins have not been fully understood, since they show wide-range variation in styles, possibly reflecting various plate tectonic conditions at the plate subduction zone. As well-documented textbooks, Dickinson and Seely [2] and Dickinson [1] compiled and summarized the general outline of the forearc basin architecture and basin-filling sediments with explanations about some actual ancient and modern example forearc basins. The major contributions of these comprehensive textbooks include not only the presentation of various forearc basin styles, but also the explanation of related elements characterizing the forearc basin styles, such as dimension, subsidence, basin filling patterns, accretionary sill conditions and trench slope break development. Among these forearc basin elements, Dickinson [1] especially picked up two major elements: basin filling conditions and sectional basin configuration controlled by the relative height of trench slope break, to determine the morphological classification of forearc basins.

This chapter attempts to examine these two major factors: basin filling condition and basin configuration, by diagnosing two contrasting actual forearc basin packages around Japan: the Eocene Ishikari–Sanriku-oki forearc basins along the NE Japan Arc and the Pleistocene Tokai-oki–Kumano-nada forearc basins along the SW Japan Arc (Figure 2). To delineate the basin filling condition, we examine sedimentological characteristics including depositional systems, sequence stratigraphic contexts and related controlling factors. Regional seismic survey sections are used to manifest the basin configuration and to discuss controlling factors on forearc sedimentation for the two example forearc basins. In addition to these major factors, we discuss the role of strike-slip tectonics on the forearc basins, as it is reported in a former literature that strike-slip movement related to oblique plate subduction may affect the forearc basin tectonics and sedimentation.

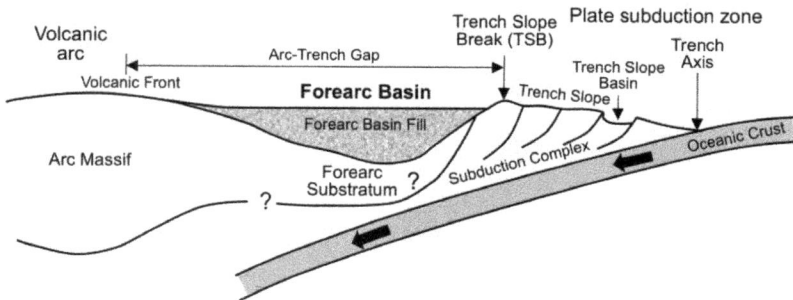

Figure 1. Schematic cross section of a forearc zone including a forearc basin, showing the basic terms used in this chapter. Modified after [1].

2. Eocene Ishikari–Sanriku-oki forearc basins

2.1. Geologic setting

The Eocene Ishikari–Sanriku-oki forearc basins were developed in the forearc zone along the NE Japan Arc (Figure 2A), which corresponds to the N-S trending narrow zone extending from the "Sorachi–Yezo Belt [3]" in central Hokkaido to the Pacific side offshore of northeast Honshu Island (Figure 2B). Although paleogeography around the NE Japan Arc was quite different from the present because the backarc basins of the NE Japan and Kuril Arcs had not opened [e.g., 4, 5], the tectonic history along the forearc zone during the Cretaceous to Paleogene can be summarized using the geologic evidences as follows. During the Cretaceous time, the eastern plates, which were regarded as the Izanagi and Kula Plates [4, 6], subducted underneath the western volcanic arc, and the forearc basin fully developed along this zone (Yezo Forearc Basin; Figure 2B [6, 7, 8]). During the Early Paleocene time, it is believed that a ridge between the Kula and Pacific Plates passed by along this forearc zone [5], causing total extinction of the forearc basins once. This tectonic event was widely recorded as "KT gap

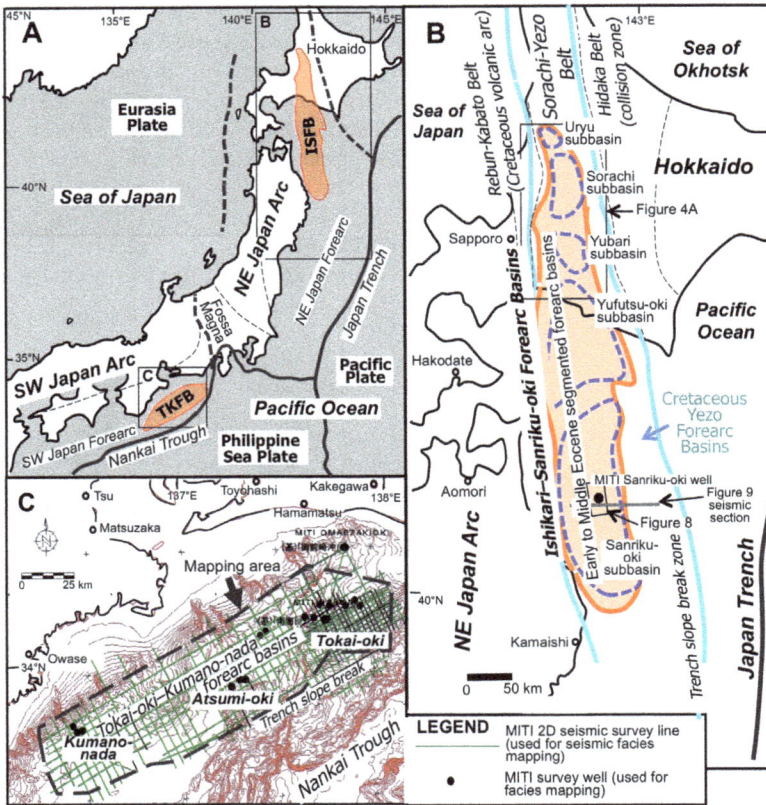

Figure 2. A) Index map showing the locations of two example forearc basins: Early to Middle Eocene Ishikari–Sanriku-oki forearc basins (ISFB) and Pleistocene Tokai-oki–kumano-nada forearc basins (TKFB). B) Close-up map showing the distribution of the Early to Middle Eocene Ishikari–Sanriku-oki forearc basins (orange lines) and the Cretaceous Yezo forearc basins (sky blue lines). The Eocene Ishikari–Sanriku-oki forearc basins are segmented into several subbasins (blue dashed lines). Compiled after [6, 8, 10, 11]. C) Close-up map of the Pleistocene Tokai-oki–Kumano-nada forearc basins, showing the mapping area and 2D seismic survey line positions used for seismic facies analysis. Modified after [23, 24].

unconformity [6, 8] (Figure 3)" seen in sedimentary successions along the Sorachi-Yezo Belt and the Ishikari–Sanriku-oki forearc zone with a minor time transgressive trend of the unconformity development possibly related to the ridge passage [8, 9]. After this tectonic event, fragmented small basins sporadically developed along the Ishikari–Sanriku-oki forearc zone. The Eocene was a relatively widespread phase of forearc basins, extending from Sanriku-oki to central Hokkaido (Figures 2B, 3). These Eocene forearc basins were segmented into several subbasins: Sanriku-oki, Yufutsu-oki, Yubari, Sorachi and Uryu subbasins [10, 11] (Figure 2B).

This section picks up the Sorachi, Yubari and Sanriku-oki subbasins for examining the basin filling condition and basin configuration.

2.2. Sorachi and Yubari subbasins (Ishikari Group)

2.2.1. Stratigraphic framework

The Sorachi and Yubari subbasins are located in central Hokkaido (Figures 4), and situated near the northern end of the Eocene Ishikari–Sanriku-oki forearc basins (Figure 2B). The Sorachi and Yubari subbasins developed and started sedimentation at the early Middle Eocene time, and continued until the Early Oligocene time with some short breaks by unconformities [12] (Figure 3). This section focuses on the Middle Eocene Ishikari Group (Figure 3), which constitutes the major part of the Sorachi and Yubari subbasin fill.

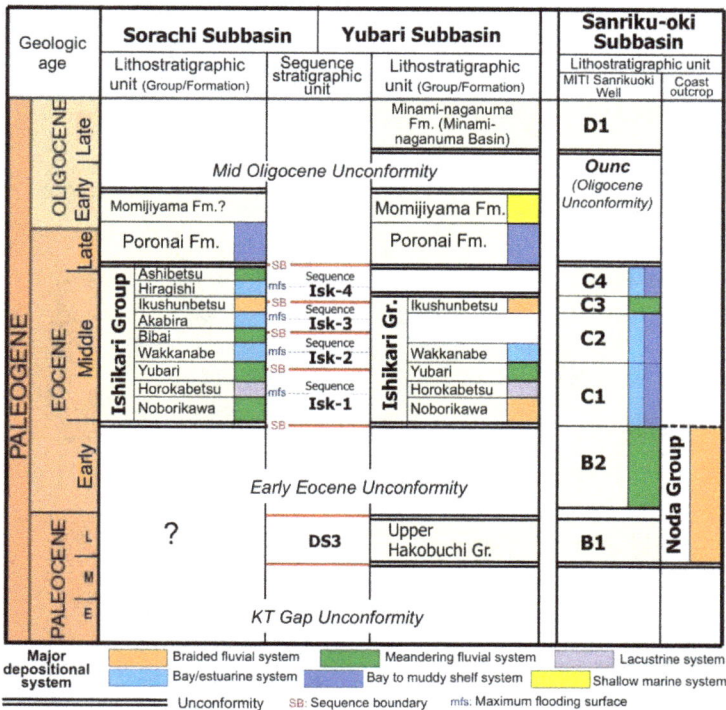

Figure 3. Generalized stratigraphic framework of the Paleocene, Eocene and Lower Oligocene in the Sorachi, Yubari and Sanriku-oki subbasins of the Ishikari–Sanriku-oki forearc zone. Colored columns beside the stratigraphic unit names denote the major depositional systems. This chapter mainly targets the Early to Middle Eocene basin fills. Compiled after [10–13, 15].

The Ishikari Group is divided into nine lithostratigraphic units: the Noborikawa, Horokabetsu, Yubari, Wakkanabe, Bibai, Akabira, Ikushunbetsu, Hiragishi and Ashibetsu Formations. From the standpoint of sequence stratigraphy, the Ishikari Group can be divided into four 3rd-order depositional sequences: Sequence Isk-1 to -4 in ascending order, and each depositional sequence is further divided into TST (transgressive systems tract) and HST (highstand systems tract), based on transgressive/regressive trends and marine incursion beds (Figures 3, 5) [11, 13]. In the Sorachi subbasin, the nine lithostratigraphic units are all developed, whereas in the Yubari basin, the Bibai, Akabira, Hiragishi and Ashibetsu Formations are absent, suggesting that the basin filling sedimentation was not continuous but episodic in the Yubari subbasin.

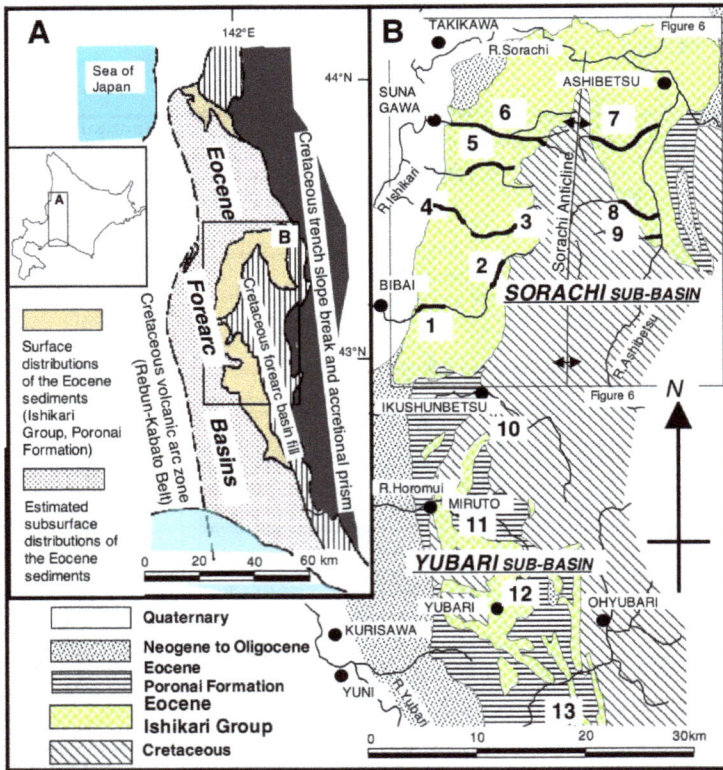

Figure 4. A) Geologic map showing the distributions of the Eocene forearc basin sediments in central Hokkaido, near the northern end of the Ishikari–Sanriku-oki forearc basins. B) Close-up geologic map showing the surface distributions of the Middle Eocene Ishikari Group in central Hokkaido. The Middle Eocene forearc basin in this area was segmented into the Sorachi subbasin on the north and the Yubari subbasin on the south. Numbers shown along rivers denote transect numbers of geologic survey, which correspond to numbers on the geologic cross section in Figures 5 and 7B. Modified after [11].

2.2.2. Depositional systems

Sedimentary facies analysis reveals that the Ishikari Group in the Sorachi and Yubari subbasins is composed of 24 sedimentary facies. These sedimentary facies are further assembled into five facies associations: braided fluvial facies association (BF), meandering fluvial facies association (MF), lacustrine facies association (LA), bay margin–estuarine facies association (ES) and bay center facies association (BA), as groups of sedimentary facies based on genetically related sedimentary environments and succession patterns [11, 13] (Figure 5). These five facies associations indicate that the Ishikari Group consists of five depositional systems: braided fluvial system, meandering fluvial system, lacustrine system, bay margin–estuarine system and bay system. Figure 5 depicts the schematic cross section showing the temporal and spatial distributions of depositional systems within the sequence stratigraphic framework in the Sorachi subbasin. As Figure 5 shows, the Ishikari Group mainly consists of a meandering fluvial system with some developments of a braided fluvial system. Lacustrine, bay margin / estuarine and bay center systems cyclically occur as marine incursion beds at around the maximum flooding surface of each 3rd-order depositional sequence.

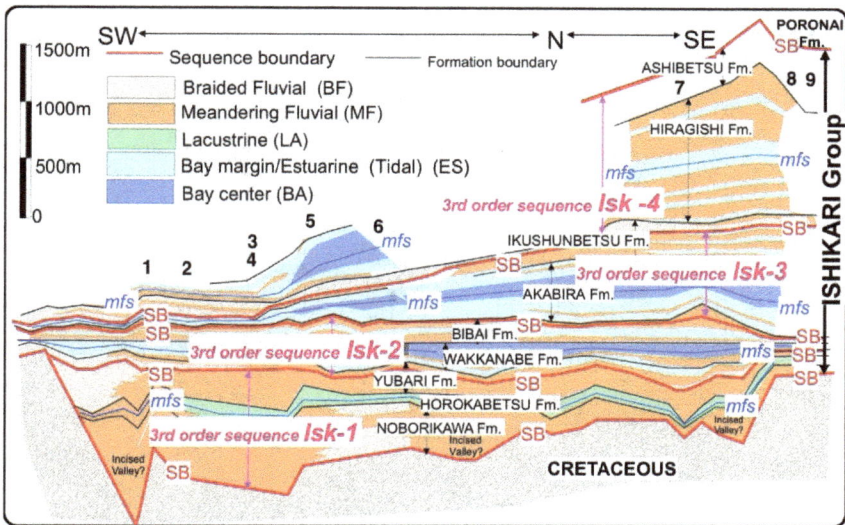

Figure 5. Schematic cross section of the Middle Eocene Ishikari Group in the Sorachi subbasin, showing sequence stratigraphic division and temporal and spatial distributions of depositional systems. Numbers above the cross section denote transect numbers of geologic survey shown in Figure 4B. Modified after [11, 13].

Figure 6 depicts facies maps showing spatial distributions of depositional systems for each systems tract of Sequences Isk-1 to -3. It is estimated that in response to relative sea level changes, the sedimentary environments in the Sorachi subbasin changed by cyclic transgression and regression. At the early phase of transgression and the late phase of regression, braided and meandering fluvial environments were dominant, whereas at the maximum flooding phase, bay center and bay margin environments were dominant. These facies maps indicate that marine influence became strong northeastward, whereas terrestrial environments such as braided / meandering river environments were dominant in the southern or south-western area of the Sorachi subbasin.

One of the notable characteristics of the depositional systems in the Ishikari Group in the Sorachi subbasin can be predominance of tidal deposits in a bay margin–estuarine system. Even though shallow marine condition periodically occurred during deposition, there are no wave-influenced sandy shallow marine facies such as foreshore and shoreface sandstone facies in the Ishikari Group. These facts indicate that the basin setting of the Sorachi subbasin was protected by wave action, and was not directly facing an open sea.

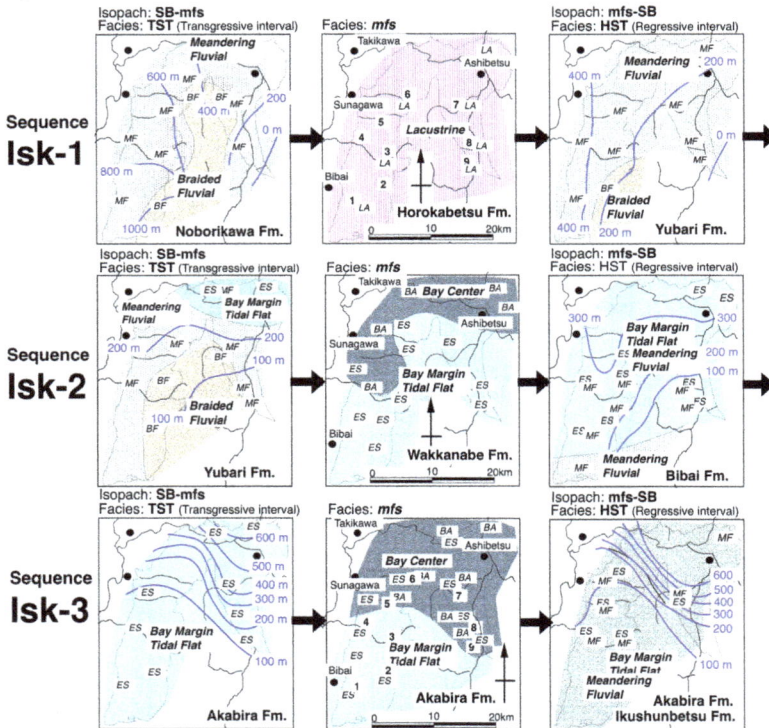

Figure 6. Spatial distribution maps of depositional systems in the Sorachi subbasin, showing paleogeographical changes for systems tracts (TST: transgressive interval; mfs: maximum transgression; HST: regressive interval) of 3rd-order Sequences Isk-1 to -3. Maps were created on the basis of facies association plots at the survey transect position. MF: meandering fluvial, BF: braided fluvial, LA: lacustrine, ES: bay margin–estuarine, BA: bay center. Blue contours denote isopach (iso-thickness) lines. Modified after [11].

2.2.3. Subsidence history

Figure 6 also depicts isopach contours for each systems tract of depositional sequences in the Sorachi subbasin. These isopach maps suggest that the thickness trend, indicating the depocenter, changed intermittently during deposition of the Ishikari Group. Since the depositional environments (altitude of deposition) in the Sorachi subbasin were more or less equivalent to a relative sea level or base level, it is regarded that the thickness trend indicates a spatial trend of total basin subsidence. Figure 7A demonstrates total subsidence curves of three different positions of the Sorachi subbasin, which were created on the basis of the thickness information. These isopach maps and total subsidence curves indicate that the western part of the subbasin rapidly subsided first. Subsequently during deposition of Sequence Isk-3 and -4, the northeastern part selectively subsided at a drastically rapid rate, and finally accumulated 3000 m-thick tidal-dominant deposits. Thus the Sorachi subbasin is characterized by a differential subsidence especially in the later half of the Ishikari Group deposition [11].

In addition to a differential subsidence within a subbasin, the subsidence patterns between subbasins show a notable difference. Figure 7B depicts the schematic cross section across the Sorachi and Yubari subbasins, showing a large thickness difference [14], possibly related to the difference in subsidence pattern as shown in Figure 7A. Accordingly, it is suggested that the segmented forearc basins in the Ishikari–Sanriku-oki forearc zone show highly variable subsidence patterns within and between subbasins.

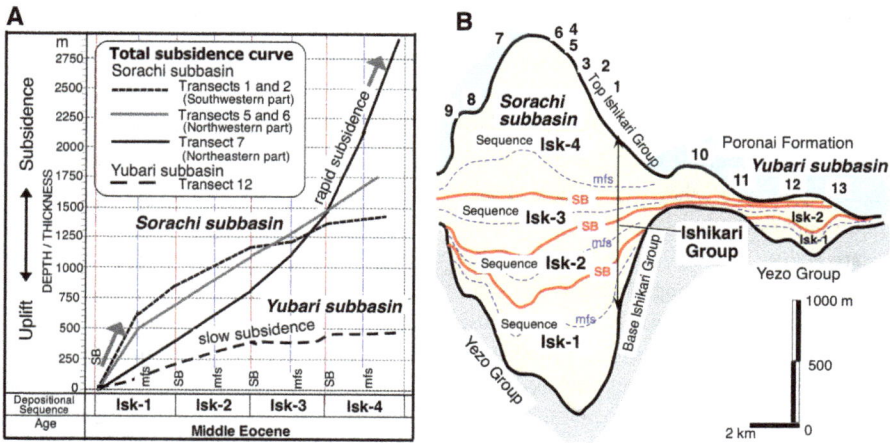

Figure 7. A) Diagram showing total subsidence histories along the selected transects during deposition of the Ishikari Group in the Sorachi and Yubari subbasins. Modified after [11]. B) Schematic sectional diagram showing thickness change of the Ishikari Group between the Sorachi and Yubari subbasins. Numbers above the Ishikari Group on the section denote transect numbers of surveys shown in Figure 4B. SB: sequence boundary, mfs: maximum flooding surface. Modified after [14].

2.3. Sanriku-oki subbasin

2.3.1. Stratigraphic framework

The Sanriku-oki subbasin is located in northeastern offshore of the Honshu Island, and situated near the southern end of the Eocene Ishikari–Sanriku-oki forearc basins (Figure 2B). After the K/T gap unconformity, the Sanriku-oki subbasin started basin-filling sedimentation in Late Paleocene time, and continued until the large-scale Oligocene unconformity (Ounc [10]) was formed (Figure 3). This section focuses on the Lower to Middle Eocene forearc basin-filling sediments, which are divided into the B2, C1, C2, C3 and C4 units [15] (Figure 3), for examining the depositional condition and basin configuration.

2.3.2. Depositional systems and basin setting

According to the MITI Sanriku-oki well report [15], the Lower to Middle Eocene succes-sions in the Sanriku-oki subbasin are mainly composed of mudstone, sandstone and coal-bearing alternating beds of sandstone and mudstone, which were deposited in terrestrial, brackish and neritic marine environments. Figure 8 demonstrates interpreted seismic facies maps of the lower and upper intervals of the Lower Eocene B2 unit in the 3D seismic surveyed area, including the MITI Sanriku-oki well location, in the central part of the Sanriku-oki subbasin (Figure 2B). These seismic facies maps, which were displayed by different colors assigned for each "seismic trace shape" class, show intricate meandering, braided or partly networked fluvial channel zones and floodplain–back mash zones.

Based on the sedimentary environment information from the MITI Sanriku-oki well and the seismic facies maps, it is interpreted that the B2 and C3 units consist mainly of a coal-bearing meandering fluvial system with minor bay center to bay margin systems as marine incursion beds, and the C1, C2 and C4 units consist mainly of bay to muddy shelf systems (Figure 3). Since all these component depositional systems resemble those of the Sorachi/Yubari subbasins, it is regarded that the Eocene Sanriku-oki subbasin was in a confined forearc setting, which was not directly facing an open sea and was protected by wave action. This basin setting during the Eocene time is supported by the basin configuration shown on a long 2D seismic section transecting the Sanriku-oki subbasin (Figure 9), in which the trench slope break prominently uplifted and eroded by Ounc (Oligocene Unconformity [10]), and the Cretaceous to Eocene basin-filling succession seems to be confined within the arcward side of the uplifted trench slope break. This confined forearc setting was terminated by the Ounc event, accompanied with seaward migration and large subsidence of the trench slope break, which finally caused transformation of the forearc basin setting from a fluvial to deep-marine slope condition as shown in the cross section in Figure 9. Consequently, it is regarded that the Sanriku-oki forearc basin setting was strongly controlled by the trench slope development.

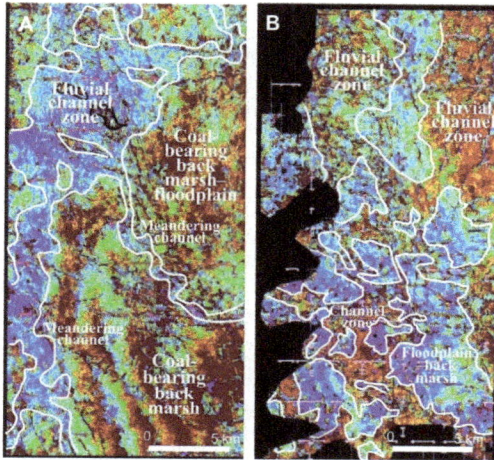

Figure 8. Seismic facies maps showing the distributions of a fluvial channel zone and a floodplain–back marsh zone in a meandering fluvial system in the central part of the Sanriku-oki subbasin. Map colors were assigned for each different seismic trace shape, which can indicate difference in sedimentary environment. A) Case of a lower horizon of the B2 unit. Bluish colors are interpreted as a channel zone on the basis of the seismic trace shape and distribution pattern. B) Case of an upper horizon of the B2 unit. Reddish colors are interpreted as a channel zone. The map location is shown in Figure 2B.

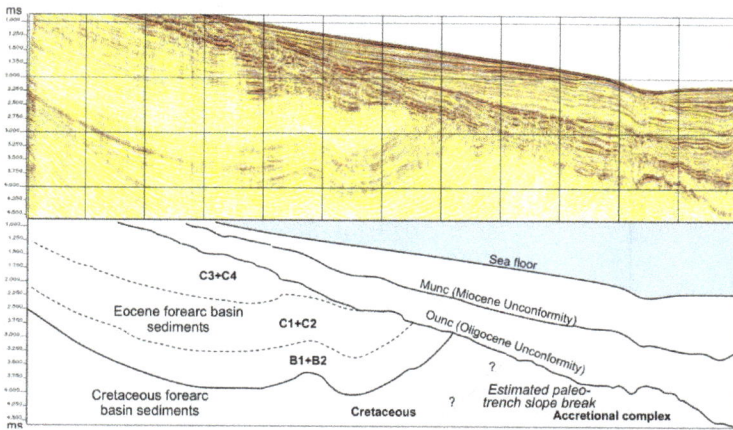

Figure 9. An E-W long 2D seismic section transecting the Sanriku-oki subbasin, showing trench slope break uplift and subbasin confinement. Although the present status of the Cretaceous to Eocene forearc basin fill and trench slope break seems to be inclined seaward, it is estimated that the trench slope break was more or less uplifted and emerged as a barrier because of the leaping-up morphology and fluvial-dominated environments in the Cretaceous to Eocene forearc basin-fill successions. The 2D seismic data were acquired in a MITI survey [16]. The seismic survey line location is shown in Figure 2B.

2.4. Forearc setting model

Based on the characteristics of depositional systems and basin configurations of the Sorachi, Yubari and Sanriku-oki subbasins, a forearc setting model of the Eocene Ishikari–Sanriku-oki forearc basins can be proposed as shown in Figure 10. The trench slope break ridge is estimated to have emerged above the sea along the eastern margin (subduction zone side) of the forearc basins, and formed a barrier to the open sea condition in the trench side of the trench slope break. This uplifted trench slope break condition is supported by previous petrography studies [17–19], which reveal that sandstones of the forearc basin fill (Ishikari Group) contain chromspinels derived from an emerged ridge of the Kamuikotan Belt. The N-S trending Kamuikotan Belt is distributed along the eastern margin of the forearc zone in Hokkaido (Sorachi–Yezo Belt), and mainly consists of serpentinite and various kinds of high pressure-type metamorphic rocks with tectonic mélanges, formed in an accretionary prism [3, 20].

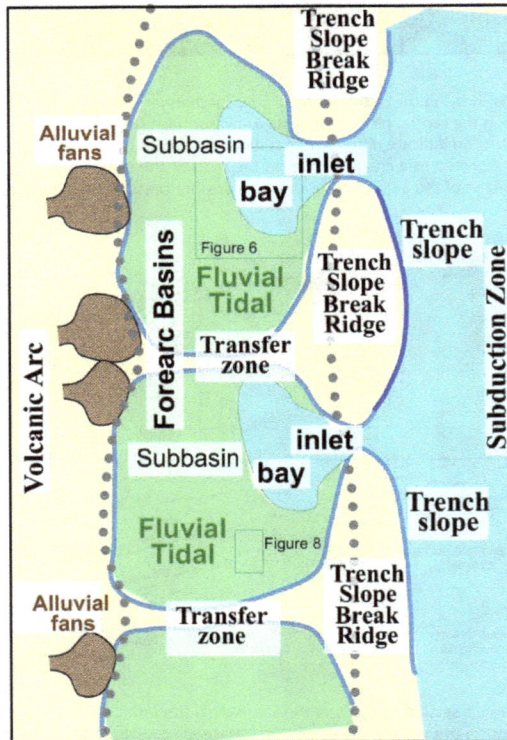

Figure 10. Schematic and conceptual forearc setting model for the Eocene Ishikari–Sanriku-oki forearc basins, including the Sorachi, Yubari and Sanriku-oki subbasins. Small rectangles inside the basin denote approximate positions of the mapped areas for the Sorachi subbasin (Figure 6) and the Sanriku-oki subbasin (Figure 8).

Inside the forearc basin, major depositional systems were bay to fluvial systems without any wave-influenced facies. In response to relative sea level changes, transgression and regression repeated, and the major depositional system alternated between a fluvial system-dominated condition and a bay system-dominated condition. Because of the existence of marine sediments, it is estimated that there were an inlet interconnecting between the open sea and the inside of the forearc basin, through which the seawater came into the inside of the forearc basin.

Our forearc setting model also demonstrates forearc basin segmentation, reflecting the fact that the Eocene Ishikari–Sanriku-oki forearc basins were segmented into 50 to 150 km long subbasins aligned along the forearc extension (Figure 2B). As described above, the segmented subbasins show a different subsidence pattern and different sediment thickness for each subbasin.

3. Pleistocene Tokai-oki–Kumano-nada forearc basins

3.1. Geologic setting and stratigraphic framework

The Pleistocene Tokai-oki–Kumano-nada forearc basins were developed in the forearc zone between the SW Japan Arc and the Nankai Trough subduction zone (Figures 2A, 2C). On the contrary to the sporadic developments of forearc basins during the late Paleogene and early Neogene time, thick sedimentary packages of the Late Pliocene to Pleistocene Tokai-oki–Kumano-nada forearc basins widely developed in this forearc zone. This section picks up the major basin-filling sediments equivalent to the Late Pliocene to Early Pleistocene Kakegawa Group (Atsumi-oki Group [21]) and Middle Pleistocene Ogasa Group (Hamamatsu-oki Group [21]; Figure 11) to examine the basin filling conditions and basin configurations. The Kakegawa Group unconformably overlies the underlying units with a certain time gap, indicating the different phase of forearc basin tectonics, and the Ogasa Group unconformably overlies the Kakegawa Group, indicating a tectonic event between depositions of the two groups. The study area is set between the Present continental slope toe and the trench slope break zone, which covers the Tokai-oki, Atsumi-oki and Kumano-nada areas (Figure 2C). From the standpoints of sequence stratigraphy and sedimentology, the targeted forearc sediments are divided into seventeen depositional sequences: Sequence Kg-a to -h and Og-a to –i, based on reflection termination patterns on the seismic sections and facies succession patterns on the well successions [22, 23](Figure 11). The major depositional system of the whole interval is a submarine fan turbidite system [22, 23].

3.2. Transformation of depositional styles

Takano et al. [22] demonstrated a series of facies maps in the Tokai-oki–Kumano-nada forearc basins for each depositional sequence unit for the interval equivalent to the Kakegawa and Ogasa Groups (Figure 12). These facies maps were created on the basis of seismic facies information plotted on the seismic survey line maps as well as some exploration well data (Figure 2C) [22–25]. These facies maps clearly show the depositional patterns of submarine fans, indicating that quite a few numbers of submarine canyons from the main land of Japanese

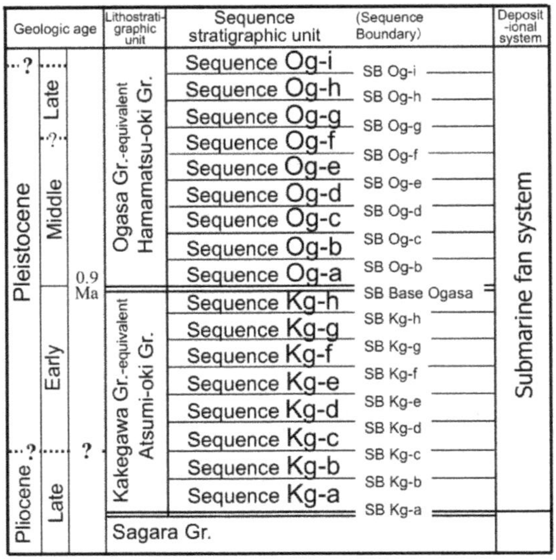

Geologic age	Lithostrati-graphic unit	Sequence stratigraphic unit	(Sequence Boundary)	Deposit-ional system
Pleistocene — Late	Ogasa Gr.-equivalent Hamamatsu-oki Gr.	Sequence Og-i	SB Og-i	
		Sequence Og-h	SB Og-h	
		Sequence Og-g	SB Og-g	
		Sequence Og-f	SB Og-f	
Pleistocene — Middle		Sequence Og-e	SB Og-e	Submarine fan system
		Sequence Og-d	SB Og-d	
		Sequence Og-c	SB Og-c	
		Sequence Og-b	SB Og-b	
0.9 Ma		Sequence Og-a	SB Og-a	
Pleistocene — Early	Kakegawa Gr.-equivalent Atsumi-oki Gr.	Sequence Kg-h	SB Base Ogasa	
		Sequence Kg-g	SB Kg-h	
		Sequence Kg-f	SB Kg-g	
		Sequence Kg-e	SB Kg-f	
		Sequence Kg-d	SB Kg-e	
		Sequence Kg-c	SB Kg-d	
?		Sequence Kg-b	SB Kg-c	
Pliocene — Late		Sequence Kg-a	SB Kg-b	
	Sagara Gr.		SB Kg-a	

Figure 11. Litho- and sequence stratigraphic framework of the latest Pliocene to Pleistocene successions in the Tokai-oki–Kumano-nada forearc basins. Modified after [22, 23].

Islands functioned as fixed feeder systems, along which submarine fans were formed in the forearc basins (Figure 12). These facies maps also suggest that submarine-fan architecture was intermittently transformed through time (Figure 12)[22, 26]; from a braided channel-dominated condition (Stage 1 represented by the map of Sequence Kg-a), through a small fan-dominated condition with shrinking separated small basins (Stage 2 represented by the map of Sequence Kg-e), and a trough-fill turbidite-dominated condition (Stage 3 represented by the map of Sequence Og-e), to a channel-levee system-dominated condition (Stage 4 represented by the map of Sequence Og-h). Although the submarine-fan architecture was transformed temporary, some spatial differences in depositional patterns between the Tokai-oki, Atsumi-oki and Kumano-nada areas can also be recognized (Figure 12), possibly resulting from forearc basin segmentation and sediment supply variation.

3.3. Transformation of basin configuration and background tectonics

To examine the relationships between the changes in the submarine-fan depositional styles and basin configuration, which may be indicating the background tectonics, we investigated seismic sections transecting the Tokai-oki–Kumano-nada forearc basins. Figure 13 depicts the interpreted cross sections with depositional stage division characterized by the different submarine-fan depositional styles as mentioned above. The interrelationships between the geologic structures and sediment thickness change shown on these cross sections reveal that the depositional stage division can be connected with tectonic phases that created specific

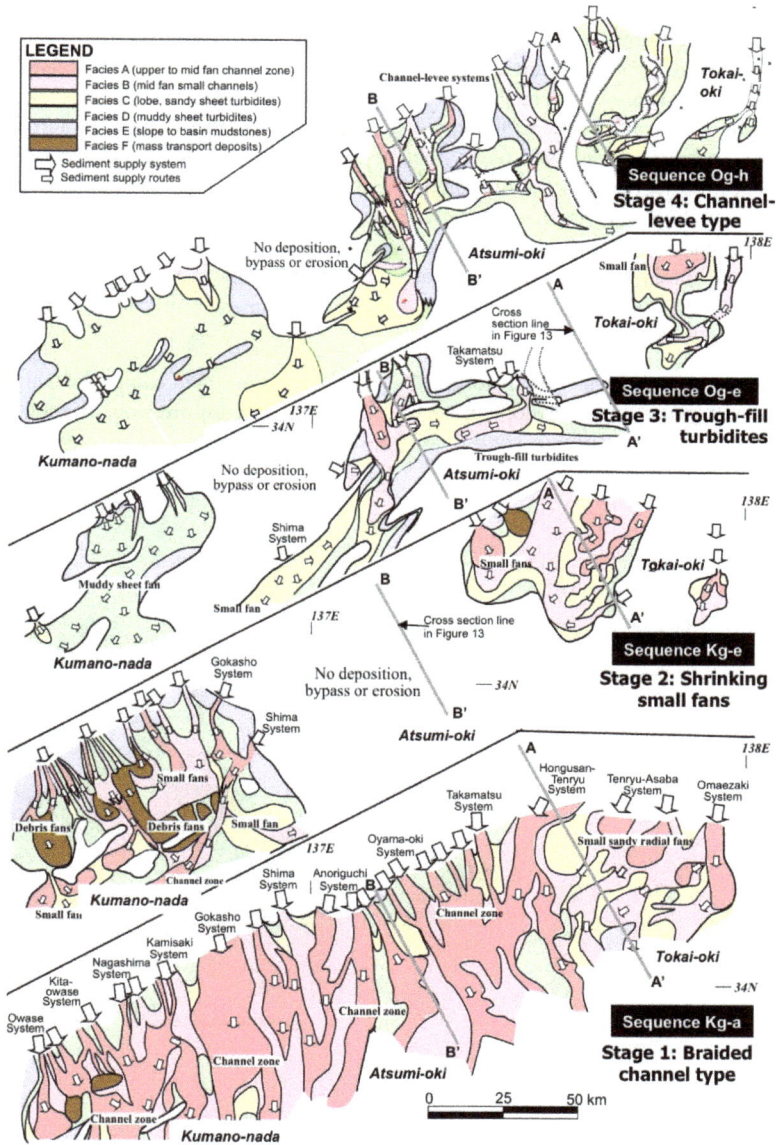

Figure 12. Facies maps of Sequence Kg-a, -e, Og-e and -h in the Tokai-oki–Kumano-nada forearc basins, showing the transformation of submarine-fan morphology and distributions. Modified after [22]. The mapping area is shown in Figure 2C.

geologic structures related to basin configuration (Figures 13, 14). Since the Stage 1 sediments show a mostly uniform thickness and a braided channel-dominated condition, the forearc basin during Stage 1 (Late Pliocene to earliest Pleistocene) is interpreted to have been a gently inclined, sloped basin without major topographic undulation, which is characteristic of an incipient phase of forearc basin development [27]. Stage 2 (Early Pleistocene) is interpreted as a compressional stress stage with trench slope break uplift, since only limited synclinal areas contain thick sediments, and the depositional areas shrunk continuously. Stage 3 (Middle Pleistocene) can be a relaxing phase, which induced subsidence of folded forearc basins, since the sedimentation is characterized by trough-fill (syncline-fill) turbidite systems and the depositional territory became wider gradually. Stage 4 (Middle to Late Pleistocene) can be a compressional stress stage again, as trench slope break prominently uplifted as shown on the section B–B' in Figure 13.

Consequently, it is suggested that during the Pleistocene time, two compressional phases occurred in response to trench slope break uplift and arcward suppression, and the forearc depositional styles were strongly controlled by these tectonic events.

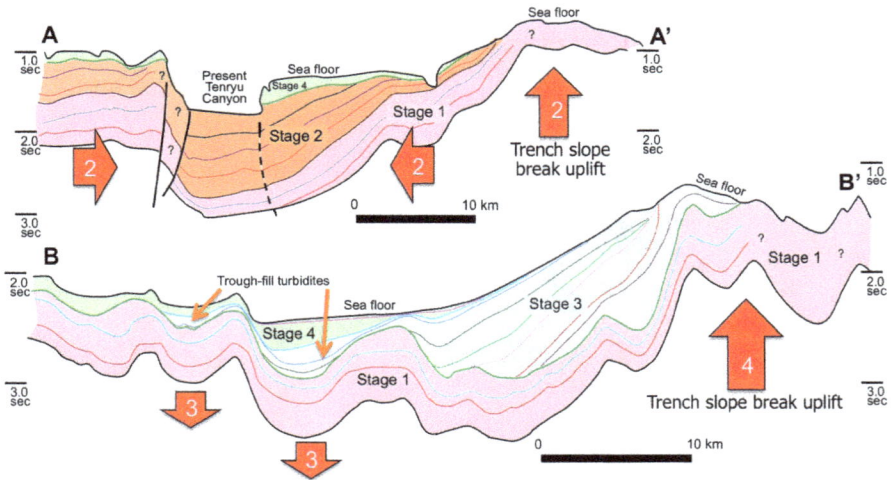

Figure 13. Cross sections based on the interpreted seismic sections transecting the Tokai-oki–Kumano-nada forearc basins, showing the basin deformation and background tectonics during Stages 2, 3 and 4. Traced lines on the cross sections denote sequence boundary (SB) horizons corresponding to those of Figure 14 in line colors. The section locations (seismic survey lines) are shown on the maps in Figure 12. Seismic sections were acquired in a MITI (Ministry of International Trade and Industry of Japan) survey [24]. Large red arrows denote compression and uplift during Stage 2, subsidence during Stage 3 and uplift during Stage 4.

Age	Sequence stratigraphy	Tectono-Sedimentary History	Stage	Submarine fan type
Pleistocene — Late (Ogasa Fm. equiv.)	Sequence Og-i	TSB uplift — Deformation — Compression	Stage 4	Channel Levee System
	Sequence Og-h			
	Sequence Og-g			
	Sequence Og-f	Cessation of compression / subsidence — Expanding deposition area	Stage 3	Trough-fill small fan / Sheet turbidites
Middle	Sequence Og-e			
	Sequence Og-d			
	Sequence Og-c			
	Sequence Og-b			
0.9 Ma	Sequence Og-a			
Early (Kakegawa Fm. equiv.)	Sequence Kg-g	TSB uplift / arcward suppression — Compression — Shrinking deposition area	Stage 2	Small radial fan type
	Sequence Kg-f			
	Sequence Kg-e			
	Sequence Kg-d	Folding		
? Pliocene?	Sequence Kg-c	Incipient slope basin?	Stage 1	Braided channel type
	Sequence Kg-b			
	Sequence Kg-a			

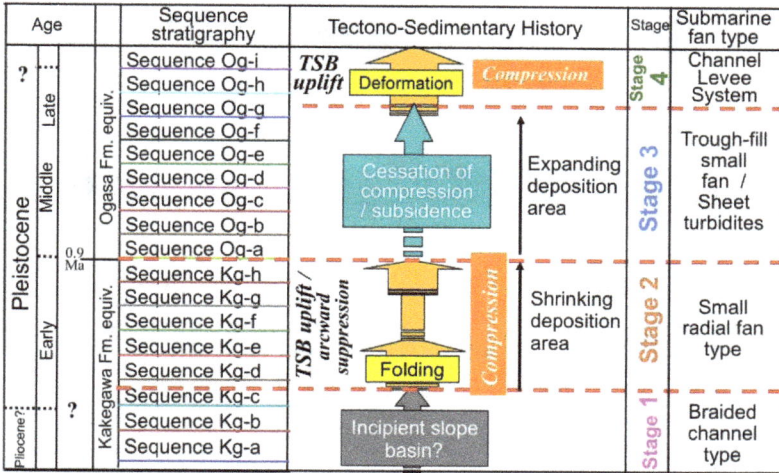

Figure 14. Generalized summary chart showing the transformation of the tectono-sedimentary conditions and submarine-fan types of the Pleistocene Tokai-oki–Kumano-nada forearc basin fill. Compiled and modified after [23, 26].

4. Discussion

4.1. Comparison with Dickinson's forearc basin classification scheme

Dickinson's simple classification scheme for forearc basin morphology [1] is based on the basin filling conditions and sectional basin configurations basically controlled by trench slope break height (Figure 15). Since the basin filling condition comprises two classes: underfilled and overfilled, and the sectional basin configuration comprises four classes: sloped, ridged/terraced, ridged/shelved and ridged/benched, depending on the trench slope break height, forearc basins can be classified into eight different types in the Dickinson's classification scheme [1] (Figure 15). According to our analysis results, the Eocene Ishikari–Sanriku-oki forearc basins can be categorized into the "emergent ridged", "overfilled shelved" to "benched" types (Figure 15), as it is interpreted that the trench slope break was uplifted and emerged, and the major sedimentary environments were mostly near the sea level except partly developed braided rivers in an elevated setting. On the contrary, the Pleistocene Tokai-oki–Kumano-nada forearc basins can be categorized into the "overfilled sloped", "underfilled submerged ridged" to "overfilled deep marine terraced" types (Figure 15), as the estimated trench slope break was submerged and low, and the major sedimentary environments were submarine fans and muddy slope to basin floor.

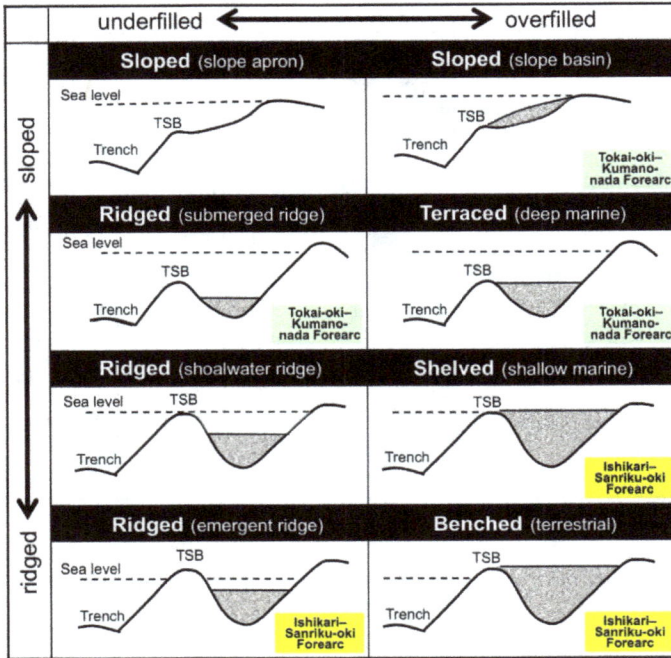

Figure 15. Dickinson's forearc basin classification chart on the basis of basin filling conditions and sectional basin configuration. Modified after [1]. TSB: trench slope break.

4.2. Controlling factors on the variation in forearc basin styles

This section attempts to discuss major controlling factors on the variation in forearc basin configurations and depositional systems on the basis of the results of the examinations above (Figures 16, 17).

4.2.1. Trench slope break development

Trench slope break is a topographic high bounding the forearc basin to a trench slope steeply dipping to the subduction zone (Figure 1). As the Dickinson's forearc basin classification places great importance [1] (Figure 15), the results of our examination also indicate that the development condition of a trench slope break is the most principal factor to control the forearc basin configurations and basin filling depositional systems. In case the trench slope break development is minor or moderate as seen in the Tokai-oki–Kumano-nada forearc basins, the trench slope break ridge is submerged, and the basin filling sediments tend to be deeper marine shales or turbidites. On the other hand, in case the trench slope break prominently develops as seen in the Ishikari–Sanriku-oki forearc basins, the trench slope break ridge is emerged, and the basin filling depositional systems tend to be fluvial to bay systems if sediment supply is enough. Dickinson [1] suggests that the trench slope break development is strongly related to differences in plate subduction conditions such as accretional prism formation and tectonic erosion.

In addition to the height of a trench slope break, related arcward suppression accompanied with the trench slope break uplift is also regarded as an important factor to control basin deformation as seen in the Tokai-oki–Kumano-nada forearc basin (Figure 13).

4.2.2. Balance between basin accommodation and sediment supply

Even in a fully uplifted trench slope break setting, a condition under minor sediment supply or relatively rapid subsidence causes a deeper marine forearc basin. The Ishikari–Sanriku-oki forearc basins maintained a balanced condition between the amount of sediment supply and the basin accommodation space, causing a thick accumulation of fluvial to bay sediments. Accordingly, it is suggested that the balance between sediment supply and forearc basin accommodation created by a trench slope break barrier and basin subsidence [28] (total subsidence) can be a crucial controlling factor not only on the forearc basin filling conditions such as underfilled and overfilled conditions but also on the variation of depositional systems. Dickinson [1] suggests that the underfilled types mostly occur along an island arc with a small amount of sediment supply, whereas the overfilled types mostly occur along a continental arc with a large amount of sediment supply.

4.2.3. Strike-slip movement related to basin segmentation

Our examination results suggest that a forearc zone is commonly segmented into subbasins. The Ishikari–Sanriku-oki forearc basins were segmented into 50 to 150 km long subbasins aligned along the forearc extension (Figure 2B). The Tokai-oki–Kumano-nada forearc basins are also possible to have been segmented as suggested by Sasaki et al. [29] and as seen in the facies maps (Figure 12), in which the sedimentary packages tend to be segmented into the Tokai-oki, Atsumi-oki and Kumano-nada possible subbasins. As described above, the segmented subbasins show a different subsidence pattern and sediment thickness for each subbasin (Figures 7, 13), and differential subsidence within a subbasin is characteristically observed (Figures 6, 7).

As a possible formation mechanism of forearc segmentation, Dickinson [1] suggests strike slip tectonics along a forearc zone, induced by oblique plate subduction underneath a forearc zone. As many of plate subduction direction at the convergent margin tend to be not complete normal direction to the subduction trench, oblique plate subduction is quite common. The oblique subduction may induce a strike-slip motion of forearc sliver and basin segmentation as seen in the Sumatra forearc and Aleutian forearc [1].

To examine the effect of strike-slip motion on the forearc basin segmentation, Kusumoto et al. [30] conducted dislocation modeling for basin segmentation, using the Sorachi and Yubari subbasins as examples. Dislocation modeling is to simulate basin dislocation by fault movement with the assumption of a homogeneous elastic body. Kusumoto et al. [30] picked up fault arrays around the subbasins, which indicate a strike-slip fault system consisting of main faults and splay faults, and set them in the model. When right-lateral motion occurred along the main faults, then the subbasins corresponding to the Sorachi and Yubari subbasins were properly simulated in the modeling [30]. This result suggests that the forearc basin segmentation was caused by strike-slip tectonics along the forearc zone.

Consequently, strike-slip tectonics is also one of the crucial factors to determine basin config-uration and depositional system distributions in a forearc zone (Figures 16, 17). Figure 17 demonstrates schematic diagram showing type variations of forearc basins as functions of trench slope break development, arcward compression and strike-slip movement. In addition to the Dickinson's forearc basin classification scheme (Figure 15), this study delineates that both arcward compression and strike-slip movement are crucial factors in forearc basin classification. In case arcward compression is intense due to trench slope break evolution, a confined shrinking or trough-fill type forearc basin can be formed, as seen in Stages 2 and 3 in the Tokai-oki–Kumano-nada forearc basins (Figures 12, 13, 14). In case strike-slip movement is dominant, a segmented marine or non-marine forearc basins can be formed, as seen in the Sorachi and Yubari subbasins (Figures 5, 10). When strike-slip movement is intense, the forearc basin can be transformed into a fragmented strike-slip basin.

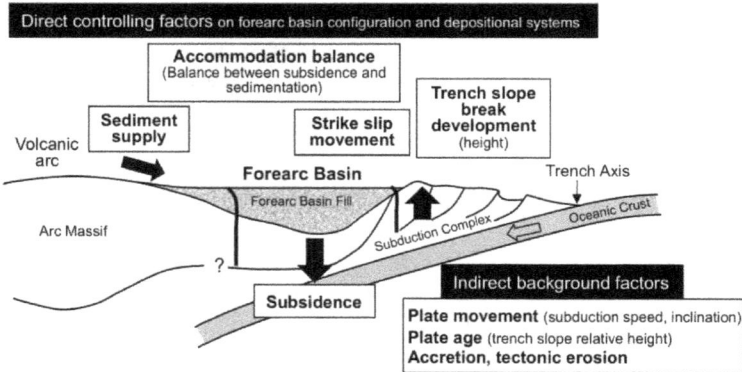

Figure 16. Controlling factors on variation in forearc basin configuration and depositional systems.

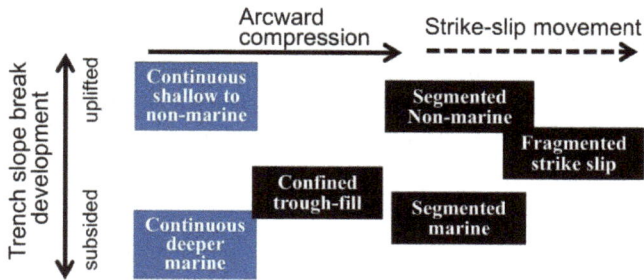

Figure 17. Schematic diagram showing type variations of forearc basins as functions of trench slope break development, arcward compression and strike-slip movement. Arrow direction denotes intensity of each factor.

5. Conclusions

To elucidate forearc basin variation and its controlling factors, the basin configurations and
basin-filling depositional systems were examined for actual examples from the Eocene
Ishikari–Sanriku-oki forearc basins and the Pleistocene Tokai-oki–Kumano-nada forearc
basins. As the results, the following points were revealed.

1. The Ishikari–Sanriku-oki forearc basins are filled with aggradational sediments consisting
    of bay to fluvial systems. Since the trench slope break is estimated to have uplifted and
    emerged to form a barrier to an open sea condition, the Ishikari–Sanriku-oki forearc basins
    can be categorized into the "emergent ridged", "overfilled shelved" to "benched" types
    of Dickinson's forearc basin classification [1]. Basin segmentation is commonly observed,
    and the subsidence pattern is different between subbasins.

2. The Tokai-oki–Kumano-nada forearc basins are filled with continuously changing
    submarine-fan systems. Since the trench slope break is estimated to have submerged, the
    Tokai-oki–Kumano-nada forearc basins can be categorized into the "overfilled sloped",
    "underfilled submerged ridged" to "overfilled deep marine terraced" types [1].

3. Our examination results suggest that the major controlling factors on the forearc basin
    configurations and depositional systems include a) the trench slope break condition such
    as development height and arcward suppression, b) the balance between basin accom-
    modation and sediment supply, c) and the strike-slip movement of forearc sliver, inducing
    forearc basin segmentation. Although the Dickinson's forearc basin classification [1] is
    effective, two factors of arcward compression and lateral-slip movement should be added
    for useful classification (Figure 17).

Acknowledgements

The authors are grateful to Drs. Ray Ingersoll, Cathy Busby and Paul Heller for useful suggestions on the tectonics and sedimentation of forearc basins. JAPEX, JX, JOGMEC, METI and MH21 Research Consortium kindly provided permission for data publication. This study was partly conducted under the MH21 Research Consortium.

Author details

Osamu Takano[1], Yasuto Itoh[2] and Shigekazu Kusumoto[3]

*Address all correspondence to: osamu.takano@japex.co.jp

1 JAPEX Research Center, Japan Petroleum Exploration, Japan

2 Graduate School of Science, Osaka Prefecture University, Japan

3 Graduate School of Science and Technology for Research, University of Toyama, Japan

References

[1] Dickinson WR. Forearc basins. In: Busby C, Ingersoll RV. (eds.) Tectonics of Sedimentary Basins. Oxford: Blackwell; 1995. p221-261.

[2] Dickinson WR, Seely DR. Structure and stratigraphy of forearc regions. American Association of Petroleum Geologists Bulletin 1979; 63: 2-31.

[3] Kiminami K, Kontani Y. Mesozoic arc–trench systems in Hokkaido, Japan. In: Hashimoto M, Uyeda S. (eds.) Accretion Tectonics in the Circum–Pacific Regions. Tokyo: Terra Publication; 1983. P107-122.

[4] Kimura G. Cretaceous episodic growth of the Japanese Islands. Island Arc 1997; 6: 52-68.

[5] Kiminami K. Cretaceous to Paleogene convergent margin. In: Niida K, Arita K, Kato M. (eds.) Regional Geology of Japan 1. Hokkaido. Tokyo: Asakura; 2010. p526-528. (in Japanese)

[6] Ando H. Stratigraphic correlation of Upper Cretaceous to Paleocene forearc basin sediments in Northeast Japan: cyclic sedimentation and basin evolution. Journal of Asian Earth Sciences 2003; 21: 919-933.

[7]   Ando H. Upper Cretaceous fore-arc basin sequences in Northeast Japan: large-scale controlling factors as eustasy, volcanism and relative plate motion. Journal of the Geological Society of Thailand 2004; 1: 35-44.

[8]   Ando H. Geologic setting and stratigraphic correlation of the Cretaceous to Paleocene Yezo forearc basin in Northeast Japan. Journal of Japanese Association for Petroleum Technology 2005; 70: 24-36. (in Japanese with English abstract)

[9]   Ando H, Tomosugi, T. Unconformity between the Upper Maastrichtian and Upper Paleocene in the Hakobuchi Formation, north Hokkaido, Japan: a major time gap within the Yezo forearc basin sediments. Cretaceous Research 2005; 26: 85-95.

[10] Osawa M, Nakanishi S, Tanahashi M, Oda H. Structure, tectonic evolution and gas exploration potential of offshore Sanriku and Hidaka provinces, Pacific Ocean, off northern Honshu and Hokkaido, Japan. Journal of Japanese Association for Petroleum Technology 2002; 67: 38-49. (in Japanese with English abstract)

[11] Takano O, Waseda A. Sequence stratigraphic architecture of a differentially subsiding bay to fluvial basin: the Eocene Ishikari Group in the Ishikari Coal Field, central Hokkaido, Japan. Sedimentary Geology 2003; 160 (1-3): 131-158.

[12] Kurita H, Hoyanagi K. Outline (Paleogene) In: Niida K, Arita K, Kato M. (eds.) Regional Geology of Japan, 1. Hokkaido. Tokyo: Asakura; 2010. P97-99. (in Japanese)

[13] Takano O, Waseda A, Nishita H, Ichinoseki T, Yokoi K. Fluvial to bay-estuarine system and depositional sequences of the Eocene Ishikari Group, central Hokkaido. Journal of the Sedimentological Society of Japan 1998; 47: 33-53. (in Japanese with English abstract)

[14] Matsui M. Paleogene. In: Editorial Committee of Hokkaido, Part 1 of Regional Geology of Japan. Tokyo: Kyoritsu Shuppan; 1990. p46-61.

[15] Japan National Oil Corporation. Report for the geological study of the MITI Sanriku-oki well. Fiscal Year 1998. Tokyo: Japan National Oil Corporation; 2000. 48p.

[16] Japan National Oil Corporation. Report on the geophysical survey "Shimokita–Kitakami". Fiscal Year 1973. Tokyo: Japan National Oil Corporation; 1974.

[17] Iijima A. On relationship between the provenances and the depositional basins of the Upper Cretaceous and Tertiary formations in Hokkaido, Japan. Journal of Faculty of Science, University of Tokyo Section 2 1959; 11: 339-385.

[18] Iijima A. The chromian distribution in the Paleogene deposits of Hokkaido and its bearing on ultrabasic rock belts, Japan. Japanese Journal of Geology and Geography 1964; 35: 17-42.

[19] IIjima A. Evolution of the Paleogene sedimentary basins in Hokkaido. Journal of Geography 1996; 105: 178-197.

[20] Maekawa H. Processing formation of the Kamuikotan metamorphic rocks from the Bibai and surrounding areas, central Hokkaido. Monograph of Association for Geological Collaboration of Japan 1986; 31: 107-117.

[21] Arai K, Okamura Y, Ikehara K, Ashi J, Soh W, Kinoshita M. Active faults and tectonics on the upper forearc slope off Hamamatsu City, central Japan. Journal of the Geological Society of Japan 2006: 112; 749–759. (in Japanese with English abstract)

[22] Takano O, Nishimura M, Fujii T, Saeki T. Sequence stratigraphic distribution analysis of methane-hydrate-bearing submarine-fan turbidite sandstones in the Eastern Nankai Trough area: relationship between turbidite facies distributions and BSR occurrence. Journal of Geography 2009; 118: 776-792. (in Japanese with English abstract)

[23] Takano O, Fujii T, Saeki T, Shimoda N, Noguchi S, Nishimura M, Takayama T, Tsuji T. Application of sedimentological methodology to the methane-hydrate exploration project in the eastern Nankai Trough area. Journal of Japanese Association of Petroleum Technology 2010; 75: 9-19. (in Japanese with English abstract)

[24] Japan National Oil Corporation. Report for the MITI offshore geophysical survey Tokai-oki–Kumano-nada. Fiscal Year 2001. Tokyo: Japan National Oil Corporation; 2003.

[25] Japan National Oil Corporation. Report for the geological study of the MITI Tokai-oki–Kumano-nada wells. Fiscal Year 2003. Tokyo: Japan National Oil Corporation; 2004.

[26] Noguchi S, Shimoda N, Takano O, Oikawa N, Inamori T, Saeki T, Fujii T. 3-D internal architecture of methane hydrate bearing turbidite channels in the eastern Nankai Trough, Japan. Marine and Petroleum Geology 2011; 28: 1817-1828.

[27] Shimamura K. Where are the terrigenous sediments going in the Japanese Island Arc? In: Tokuyama H, Shcheka SA, Isezaki N. (eds.) Geology and Geophysics of the Philippine Sea. Tokyo: Terra Scientific Publishing; 1995. p241-249.

[28] Xie X, Heller PL. Plate tectonic settings and basin subsidence history. Geological Society of America Bulletin 2009; 121: 55-64.

[29] Sasaki K, Okuda Y, Kato S. Petroleum geology of the Pacific side of Southwest Japan. In: Association of Natural Gas Mining and Association for Offshore Petroleum Exploration (eds.) Petroleum and Natural Gas Resources of Japan. Tokyo: Association of Natural Gas Mining and Association for Offshore Petroleum Exploration; 1992. P225-246.

[30] Kusumoto S, Itoh Y, Takano O, Tamaki M. Numerical modeling of sedimentary basin formation at the termination of lateral faults in a tectonic region where fault propagation has occurred. In: Itoh Y. (ed.) Mechanism of Sedimentary Basin Formation –Multidisciplinary Approach on Active Plate Margins. Rejeka: InTech; 2013. P…..

Basin Formation in Peripherals of Continental Plate: Orogeny, Rifting & Collision

Early Continental Rift Basin Stratigraphy, Depositional Facies and Tectonics in Volcaniclastic System: Examples from the Miocene Successions Along the Japan Sea and in the East African Rift Valley (Kenya)

Tetsuya Sakai, Mototaka Saneyoshi,
Yoshihiro Sawada, Masato Nakatsukasa,
Yutaka Kunimtatsu and Emma Mbua

Additional information is available at the end of the chapter

1. Introduction

There are numerous stratigraphic studies regarding rift valley fill successions. The major understanding of the rift basins' filling process was obtained from the Basin and Range regions and Rio Grande Rift, USA (for example, [1-4]); East African Rift Valley (for example, [5-7]); Suez Rift, Egypt (for example, [8-10]); Corinth Basin, Greece (for example, [11-13]) and so on. The initial stage of the rift basin evolution is characterized by the development of a series of small half-grabens. Basins become larger through the linkage of border faults of individual half-grabens [14]. Even after basin mergers, topographic lows of footwall among basins (accommodation zones before basin mergers) play an important role for sediment supply. The relay ramp developed between two normal faults dipping in the same direction (Figure 1), and its evolution, is crucial because it acts as the major entry point of the water and sediments to the basins [14-15]. The manner of sediment entry to the basins and the subsidence pattern strongly affect the architecture of basin-fills (for example, [16-17]), resulting in the formation of different systems tracts in different places within a basin at a given time [18]. In case of continental rift basins with lakes, the strata formation is much more complicated than in marine basins (for example [19-20]). The differences in sedimentation process between lake and marine basins are summarized in [21], suggesting that the terrestrial basin fills are not miniature marine basins because there were different amplitude base-level variations, linkage of climate and sediment supply and so on. Pre-rift basement structures also affect the evolution of the basin as well as fills of the early rift basins (for example, [13, 22-23]).

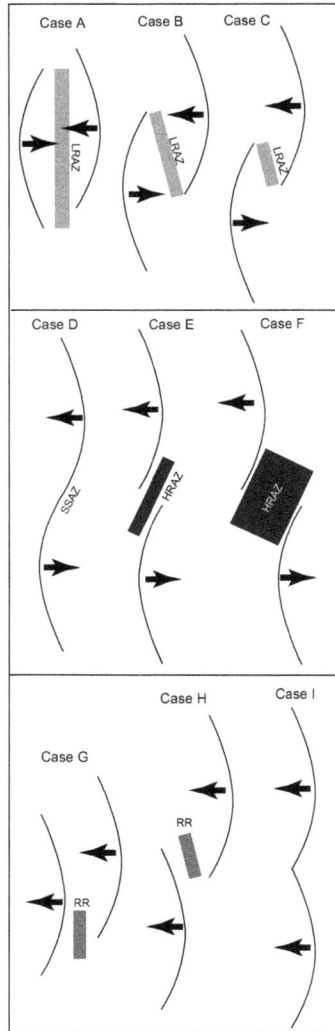

Figure 1. Various types of linked half-grabens and characteristics of accommodation zones modified from Figure 6a-c in Rosendhal (1987). Arrows indicate the direction of normal fault displacement. LRAZ: Low relief accommodation zone, HRAZ: High relief accommodation zone, SSAZ: Strike-slip accommodation zone, RR: Relay ramp.

However, studies of the rift basin fills with active volcanism have been limited and their basin-fill processes are poorly understood (see [24]). Pyroclastic fall may supply sediments from the air nearly evenly within a basin if the basin size is small relative to the pyroclastic fall area. The reworked volcaniclastics (mainly ash) supplied via rivers can be more widely spread in the lake than is the case of the siliciclastic system. This is because of smaller grain density (for example [25]), resulting in faster sedimentation even in parts of a basin starved of sediments transported by streams. Such faster sedimentation may provide opportunities to decode the high-resolution tectonic and basin-fill history through the reconstruction of environmental changes. Some examples of the basin-fill successions affected strongly by sediment supply via pyroclastic fall are, therefore, shown to discuss the evolution of the early rift basin fills. Early rift basins are expected to experience a complicated history in association with merging small basins when border fault tips propagate laterally to the next basin [14] or when one basin is filled out and sediments and water spill over to the next basins beyond the accommodation zone [16]. Examples of studies discussing such events are also limited to a small number [4]. The basins filled rapidly with pyroclastic fall are suitable for detecting such basin-merging events as well as another type of tectonic events such as subsidence.

The Miocene successions exposed along the Japan Sea on the Japanese main island contain the early rift basin fills, which were formed when the Japan Sea was opened; they are now exposed on the land because of tectonic inversion (for example, [26]). One of the basin fills was targeted in this study—the Miocene Koura Formation, exposed in SW Japan. Other targets here were the Miocene half-graben fills in Kenya (Namurungule Formation in Samburu Hills and Nakali Formation in Nakali, northern and central Kenya, respectively). The basin fills adjacent to the volcanoes must be strongly affected by the supply of volcaniclastics and lava flow, as well as subsidence/uplift related to volcanism (for example [24, 27]). However, the local volcano-related tectonics (such as caldera formation) are excluded in this study for simple discussions. Because the centre of the eruption or intrusion of magma during the deposition of the Na-murungle and Nakali Formations has not been discovered around the target basins, it is considered that the tectonic subsidence or uplift induced by local volcanisms (see [24, 28]) can be ignored for these cases. The Koura Formation example is unclear for the strong tectonic control from local volcanoes, but its effect can be ignored as well, because lava has not been found and only pyroclastic fall or flow deposits have been described from the formation.

2. Case studies of the early rift basin fills

2.1. Koura formation

The Miocene successions associated with the Japan Sea opening widely spread along the coastal region of the Japan Sea. The Miocene Koura Formation is exposed in the Shimane Peninsula, on the western part of the main island of Japan (Honshu Island) (Figure 2). The Koura Formation distribution is elongated E–W—which is almost parallel to faults in and around the western Japan Sea (ENE–WSW)—and dips mainly to the north, allowing observation of axial facies changes in the basin (Figure 2). The basin fill thickness exceeds 600 m [29-30].

The basement rock of the basin has not been confirmed yet, but granitic or metamorphic rocks are inferred to be the basement on the basis of the gravels contained in the formation [31].

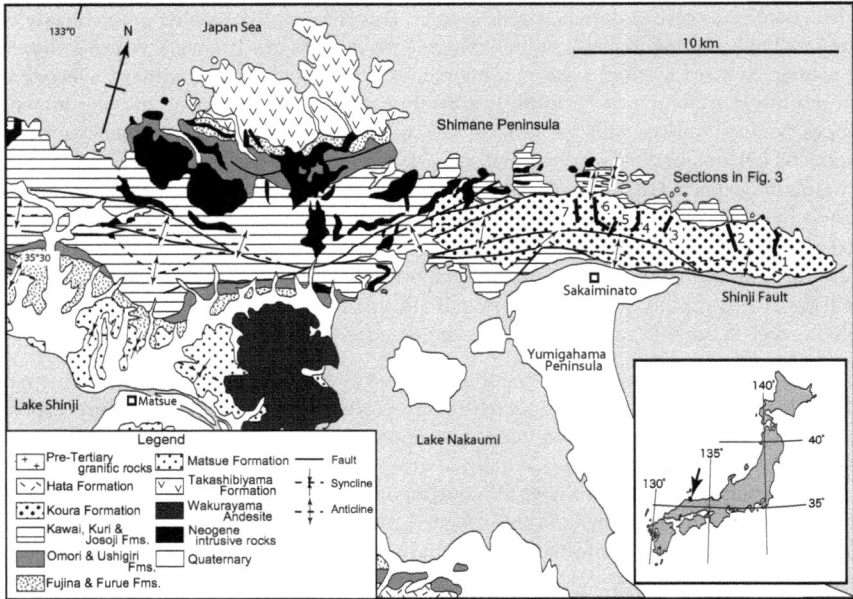

Figure 2. Location and geologic map of eastern Shimane. 1–7 in Figure 4 show the sections.

It is difficult to discuss the basin morphology at the time of deposition because of the limited extent of exposure. However, the seismic cross-sections of this area show the presence of a series of half-grabens under the bottom of the Japan Sea [32], suggesting that the Koura Basin fills a half-graben. Although the border fault of this basin has not been confirmed either, one of the major faults of this region, the Shinji Fault (or Kashima Fault), running just south of the distribution area of the Koura Formation (Figure 2) and acting presently as a right-lateral strike-slip fault [33], is most probably the border fault of the basin.

The Koura Formation consists of three members (Sakai et al., 2013). For simplicity, these three members are referred to as "lower", "middle" and "upper" formations. The lower formation consists of conglomeratic sandstone beds (alluvial fan origin [30]) and the overlying alternation of the sandstone and mudstone beds (meandering and braided streams, floodplain, marsh and shallow lake origin [34]) (Figures 3 and 4). The middle formation consists of andesitic volcaniclastics deposited in a shallow (probably fresh) lake and floodplain (Figures. 3 and 4). The sediments in this interval are predominated by those from pyroclastic fall and small-scale gravity flows (Kano, 1991; Sakai et al., 2013). The upper formation is characterized by alternations of tuffaceous sandstone and mudstone beds (fan delta deposit; Figures 3 and 4) and conglomerate interbeds filling small sublacustrine channels developed on the fan delta slope — which were deposited in a blackish lake, as suggested by the presence of *Ostrea* and *Corbicula* fossils and burrows by *Teredo* sp. as well as geochemical data [29, 34, 36-37]. The upper Koura Formation contains hummocky cross-stratified (HCS) sandstone beds, implying that the lake size became wide enough for generating large waves. The sediment supply to the basin was mainly from the east during the deposition of the upper formation [30, 35]. Three thick lapilli tuff beds (T3–T5) are strong tools for correlation within the basin (Figure 3). The chronology of the Koura Formation has been insufficient. Ages of 16–24 Ma were obtained from the upper formation through fission-track dating [38], and the age of the base of the overlying Josoji Formation (i.e. the top of the Koura Formation) was estimated to be 20–18 Ma or younger [39].

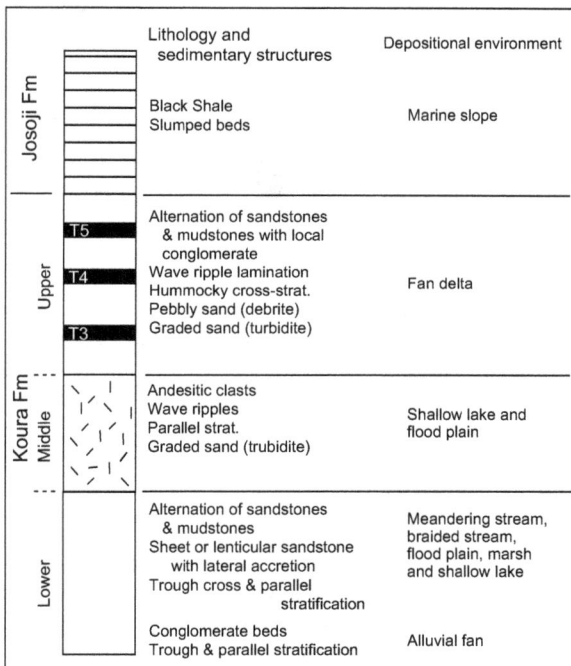

Figure 3. Lithostratigraphy of the Koura and lowest Josoji Formations. T3–T5 indicate lapilli tuff beds in the upper Koura Formation.

The boundaries of the lower-middle and middle-upper formations are marked by a surface that is then overlain by an up to 10 m sediment interval consisting of cross-stratified sandstone or conglomerate beds (Figure 5). It is interpreted that each cross-stratified interval was deposited from a basin-wide flood-flow incoming from another basin, and subsequent lake-level rise occurred when this and the other basin were merged [34], on the basis of the following

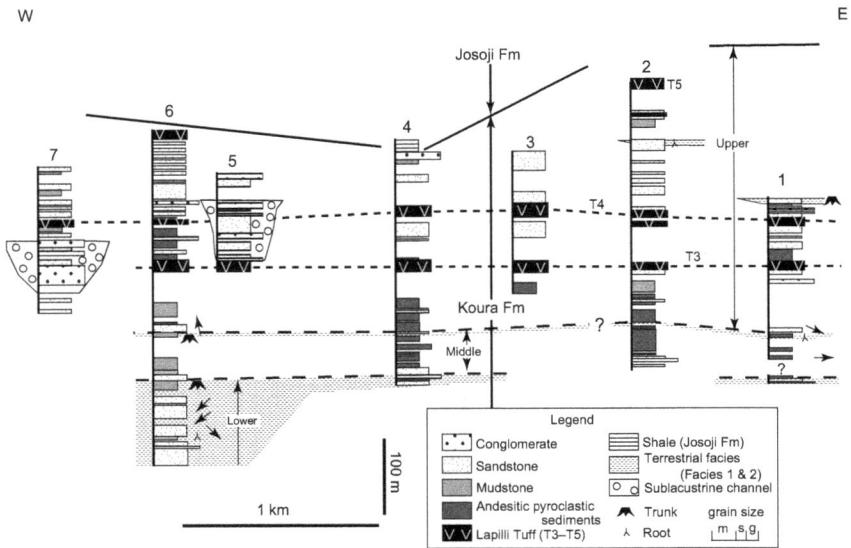

Figure 4. Columnar cross-sections of the Koura Formation (modified from [34]). m: mud, s: sand, g: gravel.

reasons: (1) absence of a major erosion surface within both cross-stratified intervals and homogeneous lithology imply their deposition within a short period; (2) both the cross-stratified intervals cover terrestrial deposits with tree trunks, and change upward into the alternations of sandstone and mudstone bed and the andesitic volcaniclastic beds containing both pyroclastic fall, turbidite and beds with small-scale slump structures (lake deposit) (Figure 5). The rapid lake-level rise suggests the merger with a basin having a base level higher than that of the Koura Basin. The second event may record the merger of this basin with a marine one to become a blackish lake basin.

The top of the Koura Formation is marked by a major flooding surface below the black marine shale of the Josoji Formation. The Josoji Formation is interpreted to consist of sediments of the climax phase of the Japan Sea opening. In the context of sequence stratigraphy, the lower and middle formations are interpreted to be the lowstand systems tract (LST). The base of the upper formation is interpreted to be the first flooding surface and the upper formation is interpreted to be the transgressive systems tract (TST) together with a part of the overlying Josoji Formation.

The upper Koura Formation hosts sediment cycles (Figure 6). The thickness of each cycle ranges from 5 to 20 m. Some of the cycles are bounded by flooding surfaces (see [40]) (Figure 6A). Such cycles mainly consist of sediments with an upward-shallowing trend. The basal flooding surface is covered with a massive mudstone bed (outer shelf equivalent deposit of the fan delta) or an alternation of HCS sandstone and mudstone beds of inner shelf equivalent

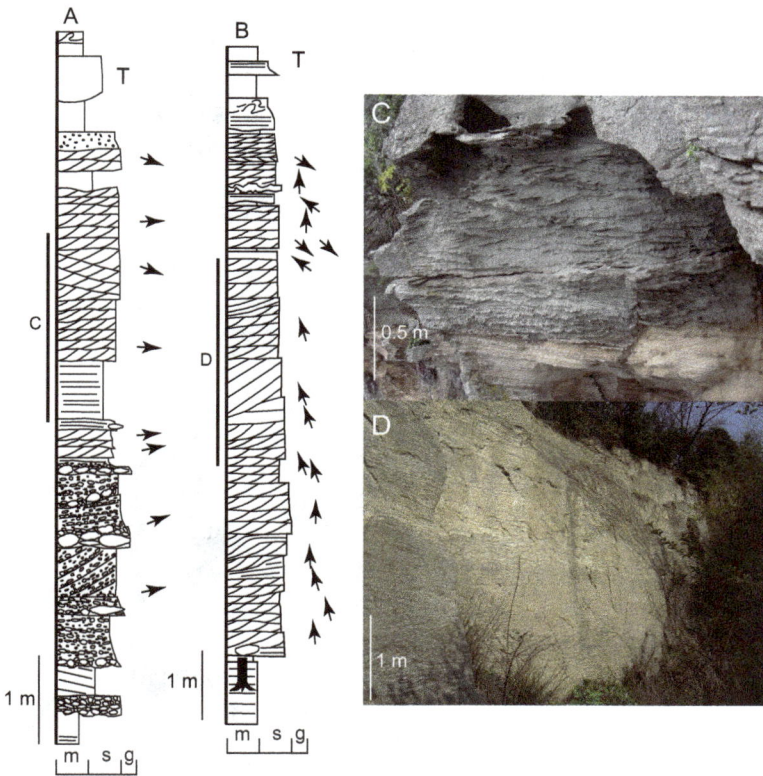

Figure 5. Columnar cross-sections of the event beds. Each event bed is interpreted as having been deposited from an outburst flood associated with a basin merger, followed by a lake-level rise (modified from [34]). A: event beds at the base of the middle Koura Formation. Arrows indicate the palaeoflow direction (up = north). B: event beds at the base of the upper Koura Formation. Both horizons grade upwards into the lake deposits containing turbdite (T) and slumped deposit. See Figure 7 for legend. C: outcrop photograph of a part of the event beds (cross-stratified beds) at the base of the middle formation. D: photograph showing the wide view of the event beds at the base of the upper Koura Formation.

deposits of the fan delta [34]. On the other hand, most of the cycle bases are represented by a surface covering a slumped deposit (Figures 6B–D). Each surface is undulating (i.e. erosional) (Figures 6B and 6D), and is then covered with facies beds shallower than those below the surface (Figure 6). Parts of the sediments just above the surface are also dragged into the slumped deposits in some places (Figure 6C), indicating that sediment accumulation above the surface occurred almost simultaneously with slumping. The sediment overlying the surface is then punctuated by the flooding surface (Figure 6), covered by a shallowing-upward succession. The cycle boundaries are, therefore, interpreted to have been formed by relative uplift of this area at the time of deposition.

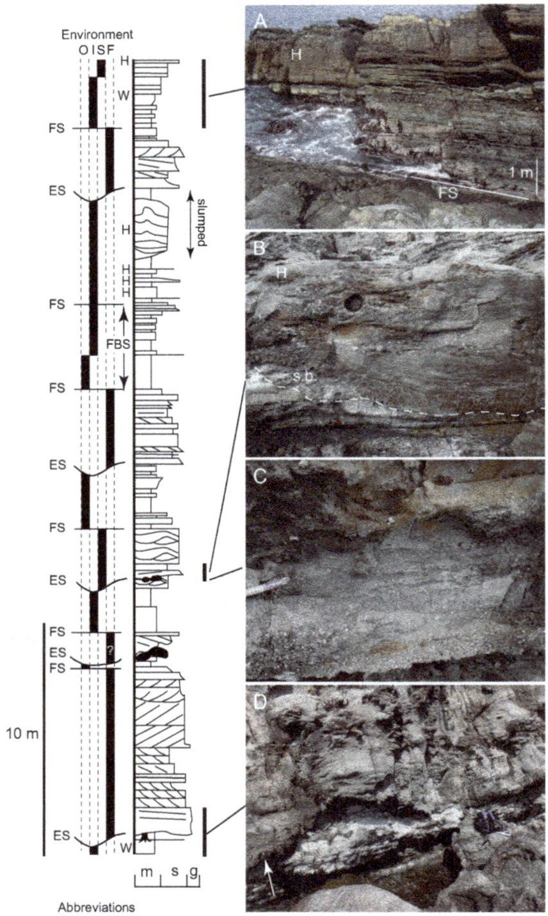

Abbreviations

ES: erosion surface, FS: flooding surface, FBS: flooding-surface bounded cycle,

H: hummocky cross-straification, W: wave ripple lamination,

O: fan delta slope environment equivalent to outer shelf,

I: fan delta slope environment equivalent to inner shelf, S: shoreface, F: fluvial.

Figure 6. Columnar cross-section of a part of the upper Koura Formation and changes in depositional environment. A: a sediment cycle showing an upward-shallowing trend. The cycle base is represented by a flooding surface. B: an erosion surface truncating the inner shelf equivalent deposit, and is then covered with HCS sandstone beds of the shoreface origin. s.b.: slump block. Note the hammer for the scale. C: a close-up photograph of the sediment just below the erosion surface (cycle boundary). There are several slip surfaces of the slump in the sediments. Coarse-sediment grains, which can be found only above the surface, are also incorporated (probably dragged) into the slumped horizon. The scale is 0.2 m long. D: a basal erosion surface of the cycle, truncating the inner-shelf-equivalent deposit and being covered with fluvial channel deposit. The white arrow indicates the surface. See Figure 7 for the legend of the columnar section.

Such sediment cycles were not identified in the lower and middle formations. The detailed outcrop observations in the lower formation revealed that either the top or the base of the sandstone intervals (fluvial channel facies) is marked by a surface associated with minor sliding (Figure 7); the former case is the most common (Figure 7). The surface is then covered with a thin poorly sorted silty sandstone bed (up to 0.1 m thick)(Figure 7A). Some of the silty sandstone beds contain pebble-sized sandstone or mudstone clasts of the underlying beds (Figure 7). In some places, the very small syndepositional faults extending almost parallel to the bedding plane are recognized below the silty sandstone beds. The silty sandstone beds are then covered with a massive or a laminated mudstone bed of a small and shallow lake origin (Figure 7B), showing a lake-level rise immediately after the sliding event. The slide may have been associated with subsidence of the basin.

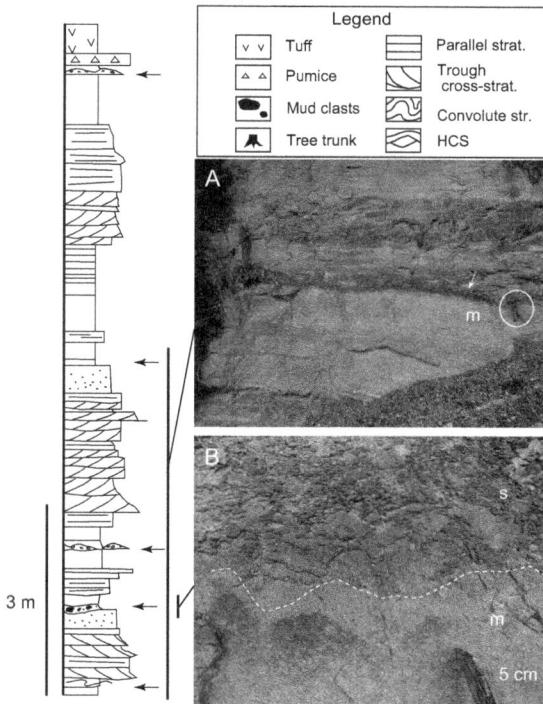

Figure 7. Example of the columnar cross-section of the lower Koura Formation taken along section 6. The arrows indicate the horizons showing evidence of a small-scale slide. A: outcrop view of the fluvial deposit. The arrow indicates the horizon of Figure 7B. The scale (hammer) is 0.3 m long. B: close-up photograph of the top of the fluvial channel fill sandstone beds. The dotted white line indicates the surface of the slide, which is then overlain by a silty sandstone bed with abundant sand clasts originating from the underlying sandstone bed. The scale (a part of the hammer head) is 0.05 m long. m: massive sandstone bed, s: shallow lake deposit, HCS: hummocky cross-stratification.

2.2. Examples from the East Africa rift valley

In the Kenya Rift, the Miocene rift basin-fill successions are exposed (Figure 8). The activity of the rift system started in the Oligocene and attained its maximum in the middle to late Miocene [41]. We targeted the half-graben fills exposed in the Samburu Hills, northern Kenya [42-44], and Nakali, central Kenya [45]. The target sediment successions of both areas (Namurungule and Nakali Formations) have not been classified into members based on the international stratigraphic nomenclature, although each formation can be divided into three units. Therefore the terms, the lower, middle and upper formations, are used for three units of each formation in the present study.

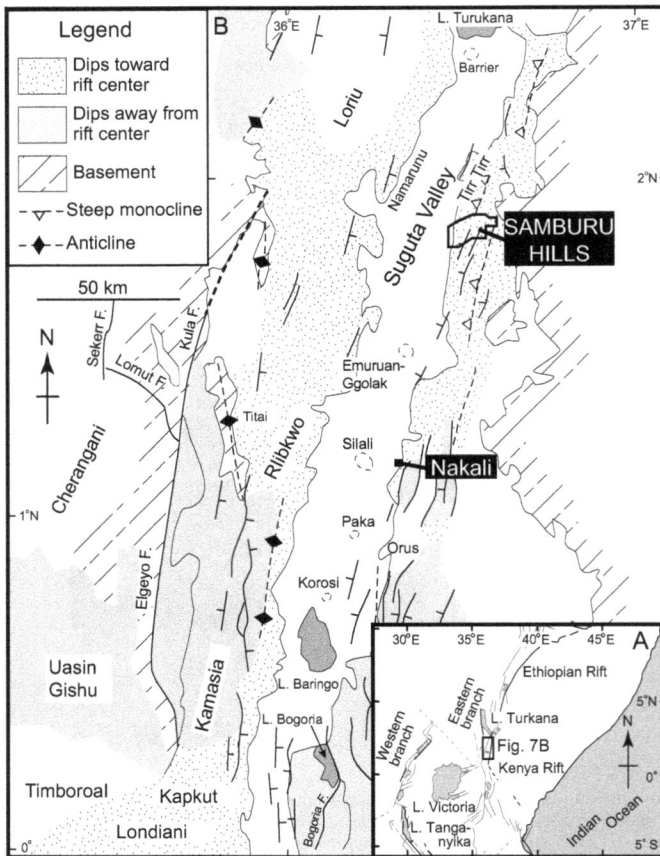

Figure 8. Location map of Samburu Hills and Nakali in central and northern Kenya.

a.   Samburu Hills

Samburu Hills are located in the eastern shoulder of the eastern branch of the East African Rift Valley system, northern Kenya (Figure 8). The Nachola, Aka Aiteputh, Namurungule and Kongia Formations (ca. 20–5.3 Ma [43]) make up the Miocene succession, which covers the Precambrian Mozambique Belt rocks (gneiss and granitic rocks)(Figure. 9). The upper Aka Aiteputh to the Namurungule Formations' phase (ca. 10–9.3 Ma) was one of the major rifting periods in this area, as suggested by the development of a series of small half-grabens, which is indicated in the geologic map as the scattered distribution of the Namurungule Formation [42] (Figure 9). Each formation body has a lenticular plan view and one or both sides of the body are punctuated by faults (Figure 9).

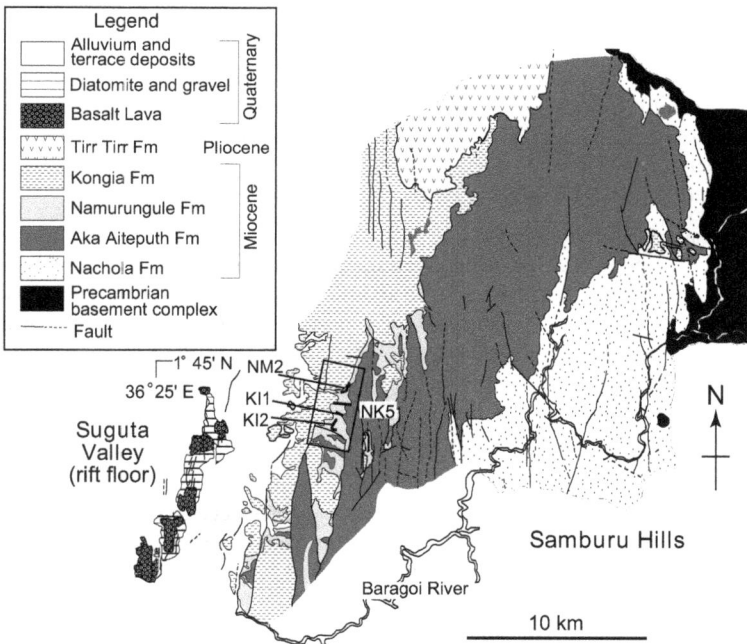

Figure 9. Geologic map of the Miocene in Samburu Hills. The enclosed part is the studied area. KI1, KI2, NM2 and NK5 are locations of columnar cross-sections in Figure 10.

The target basin has a lenticular shape extending N–S (Figure 9). Although the western margin of the basin is truncated by the overlying Kongia Formation— which is interpreted as having been deposited during the rejuvenated phase of the rift after 7 Ma (see [43])—the border fault of the basin runs in the western margin, as suggested by the Namurungule sediments thickening to the west [42]. There is a gap in fault location in the northern and southern halves of this basin. In the earliest phase of basin evolution, there may have been an accommodation zone in the boundary between the northern and southern halves of this basin.

The northern half of the basin, where spectacularly well exposures allow sediment correlation among outcrops, was targeted in the present study. A half-graben fill consists of the upper Aka Aiteputh Formation, which is characterized by red soil beds with abundant calcrete layers and basalt lavas with basalt conglomerate layers [44]. The overlying Namurungule Formation consists of four parts: the basal conglomerate beds of alluvial fan origin, the alternations of tuffaceous mudstone and sandstone beds (mudstone-dominated) of the lower part (both parts form the lower formation) and an about 20-m-thick lahar deposit of the middle formation (Figure 10). The upper formation is represented by a pile of sediment cycles, each of which consists of a sandstone-dominated and an overlying mudstone-dominated interval, as mentioned below. The age of the Namurungule Formation ranges from 9.6 to 9.3 Ma [43], and the rapid sedimentation rate was estimated to be 1.52 m/ky for the lower formation and 0.24 m/ky for the upper formation [42].

From the viewpoint of sequence stratigraphy, the red soil beds and basalt lava interval of the upper Aka Aiteputh Formation and alluvial fan interval of the basal Namurungule Formation are interpreted as the LST; most of the lower Namurungule Formation, except for its basal and uppermost part, is the TST showing retrogradational succession. The remaining part is the highstand systems tract (HST) (Figure 10: see also [44]).

The TST is characterized by a rapid lateral facies change from the thick lake facies in the southern part to the terrestrial facies represented by the alternations of root-bearing mudstone and sandstone beds in the northern part. The up to 20-m-thick terrestrial sediments in the TST contain a few stream deposits represented by an up to 0.5 m of sheet sandstone beds with parallel and trough cross-stratification. Other sandstone beds in the terrestrial deposits are associated with temporary lake expansions, as suggested by the wave-generated sedimentary structures (wave ripple lamination and small hummocky cross-stratification) in sandstone beds [42] (Figure 11). There are local slide deposits, represented by pebble- and cobble-sized mudstone breccia in the succession (Figure 11).

On the other hand, the HST is represented by a pile of sediment cycles [42]. Each cycle consists, from the base to the top, of the conglomeratic sandstone beds of fluvial channel fill origin, root-bearing mudstone beds of floodplain origin, laminated mudstone beds of lake origin and tabular cross-stratified sandstone beds of delta front origin (Figure 12). The cycle boundaries are sharp and undulating, and truncate the underlying sediments (delta front and lake deposit) (Figure 12).

Individual cycles tend to thicken to the west, which are interpreted to be owing to tectonic subsidence in the western part of the basin. The basal surface truncates the underlying sediments in each cycle, indicating a lake-level fall probably because of the migration of lake water to the basin centre when the basin subsidence occurred (for example, [2]).

Petrographical analysis indicates that the sediments (feldspar and rock fragments, mainly basalt grains) were supplied only from the adjacent area during the deposition of the upper Aka Aiteputh and lower Namurungule Formations (Figure 10), showing the poor develop-

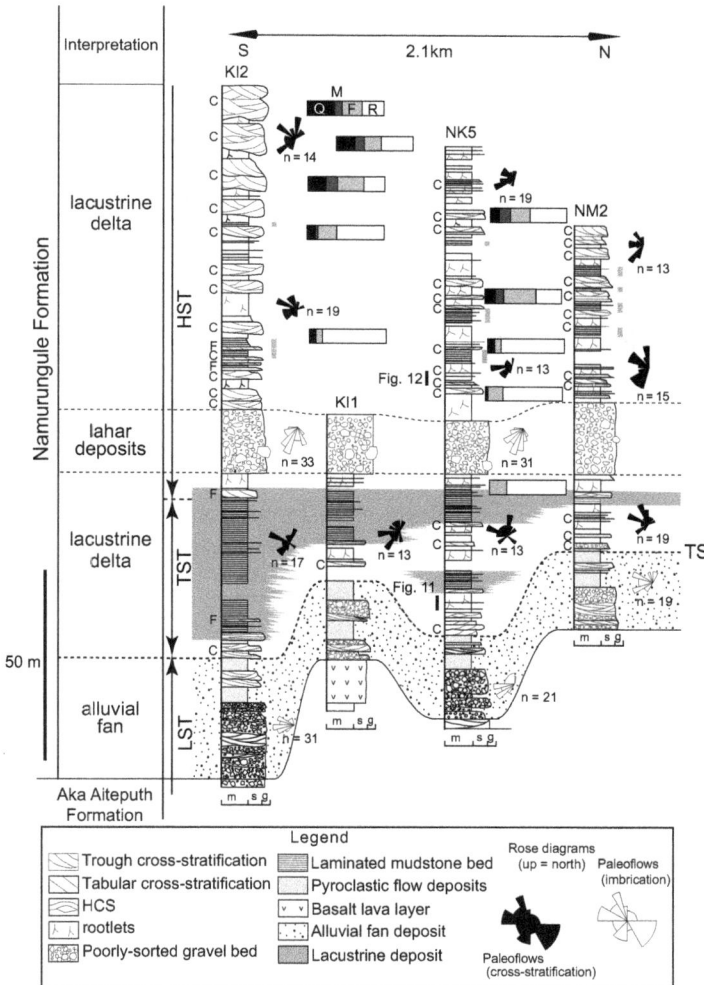

Figure 10. Columnar cross-sections of the Namurungule Formation (modified from [44]). Graphs indicate the petrographical analysis results. Grey bars indicate the horizons of lake deposits. C: fluvial channel fill, F: delta front deposit. Q: quartz, M: microcline, F: alkali-feldspar, R: rock fragment.

ment of the drainage system. On the other hand, the sediment grains in the upper formation supplied from the basement (Mozambique Belt), such as quartz and microcline, suggest that the drainage basin became wider through time [46]. This trend, later appearance of the grains originating from basement rocks, has been reported from other rift basins (for example, [46]).

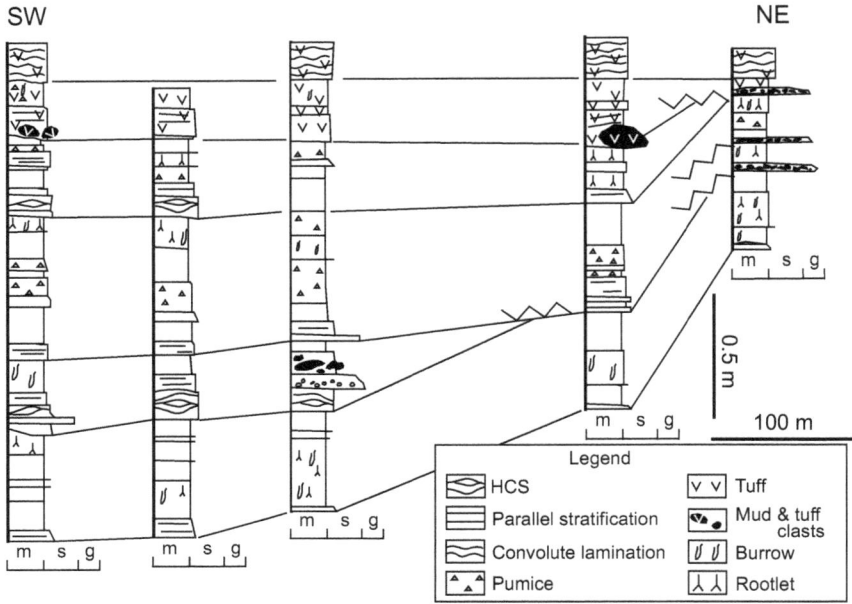

Figure 11. Example of the columnar cross-sections from the lower Namurungule Formation (modified from [42]). The zigzag line indicates an erosion surface that can be seen only in the northern part of the measured area.

b. Miocene Nakali Formation

Nakali is located about 50 km south of the Samburu Hills (Figure 8). The Miocene Nakali and Nasorut Formations are distributed in this area [45] (Figure 13). The lower part of the lower Nakali Formation is characterized by the alternations of tuffaceous sandstone and mudstone beds, which are interpreted to be turbidite and slumped deposits (delta front deposit), and the overlying thick lapilli tuff beds that bury the lake (Figure 14). The fluvial channel fill, floodplain and shallow lake deposits (conglomerate, tuffaceous sandstone and mudstone beds) characterize the upper part of the lower formation. The middle formation is represented by thick pyroclastic flow deposits (ca. 40 m). The lower part of the upper formation shows sediment characteristics similar to the upper part of the lower formation. The upper part of the upper formation is represented by tuffaceous mudstone beds and conglomerate and sandstone interbeds with slump structures. The slumped deposits in this interval indicate that this is of the lake slope origin (Figure 14). One of the important hominoid fossils, *Nakalipithecus* [45], was discovered near the top of this interval. In terms of sequence stratigraphy, most of the formation forms the LST—except for the upper part of the upper formation, where deeper lake mudstone facies predominates. This part is interpreted as the TST. The Nakali Formation is then overlain by trachitic or basaltic lava and volcaniclastics of the Nasorut Formation (Figure 14). The magnetostratigraphic reversed polarity zone in the middle portion of the Nakali

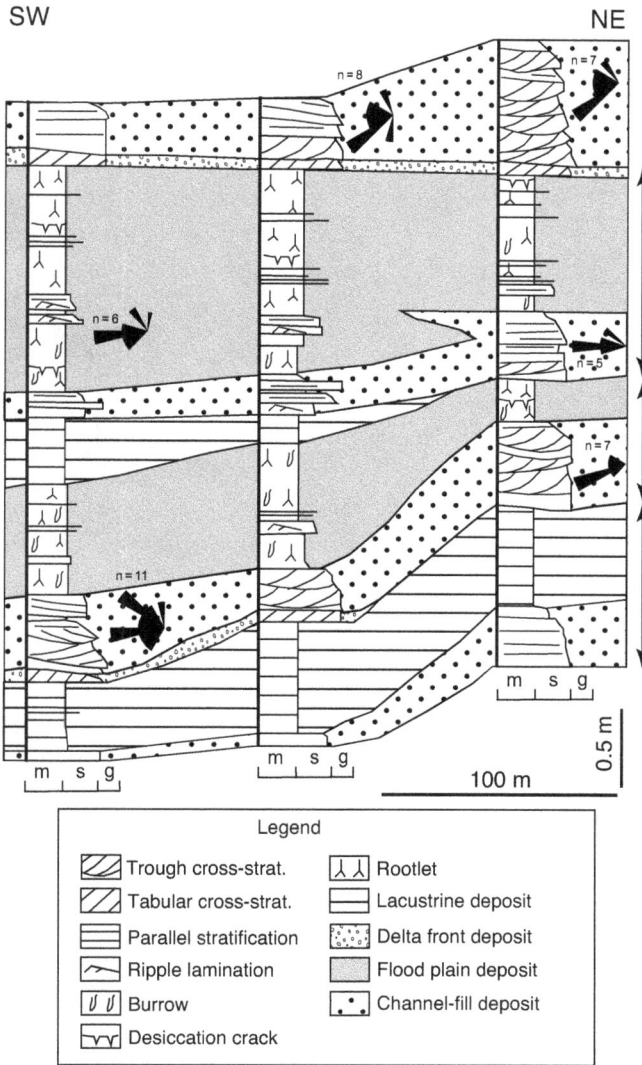

Figure 12. Example of the columnar cross-sections from the upper Namurungule Formation (modified from [42]). The arrowed intervals show individual cycles. The rose diagrams indicate palaeoflow directions shown by cross-stratification (up = north).

Formation was correlated to the Chron C5n.1n (9.88–9.74 Ma [46]) on the basis of the Ar–Ar ages [45].

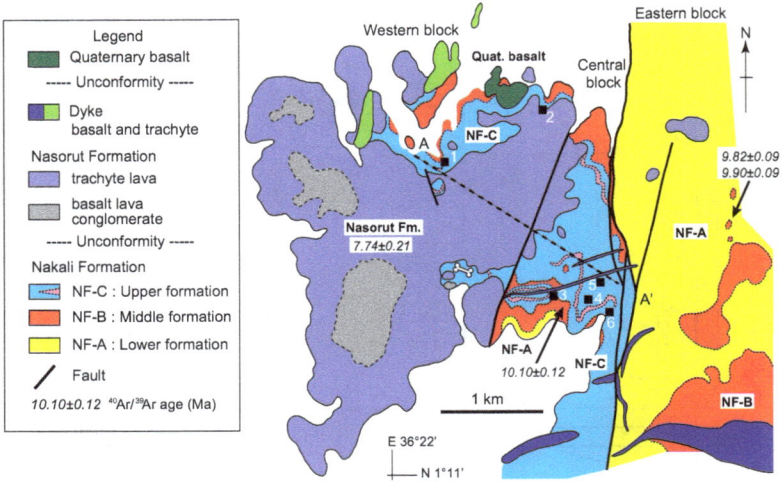

Figure 13. Geologic map of the Nakali Formation (modified from [45]). A-A' is the trend of the columnar cross-sections obtained from locations 1–6 in Figure 15.

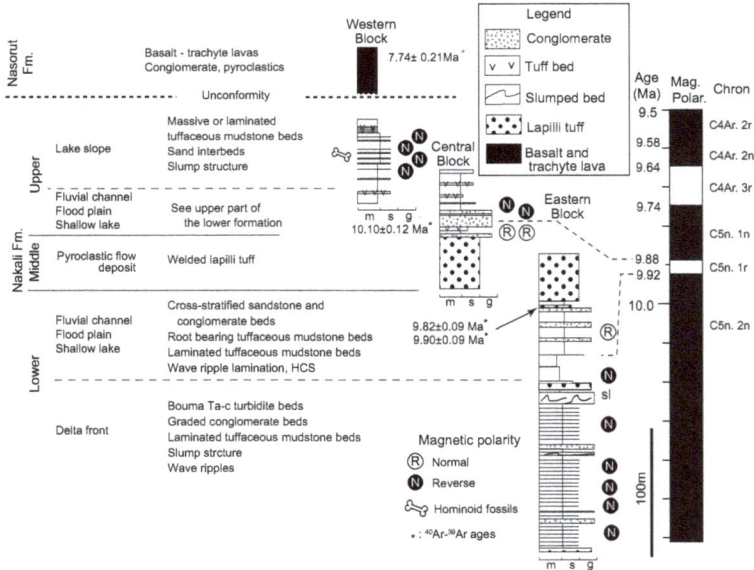

Figure 14. The generalized litho- and chronostratigraphy of the Nakali and Nasorut Formations (modified from [45]) and facies description and interpreted depositional environments.

Two normal faults extending N–S separate the rocks of this formation into three blocks (eastern, central and western) (Figure 13). The displacement of the eastern fault is larger than those of the others and is estimated to be about 200 m on the basis of the altitude gap of the middle formation between the eastern and central blocks.

c.   Subsidence history of the upper Nakali formation

The good exposures of the lower part of the upper formation and the frequently interbedded tuff beds allow observation of lateral facies changes in the field within and among blocks (Figure 15). Six tuff beds were identified in this horizon, and are named Twin (two white tuff beds), Exo (white tuff bed containing abundant trachyte fragments), Pum (pumice tuff), Fu (poorly sorted pumice tuff bed), Mfu (poorly sorted pumice and accretionary lapilli tuff bed) and Ma (white tuff bed containing accretionary lapillis) (Figure 15). The thick cemented beds with weakly weathered soil beds (termed 'terrace forming bed' in Figure 15, because this bed forms a wide terrace in this place); the White beds, represented by the sandstone and conglomerate beds rich in small breccia of white tuff, and the Red beds, characterized by red conglomerate and sandstone beds of fluvial channel fill and floodplain origin (Red beds), can be used for the correlation as well. The bases of the lake deposits, flooding surfaces, were also used as one of key horizons (Figure 15).

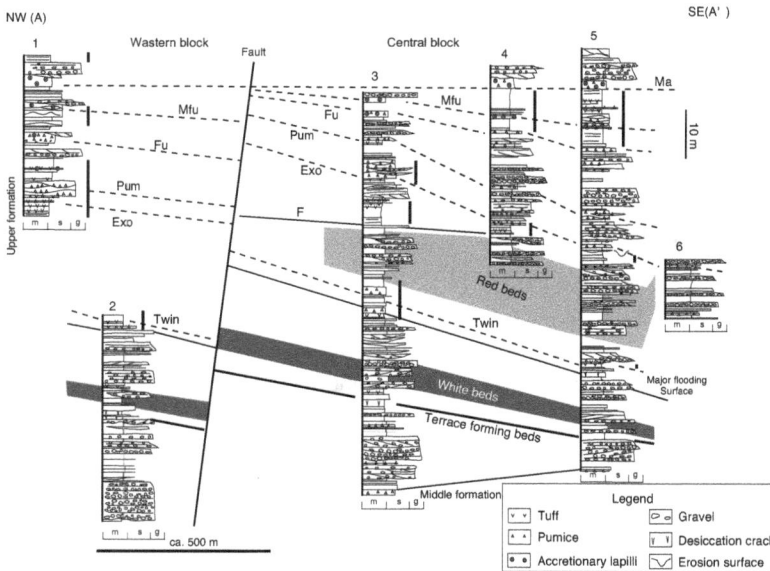

Figure 15. Columnar cross-sections of the lower part of the upper Nakali Formation. The black bar indicates the horizon of lake deposit. The dotted and solid lines indicate correlated tuff beds and flooding surface, respectively. A bold line indicates "the terrace forming bed". See text for tuff names. F: local flooding surface.

The correlation results (Figure 15) show thicker sediments in the western part of the central block below the Twin Tuff bed. The thickness of sediments between the base of the upper formation and Twin Tuff bed consistently increases from section 5 to section 2, even though section 2 is located in the west of the fault separating the western and central blocks. This may show that the fault was inactive before the Twin Tuff deposition, and another fault, which is not indicated on the geologic map and is running west of the study area, was active instead.

Thickness of the sediments between the Twin and Pum Tuff beds is almost constant in the central block. However, the Red beds tend to be thicker to the east, and the sediments between the local flooding surface (F in Figure 15) and Exo Tuff bed tend to be thicker to the west, suggesting a temporary seesaw subsidence during the deposition between the Twin and Pum Tuff beds (Figure 15).

The sediments above the Pum Tuff bed (Figure 15) tend to be thicker in the eastern part of the central block. This thickness change and the appearance of the thicker lake deposits in the eastern part clearly indicate that the depocentre was shifted in the eastern part of the basin. The seesaw subsidence seems to have ceased just before the Pum Tuff deposition. The thicker sediments to the east indicate that the fault separating the central and eastern blocks may have been formed in this phase. Note the thickness variation between the Pum and Ma Tuff beds—which tend to be thicker from section 3 to 5, but have thicker sediments in the same horizon in section 1, indicating larger subsidence around section 1 than section 3. This suggests that the fault separating the central and western blocks also became active after the Pum Tuff bed deposition.

Such a seesaw subsidence pattern suggests that the study area was located on the accommodation zone [3, 5] during the deposition of the lower part of the upper Nakali Formation. The Case C fault linkage and accommodation zone proposed in [5] (Figure 1) is inferred for this case.

Seesaw subsidence was reported from the Santo Domingo Basin in the Rio Grande Rift system [3], which has been long lived from the Oligocene to Pleistocene, because of changes in the shift of the active part of the faults forming the accommodation zone. This study showed a gradual facies shift because of such long-term seesaw subsidence (Figure. 8 in [3]). In case of the Nakali Formation, the movement's scale is much smaller and shorter than the case in [3]. This seesaw subsidence may have been related to the development of the block-bounding faults, which propagated either from the south or north. Such a temporary seesaw subsidence pattern may be the typical subsidence pattern of the Case C accommodation zone (Figure 1) when the zone is incorporated into a larger basin because of the merger of smaller basins. This result additionally suggests that the constant thickness sediments within a half-graben fill could be the consequence of the seesaw subsidence happening in a short period.

3. Discussion

3.1. Effects of supply mainly by pyroclastic fall on stratigraphic architecture

Samburu Hills provide a good example of a basin that was strongly controlled by sediment supply from pyroclastic fall. The target basin did not seem to experience a complicated tectonic history during the Namurungule phase (interaction with another basin, such as a basin merger) like other examples, so it is a suitable place to discuss the contribution of fine volcaniclastics supplied by falls or streams on stratigraphic architectures. Because the border fault of this basin runs along the centre of the rift basin, sufficient sediment supply from the footwall slope would not have been expected, and the basin should have been starved in terms of sediment supply (particularly siliciclastic sediments). However, the supply by pyroclastic fall or by streams that transported reworked pyroclastic fall sediments to the lake contributed to the high rate of sedimentation. The total thickness of the lake deposit (TST) at the southern end of the study area becomes almost double that at the northern end of the basin. This suggests that the newly formed accommodation space was rapidly filled even near the basin centre.

The presence of different systems tracts within a half-graben in the same period was expected on the basis of computer simulations [18]. The study simulated marine basins, but its results are also applicable to continental basins, except for a different response of the lake- or sea-level changes compared with the tectonic subsidence (see [21]). As expected in [18], a high rate of sediment supply might have resulted in a progradational stacking pattern in the northern end of the target basin, where the subsidence rate was small. The absence of the progradational unit in this place can be explained by dispersion of the eroded sediments into the basin due to the larger mobility of fine volcaniclastics. However, we need more tests to evaluate the effect of the higher mobility of volcaniclastics compared with siliciclastic sediments on the stratigraphic architecture.

Another two basin sediments (Koura and Nakali Formations) were dominated by volcaniclastics, and show high sedimentation rates [44-45]. The high-resolution tectonics related to basin evolution are discussed as follows.

3.2. Record of basin mergers

Both Koura and Nakali Formations record that terrestrial or shallow lake environments were finally changed to deep-water environments (Figures. 4 and 14) after several periods of rapid environmental change. As mentioned in Sakai et al. (2013), it is highly possible that the Koura Formation experienced at least two periods of outburst floods and subsequent lake-level rise as a result of merging basins.

The major flooding surface of the upper Nakali Formation is also interpreted as having been associated with a basin merger event. The hummocky cross-stratified beds and conglomeratic sandstone interbeds just below the flooding surface may be a record of strong waves and currents just before this basin was deeply submerged (Figure 16). Another basin merger event is expected to have occurred when the subsidence centre jumped from the western to eastern part of the central block around the deposition of the Pum Tuff bed. However, distinct evidence of basin merger cannot be found in the sediments. This was probably because of lower topographic relief in the accommodation zone (Figure 1), which was not high enough to cause the major shift in lake water when two basins were merged.

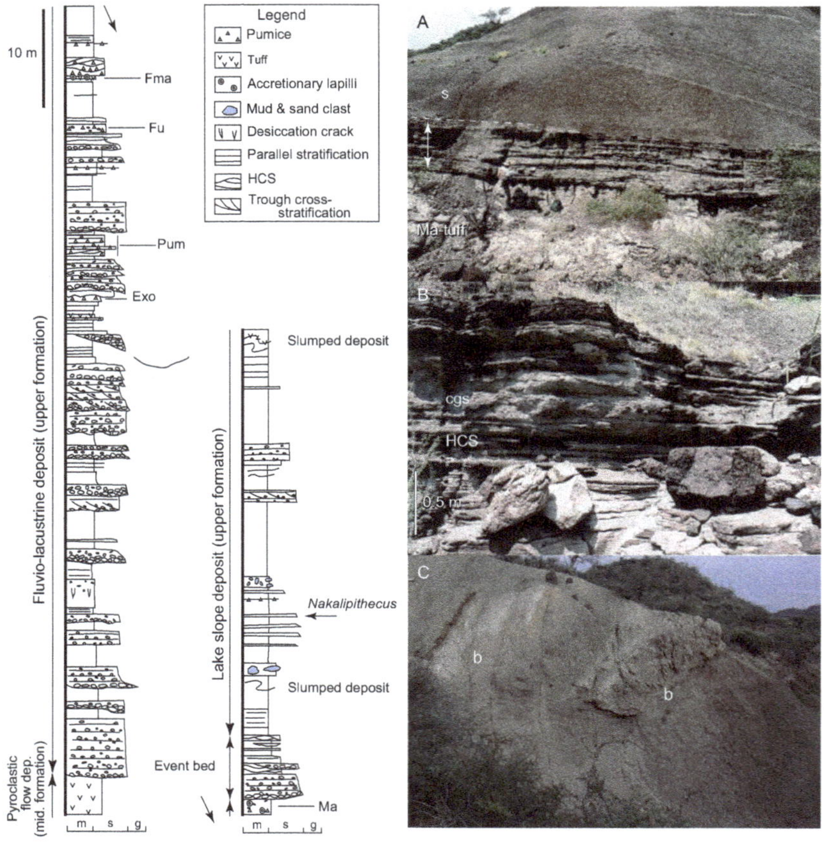

Figure 16. Example of the columnar cross-section of the upper formation, showing the boundary of the upper and lower halves of the upper formation. A concave-up solid line indicates an erosion surface. A: outcrop photograph of the boundary of the lower and upper halves of the formation. B: close-up photograph of the boundary. HCS: hummocky cross-stratification, cgs: conglomerate beds, s: slope deposit. C: an example of the slumped beds in the upper half of the upper formation. b: large slump block.

The process of the basin merger and related basin fill has been modelled in some previous studies [4, 16], which emphasized the hydraulic connection between two adjacent basins after one basin reached the over-filled condition (see [19]). In the present examples (Koura and Nakali cases), each event seems to have been related to the outburst flood and associated with a rapid deepening event. Both basins finally submerged into the Japan Sea or deep lake in short periods, implying a high subsidence rate in these basins. Therefore, the tectonic merger of the basins (i.e. connection of border faults of adjacent basins) is strongly expected for these cases. Because the Japan Sea was opened rapidly during the middle Miocene, evidence of such basin mergers is expected to be found from many basins along it.

3.3. Appearance of cycles in the upper Koura and Namurungule formations

In the Namurungule and Koura Formations, sediment cycles appear in their upper parts[34, 44]. Similar types of cycles have been reported from other areas, and some of the cycle formation was explained simply by migration of the fluvial system ([47]). Strong pulse of pyroclastic sediment supply could form small cycles as well. The Namurungule case, thickening of individual cycles to the west, indicates that the cycle formation is controlled by subsidence within the basin [42].

The Koura Formation example shown here is only a one-dimensional section, and is not enough to discuss the origin of the cycles. However, some of the erosion surface formation is clearly associated with tectonics. The shallower facies covering basal cycle surfaces without sedimentation gaps (Figures 6B and 6C) implies a lake-level fall induced by a relative uplift against the basin centre around the measured section. Although it is impossible to know the quantity of the relative uplift, the estimated uplift might be a few metres on the basis of the facies gap above and below the surface. The formation of the flooding surface and some of the cycle boundaries may be related to eustatic sea-level rise and fall.

On the contrary, both formations do not contain such cycles in their lower and middle parts. The lower parts of both formations, however, show evidence of small-scale sliding in the sediments (Figures 7 and 11). There is a small gap in the environment above and below the slide interval of the lower Koura Formation, indicating that a small-scale subsidence occurred. However, the subsidence was not of sufficient amplitude to form a cycle boundary like the case of the upper formation.

This matches with the general understanding of the rift basin evolution, where the displacement of the border fault becomes larger through the basin enlargement (for example, [14, 16]). The absence of the poorly developed drainage system also contributed to the absence of sediment cycles in case of the Namurungule Formation, because streams do not have enough strength to form an erosion surface when the relative uplift occurred. Therefore, the earliest phase of the rifting is not favourable for generating small sediment cycles related to tectonics because of smaller fault displacement.

The Nakali Formation does not contain such small sediment cycles, which indicates that the uplift or subsidence associated with fault displacement was not distinct in this place. Because the area we observed may have been situated near the accommodation zone when the upper formation was deposited, the fault displacement causing subsidence/uplift may have been smaller than that near the basin centre and was not enough to form sediment cycles.

4. Conclusions

Three examples of the early rift basin fills from the Koura Formation in SW Japan, and from the Namurungule and Nakali Formations in central and northern Kenya, have been indicated. The three basin fills consist mainly of volcaniclastics and are represented by rapid sediment accumulation. The Namurungule Formation's succession may be strongly affected by the wide dispersal of volcaniclastics in the lake resulting in a single systems tract within the basin, although a progradational unit is expected to be formed from the marginal part of the basin with a smaller subsidence rate during the TST formation in the central part of the basin. The longer-lived Koura and Nakali Basin fills may record basin merger events followed by lake-level rises probably associated with tectonic basin mergers. The appearance of the cycles only in the upper part of the Koura and Nakali Formations is interpreted to have been associated with the larger displacement of the border faults than when their lower and middle parts were deposited. Absence of the cycles in the lower part of the upper Nakali Formation can be explained by insufficient relative uplift/subsidence of the basin for cycle formation.

Acknowledgements

We thank Dr Y. Itoh of the Osaka Prefecture University and Dr O. Takano of JAPEX, who gave us the opportunity to submit this manuscript to this publication. Research in Samburu Hills and Nakali was permitted by the Government of Kenya and was supported by many local people. This study was supported by a grant in aid from the Ministry of Education, Japanese Government (17740335 for TS, 19207019 for MN and 14253006 for YS).

Author details

Tetsuya Sakai[1], Mototaka Saneyoshi[2], Yoshihiro Sawada[3], Masato Nakatsukasa[4],
Yutaka Kunimtatsu[4] and Emma Mbua[5]

1 Department of Geoscience, Shimane University, Shimane, Japan

2 Hayashibara Museum of Natural Sciences, Okayama, Japan

3 Department of Geoscience, Shimane University, Shimane, Japan

4 Department of Zoology, Graduate School of Science, Kyoto University, Kyoto, Japan

5 Department of Earth Sciences, National Museums of Kenya, Nairobi, Kenya

References

[1] Mack, G.H., William, R.S. and Kieling, J., 1994, Late Oligocene and Miocene faulting and sedimentation, and evolution of the southern Rio Grande rift, New Mexico, USA. Sedimentary Geology, 92, 79-96.

[2] Strecker, U., Steidtmann, J.R. and Smithson, S.B., 1999, A conceptual tectonostratigraphic model for seismic facies migration in a fluvio-lacustrine extentional basin. AAPG Bulletin, 83, 43-61.

[3] Smith, G.A., William, M. and Kuhle, A.J., 2001, Sedimentologic and geomorphic evidence for seesaw subsidence of the Santo Domingo accommodation-zone basin, Rio Grande rift, New Mexico. Geological Society of America Bulletin, 113, 561-574.

[4] Keighley, D., Flint, S, Howell, J. and Moscariello, A. (2003) Sequence stratigraphy in lacustrine basins: a model for part of the Green River Formation (Eocene), southwest Uinta Basin, Utah, USA. Journal of Sedimentaty Research, 73, 987-1006.

[5] Rosendahl, B.R., 1987, Architecture of continental rifts with special reference to East Africa. Annual Review of Earth and Planetary Science, 15, 445-503.

[6] Scholz, C.A., Rosendahl, B.R. and Scott, D.L., 1990, Development of coarse-grained facies in lacustrine rift basins: Examples from East Africa. Geology, 18, 140-144.

[7] Morley, C.K., and Wescott, W.A., 1999, Sedimentary environments and geometry of sedimentary bodies determined from subsurface studies in East Africa, In: Geoscience of rift systems—evolution of East Africa (Ed C.K. Morley), AAPG Studies in Geology, 44, 211-231.

[8] Gupta, S., UnderHill, J.R., Sharp, I.R. and Gawthorpe, R.L., 1999, Role of fault interaction in controlling synrift sediment dispersal paterns: Miocene, Abu Alaqa Grojup, Suez Rift, Sinai, Egypt. Basin Research, 11, 167-189.

[9] Young, M.J., Gawthorpe, R.L. and Sharp, I.R., 2000, Sedimentology and sequence stratigraphy of a transfer zone coarse-grained delta, Miocene Suez Rift, Egypt. Sedimentology, 47, 1081-1104.

[10] Strachan, L.J., Rarity, F., Gawthorpe, R.L., Wilson, P., Sharp, I. and Hodgetts, D., 2013, Submarine slope processes in rift-margin basins, Miocene Suez Rift, Egypt. Geological Society of America Bulletin, 125, 109-127.

[11] Rohais, S., Eschard, R., Ford, M., Guillocheau, F. and Moretti, I., 2007, Stratigraphic architecture of the Plio-Pleistocene infill of the Corinth Rift: Implications for its structural evolution. Tectonophysics, 440, 5-28.

[12] Bell, R.E., McNeill, L.C., Bull, J.M., Henstock, T.J., Collier, R.E.L., and Leeder, M.R., 2009, Fault architecture, basin structure and evolution of the Gulf of Corinth Rift, central Greece. Basin Research, 21, 824-855.

[13] Taylor, B., Weiss, J.R., Goodliffe, A.M., Sachpazi, M. Laigle, M. and Hirn, A., 2011, The strcture, stratigraphy and evolution of the Gulf of Corinth rift, Greece. Geophysical Journal International, 185, 1189-1219.

[14] Gawthorpe, R.L. and Leeder, M.R., 2000, Tectono-sedimentary evolution of active extentional basins. Basin Research, 12, 195-218.

[15] Athmer, W. and Luthi, S.M., 2011, The effects of relay ramp on sediment routes and deposition: A review. Sedimentary Geology, 242, 1-17.

[16] Lambiase, J.J., 1990, A model for tectonic control of lacustrine stratigraphic sequences in continental rift basin. AAPG Memoir, 50, 265-276.

[17] Howell, J.A., and Flint, S.S., 1996, A model for high resolution sequence stratigraphy within extensional basins. In: High Resolution Sequence Stratigraphy – Innovations and Applications (Eds J.A. Howell, and J.F. Aitken), Geological. Society of London, Special Publication, 104, 129-137.

[18] Gawthorpe, R.L., Hardy, S. and Ritchie, B., 2003, Numerical modelling of depositional sequences in half-graben rift basins. Sedimentology, 50 169-185.

[19] Carroll, A.R. and Bohacs, K.M., 1999, Stratigraphic classification of ancient lakes: balancing tectonic and climatic controls. Geology, 27, 99-102.

[20] Withjack, M.O., Schlische, R.W. and Olsen, P.E., 2002, Rift-basin structure and its influence on sedimentary system. SEPM Special Publication, 73, 57-81.

[21] Ilgar, A. and Nemec, W., 2005, Early Miocene lacustrine deposits and sequence stratigraphy of the Ermenek Basin, Central Taurides, Turkey. Sedimentary Geology, 173, 233-275.

[22] Smith, M. and Mosley, P., 1993, Crustal heterogeneity and basement influence on the development of the Kenya Rift, East Africa. Tectonics, 12, 591-606.

[23] Paton, D.A. and Underhill, J.R., 2004, Role of crustal anisotropy in modifying the structural and sedimentological evolution of extensional basins: the Gamtoos Basin, South Africa. Basin Research, 16, 339-359.

[24] Wei, H.H., Meng, Q.R., Wu, G.L. and Li, L., 2012, Multiple controls on rift basin sedimentation in volcanic settings: Insights from the anatomy of a small Early Cretaceous basin in the Yanshan belt, northern North China. Geological Society of America Bulletin, 124, 380-399.

[25] Königer, S. and Stollhofen, H., 2001, Environmental and tectonic controls on preservation potential of distal fallout ashes in fluvio-lacustrine settings: the Carboniferous-Permian Saar-Nahe Basin, southwest Germany. In: Volcaniclastic sedimentation in lacustrine settings (Eds J. White and N. Riggs). International Association of Sedimentologists Special Publication 30, 263-284.

[26] Okamura, Y., Watanabe, M., Morijiri, R., and Satoh, M., 1995, Rifting and basin inversion in the eastern margin of the Japan Sea. The Island Arc, 4, 166-181.

[27] Ashley, G. and Hay, R.L., 2002, Sedimentation patterns in a Plio-Pleistocene volcaniclastic rift-platform basin, Olduvai Gorge, Tanzania. SEPM Special Publication, 73, 107-122.

[28] Cas, R.A.F., Edgar, C. Allen, R.L., Bull, S. Clifford, B.A., Giordano, G. and Wright, J.V., 2001, Influence of magmatism and tectonics on sedimentation in an extensional lake basin: the Upper Devonian Bunga Beds, Boyd Volcanic Complex, South-eastern Australia. International Association of Sedimentologists, Special Publication, 30, 83-108.

[29] Takayasu, K., Yamasaki, H., Ueda, T., Akagi, S., Matsumoto, T., Nomura, R., Okada, S., Sawada, Y., Yamauchi, S. and Yoshitani, A., 1992, Miocene stratigraphy and paleogeography of the San'in district, Southwest Japan. Memoir of Geological Society of Japan, 37, 97-116.(in Japanese with English Abstract)

[30] Kano, K, and Nakano, S., 1985, Geology of the Mihonoseki District. With Geological Sheet Map at 1:50, 000, Geological Survey of Japan, 28 p.(in Japanese with English Abstract)

[31] Miura, K., 1973, A finding of gneiss gravels from the Koura formation in the Shimane Peninsula and its geological significance. Journal of Geological Society of Japan, 79, 701-702.(in Japanese with English Abstract)

[32] Tanaka, T. and Ogusa, K., 1981, Structural movement since middle Miocene in the offshore San-in sedimentary basin, the Sea of Japan. Journal of Geological Society of Japan, 87, 725-736.(in Japanese with English Abstract)

[33] Sato, T. and Nakata, T., 2002, Kashima fault as a model fault for active fault segmentation. Active Fault Research, 21, 99-110.(in Japanese with English Abstract)

[34] Sakai, T., Furukawa, A. and Kawano, S, 2013, Stepwise environmental changes in the lower Miocene Koura Formation, southwest Japan, associated with Japan Sea evolution. Journal of Geological Society of Japan, 119, 285-299.

[35] Kano, K., 1991, Volcaniclastic sedimentation in a shallow-water marginal basin: the Early Miocene Koura Formation, SW Japan. Sedimentary Geology, 74, 309-321.

[36] Yamauchi, S., Mitsunashi, N., and Yamamoto, Y., 1980, The Miocene of the Shimane Peninsula. The guidebook of the Geological Society of Japan 87th annual meeting. 39 p. (in Japanese)

[37] Kogane, N., Imaoka, Y., Ueda, Y., Sampei, Y. and Suzuki, N., 1994, Transgression of the Japan Sea in the Middle Miocene as shown by stratigraphic variations of total organic carbon and sulfur concentrations of mudstones near the boundary of Koura and Josoji Formations, eastern Shimane Peninsula, Japan. Geological reports of Shimane University, 13, 57-67.(in Japanese with English Abstract)

[38] Kano, K. and Yoshida, F., 1984, Radiometric ages of the Neogene in central eastern Shimane Prefecture, Japan and their implication in stratigraphic correlation. Bulletin of Geological Survey of Japan, 35, 159-170.(in Japanese with English Abstract)

[39] Kano, K., Yamauchi, S., Takayasu, K., Matsuura, H., and Bunno, M., 1994, Geology of the Matsue District. With Geological Sheet Map at 1:50, 000, Geological Survey of Japan, 57 p.(in Japanese with English Abstract)

[40] Van Wagoner, J.C., Posamentier, H.W., Mitchum, R.M., Vail, P.R., Sarg, J.F., Loutit, T.S. and Hardenbol, J., 1988, An overview of the fundamentals of sequence stratigraphy and key definitions. SEPM Special Publication, 42, 39-45.

[41] Baker, B.H., 1986, Tectonics and volcanism of the southern Kenya Rift Valley and its influence on rift sedimentation. Geological Society of London Special Publication, 25, 45-57.

[42] Saneyoshi, M., Nakayama, K., Sakai, T., Sawada, Y. and Ishida, H., 2006, Half-graben filling processes in the early phase of continental rifting: The Miocene Namurungule Formation of the Kenya Rift. Sedimentary Geology, 186, 111-131.

[43] Sawada, Y., Saneyhoshi, M., Nakayama, K., Sakai, T., Itaya, T., Hyodo, M., Mukyokya, Y., Pickford, M. Senut, B., Tanaka, S., Chujo, T. and Ishida, H., 2006, The ages and geological backgrounds of Miocene Nacholapithecus, Samburupithecus, and Orrorin from Kenya. In: Human Origins and Environmental Backgrounds (Eds H. Ishida, R.H. Tuttle, M. Pickford, N. Ogihara, and M. Nakatsukasa), Springer, New York, 71-96.

[44] Sakai, T., Saneyoshi, M., Tanaka, S., Sawada, Y., Nakatsukasa, M., Mbua, E. and Ishida, H., 2010, Climate shift around 10 Ma recorded in Miocene succession of Samburu Hills, northern Kenya Rift, and its significance. Geological Society of London Special Publication, 342, 109-127.

[45] Kunimatsu, Y., Nakatsukasa, M., Sawada, Y., Sakai, T., Hyodo, M., Hyodo, H., Itaya, T., Nakaya, H., Saegusa, H., Mazurier, A., Saneyoshi, M., Tsujikawa, H., Yamamoto, A. and Mbua E., 2007, A new Late Miocene great ape from Kenya and its implications for the origins of African great apes and humans. Proceedings of the National Academy of Sciences, 104, 19220-19225.

[46] Gorzáles-Acebrón, R., Arribas, J., Mas, R., 2007, Provenance of fluvial sandstones at the start of late Jurassic-Early Cretaceous in the Cameros Basin (N. Spain). Sedimentary Geology, 202, 138-157.

[47] Cande, S.C. and Kent, D.V., 1995, Revised calibration of magnetostratigraphic polarity timescale for the Late Cretaceous and Cenozoic. Journal of. Geophysical Research, 100, 6093-6095.

[48] Johnson, C.L. and Graham, S.A., 2004, Cycles in perilacustrine facies of late Mesozoic rift basins, southeastern Mongolia. Journal of Sedimentary Research, 74 786-804.

Chapter 4

East Asia-Wide Flat Slab Subduction and Jurassic Synorogenic Basin Evolution in West Korea

Kosuke Egawa

Additional information is available at the end of the chapter

1. Introduction

The interplay between oceanic plate subduction and the development of continental margins is of considerable geological interest, and of a particular interest for Asian structural geologists and petrologists is the subduction of the present and ancient Pacific plates, which triggered orogenic development and contributed to crustal evolution in the circum-Pacific regions through the Phanerozoic [1, 2]. Since the Triassic, the northwestern circum-Pacific region (also known as the East Asian continental margin) initiated the evolution of a continental arc stretching several thousand kilometers, which resulted in an East Asia-wide crustal shortening and thickening, orogenic basin formation, and landward magmatic progradation [2, 3, 4, 5, 6, 7, 8, 9]. It is noted that although the paleo-Pacific subduction along this region was also present in Paleozoic time, it did not exert a major tectonic impact on the Asian continents [10, 11, 12], and that this lack of impact was probably related to the fact that the Paleotethys Ocean lay between Laurasia and Gondwana until the Triassic period when the East Asian continental blocks had not been yet assembled [13, 14].

The Korean Peninsula, situated in the middle of the East Asian continental margin (Fig. 1), was plunged into a tectonically active phase in Mesozoic time, and three major orogenies are recorded; the Songnim, Daebo, and Bulguksa [4]. Among these, the Songnim orogeny (260–220 Ma) is represented by regional metamorphism in a close association with the final amalgamation of Chinese continental blocks in Permian–Triassic period [15, 16]. A drastic tectonic transition followed this orogeny, and the evolution of a continental-magmatic arc occurred during the Daebo (190–135 Ma) and Bulguksa (100–45 Ma) orogenies, which resulted from the flat slab subduction and subsequent slab rollback of the western paleo-Pacific plates, respectively [5, 8, 17, 18]. It is evident that the Songnim–Daebo tectonic transition led to a radical shift of the Korean sedimentary environments, from Paleozoic marine to Mesozoic nonmarine domains [12, 19]. The evolution of the continental arc ultimately produced a derivation of Korean-derived detrital sediments in the Pacific-side regions, such as in the Inner Zone of Southwest Japan [20, 21, 22, 23, 24, 25].

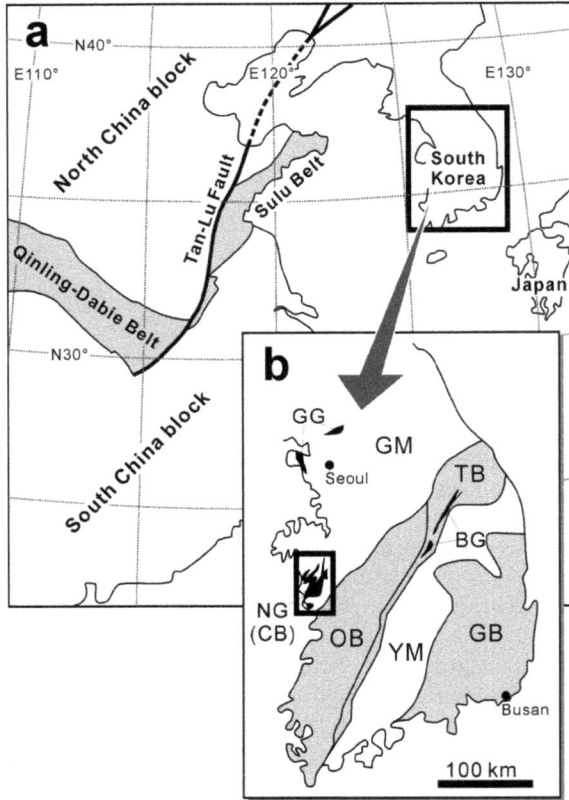

Figure 1. (a) Simplified tectonic map of East Asia, and (b) close-up of South Korea showing the major tectonic provinces with the study area (boxed) (modified after Egawa and Lee [7]). BG, Bansong Group; CB, Chungnam Basin; GB, Gyeongsang Basin; GG, Gimpo Group; GM, Gyeonggi Massif; NG, Nampo Group; OB, Okcheon Belt; TB, Taebaeksan Basin; YM, Yeongnam Massif.

The tectonism, magmatism, and sedimentation of South Korea have been systematically well reviewed and summarized by Korean geologists [4, 12, 19, 26, 27]. However, such work has included only limited description and minor discussion on Jurassic basinal evolution because of the very limited distribution and publication of research in comparison with studies related to other Phanerozoic basins (Fig. 1). In contrast, many of Jurassic structural and igneous events have been reported and detailed [4, 5, 18, 28, 29, 30, 31].

It has been conventionally interpreted that Jurassic non-marine basins are interorogenic basins, formed during the period between the Songnim and Daebo orogenies [4, 19, 32]. However, recent radiometric dating of detrital zircon [30, 33, 34] has provided an alternative view of this conventional interpretation and has shown that the depositional age of these Jurassic basins corresponds to the early phase of the Daebo orogeny; indicating a close association with the subduction-induced continental arc evolution.

From recent petrologic analyses and radiometric dating, the Chungnam region in western South Korea (known as the Hongseong Belt) has been interpreted as being an eastern extension of the Qinling–Dabie–Sulu Belt (the collisional belt between the North and South China blocks) (Fig. 1) [15, 35]. In the Chungnam region, it appears that Proterozoic to Paleozoic basement rocks were regionally metamorphosed with a high to ultra-high pressure facies during the Songnim orogeny [35, 36, 37, 38, 39]. The subsequent rapid uplift and denudation of these basement rocks then delivered their detritus into the Jurassic Chungnam Basin [34], which was followed by a structural disturbance during the late stage of the Daebo orogeny [31, 40].

The author of this paper has been studying the Chungnam Basin for several years [7, 40, 41, 42, 43, 44], and has demonstrated that the basin filling and thermal history are closely related to the Daebo continental-arc evolution. This paper presents an overview of the characteristics and mechanisms of Mesozoic flat slab subduction in East Asia, and then summarizes the sedimentary and structural evolution of the Chungnam Basin during the Daebo orogeny, with the intention of promoting a better understanding of the basin-filling processes in West Korea and also of the interplay between basinal and crustal evolution at the active continental margin of East Asia.

2. Flat slab subduction

2.1. Evidence of flat slab subduction in and around Korea

Recent igneous studies suggest that the Mesozoic continental arc evolution was triggered by the flat slab subduction of the western paleo-Pacific plates underneath the East Asian continent [6, 45, 46]. According to the observation of modern subduction zones in the Andes, there is a close relationship between flat slab subduction, crustal shortening and thickening, and inlandward-migrating magmatism [47, 48, 49, 50]. Subducted slab dip is fundamentally constrained by slab buoyancy. Therefore, a slab with oceanic plateaus or ridges is flatly subducted over a long distance, while steeper subduction occurs when such features are absent [49, 51].

Evidence for the subduction of such buoyant oceanic materials is found in the Mesozoic accretionary complexes along the eastern margin of Asia, stretching a distance of several thousand kilometers, and is seen particularly in Japan and Russian Far East [2, 52]. These complexes generally consist of oceanic plateau basalts and deep marine deposits, which were accreted and underplated underneath the Asian continental crusts during subduction [53, 54, 55]. Paleomagnetic analysis has revealed that the Japanese Islands were geologically connected to the Asian continent before the opening of the Japan Sea in Miocene epoch [6, 56], and that the Jurassic accretionary complex in Southwest Japan was situated next to South Korea during its formation [25, 29], which was initiated in, at the latest, the early Late Triassic period [57] and continued through the Jurassic period [52]. Adakitic granites, which are indicators of slab melting, intruded widely into the Korean continental crusts with an inlandward younging trend during the Jurassic period [5, 8, 18, 58], supporting the interpretation of inlandward slab migration [47, 48, 59].

2.2. Orogenic gaps between Korea and South China

The geology of South China records two major Mesozoic orogenies: the Indosinian orogeny (250–205 Ma) indicated by inlandward-migrating magmatic front with crustal thickening and shortening, and the Yanshanian orogeny (180–66 Ma) characterized by an oceanward-retrograding magmatic front with crustal thinning and stretching [6, 60]. These two orogenic events resulted from a flat slab subduction with a length of 1400 kilometers, and a subsequent slab rollback [6, 45]. The Korean Peninsula is situated just 500 km northeast of South China,

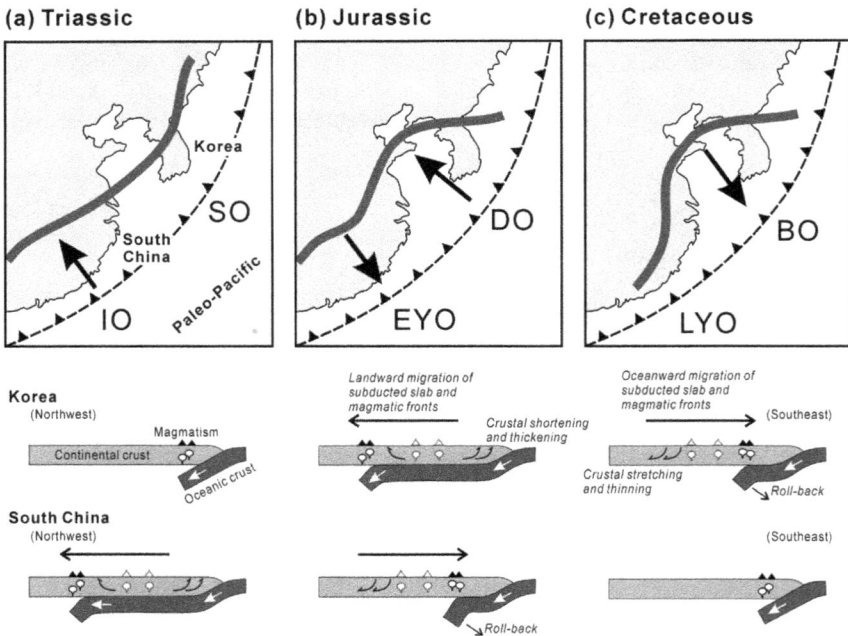

Figure 2. Landward and subsequent oceanward migration of subducted slab and magmatic fronts (direction indicated with a heavy line) in Korea and South China: (a) Triassic, (b) Jurassic, and (c) Cretaceous periods (modified after Li and Li [6], Choi et al. [8], Egawa [44], Kiminami et al. [46], and Zhou et al. [60]). Migration of magmatic front is linked to the morphology of the subducted slab of oceanic plate. BO, Bulguksa orogeny; DO, Daebo orogeny; EYO, Early Yanshanian orogeny; LYO, Late Yanshanian orogeny; IO, Indosinian orogeny; SO, Songnim orogeny.

and two peninsula-wide orogenies, the Songnim and Daebo orogenies, occurred almost contemporaneously with the Indosinian and Yanshanian orogenies, respectively. It has therefore been conventionally interpreted that these Chinese and Korean orogenies progressed under the same subduction processes [2, 4, 61].

It is necessary to reiterate here that the Songnim and Daebo orogenies are represented by a regional metamorphism related to the Chinese final assembly and by the evolution of the continental-magmatic arc associated with the paleo-Pacific subduction, respectively. Such facts therefore provide an alternative interpretation: there were distinct orogenic gaps between South China and Korea (Fig. 2) [8, 18]. This implies that when the Triassic flat slab subduction has already initiated the Indosinian orogeny in South China, the Songnim regional metamorphism in Korea was then caused by the ongoing final amalgamation of the Chinese continental blocks. The subsequent Daebo continental-magmatic arc evolution then occurred 60 m.y. later than the compressional Indosinian orogeny, and by this time South China was already in the phase of the extensional Yanshanian orogeny.

3. Synorogenic basin evolution in West Korea

The foregoing flat slab subduction then triggered and drove the Daebo orogeny in Korea, with a significant crustal shortening and thickening [4, 30]. This crustal deformation created an orogenic wedge in middle South Korea, which consists of the southeast- and northwest-vergent fold-and-thrust belts (Fig. 3) [62]. The former belt corresponds to a pro-wedge region, which includes the Okcheon Belt and the Taebaeksan Basin, and the latter-mentioned belt developed as a retro-wedge region, which includes the Chungnam region [4, 7, 30, 33]. Such wedge structures were probably formed under a NW–SE-directed compressional setting during the orogeny [63, 64].

The Chungnam Basin (consisting of several separated subbasins—the Ocheon, Oseosan, and Seongju subbasins, and other unnamed) was filled with a Jurassic nonmarine deposit, known as the Nampo Group (Fig. 4). This group unconformably covers the pre-Jurassic metamorphic basement rocks, and was structurally underlain by these rocks due to the postdepositional thrust faulting [40, 41]. The stratigraphy of the Nampo Group is subdivided into the Hajo, Amisan, Jogyeri, Baegunsa, and Seongjuri formations with decreasing age [65, 66]. Among them, the Hajo, Jogyeri, and Seongjuri formations are mainly composed of conglomerate and sandstone, whereas the Amisan and Baegunsa formations are dominated by an alternation of coal-bearing shale and sandstone. In this study, the stratigraphy of the Oseosan Subbasin (as defined by Egawa and Lee [7, 41]) is revised on the basis of the recognition of the Oseosan Thrust, which allows the structurally repetitive distribution of the Hajo and Amisan formations (Figs. 4, 5). The depositional age of the Nampo Group is inferred as being between Sinemurian and Aalenian, based on U–Pb zircon dating of regionally metamorphosed basement rocks (230–220 Ma) [35, 37, 38] and felsic lapilli tuff of the Baegunsa Formation (170 Ma) [30], which is synchronous with the magmatic event in the early stage of the Daebo orogeny (180–170 Ma; U–Pb sphene and Rb–Sr whole-rock ages) [5].

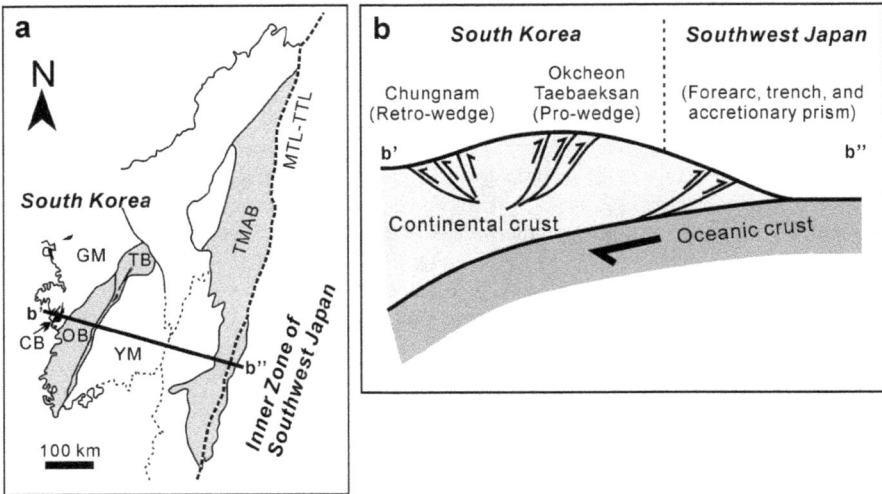

Figure 3. (a) The possible tectonic arrangement of South Korea and the Inner Zone of Southwest Japan during the Jurassic period (modified after Egawa and Lee [7]). CB, Chungnam Basin; GM, Gyeonggi Massif; MTL-TTL, Median Tectonic Line–Tanakura Tectonic Line; OB, Okcheon Belt; TB, Taebaeksan Basin; TMAB, Tanba–Mino–Ashio Belt (Jurassic accretionary complex); YM, Yeongnam Massif. (b) Schematic cross section along the b'–b'' section in (a) showing the possible evolution of a continental arc (not to scale).

3.1. Basin filling controlled by a tectonic cycle

Egawa and Lee [7] detailed and classified the nonmarine sedimentary characteristics of the Nampo Group into seven sedimentary facies associations: colluvial fan, alluvial fan, braid-plain, delta plain, delta front, offshore lacustrine, and volcaniclastic plain (Fig. 5). A combination of these facies associations reveals a vertical cyclic pattern presented by the fining- to coarsening-upward lower and upper sequences of the alluvio-lacustrine system in the Ocheon, Oseosan, and Seongju subbasins. These depositional cycles are subdivided by the thick, progressive colluvial/alluvial fan deposits of the Jogyeri Formation, along with strong interformational unconformities occurring between the Amisan and Lower Jogyeri formations (U1 unconformity) and between the Lower and Upper Jogyeri formations (U2 unconformity).

Such stratigraphic features correspond to typical alluvial basin-filling patterns, and are attributable to tectonically-driven sediment flux or climate-driven diffusivity occurring over a relatively short time-scale [67, 68]. The lack of stratigraphic or temporal variations in the degree of chemical weathering [69], along with the presence of coal deposits [70, 71], indicates little or no climate fluctuation at the time of basin filling. This illustrates that a process of tectonically-driven sediment flux is most likely to have occurred. As variation of sediment flux is an index of tectonic activity, the remarkable gravel progradation of the Jogyeri Formation probably records a time of low sediment flux and quiescent tectonism (Fig. 6) [67, 72, 73]. Under this assumption, therefore, the fine-grained sediments in the

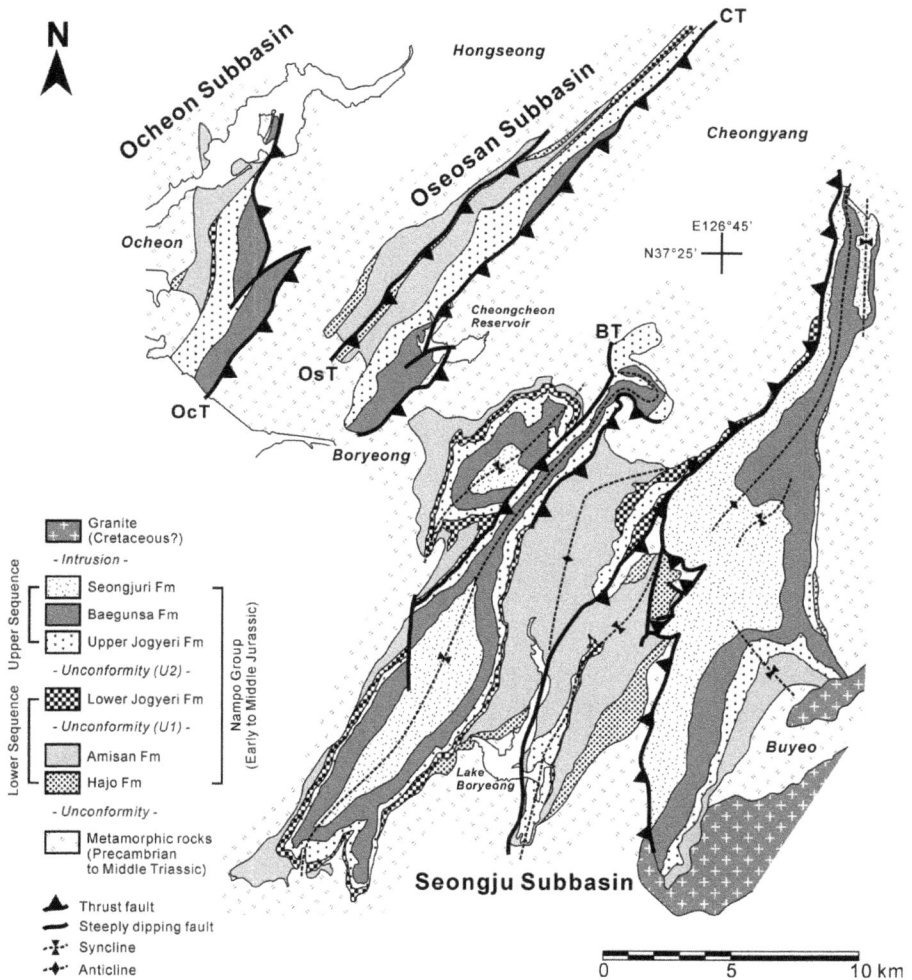

Figure 4. Geological map of the Chungnam Basin (which consists of the Ocheon, Oseosan, and Seongju subbasins) filled with the Jurassic Nampo Group (modified after Egawa and Lee [7]). BT, Baegunsa Thrust; CT, Cheongla Thrust; OcT, Ocheon Thrust; OsT, Oseosan Thrust.

other four formations are interpreted to have been deposited under active tectonism. It is assumed that the phase of Jogyeri gravel progradation reflected the progressive encroachment of deformation into the foreland [74, 75] due to the subduction-induced crustal shortening. These relationships permit a possible interpretation of the Chungnam Basin as being a piggyback or wedge-top basin [76, 77].

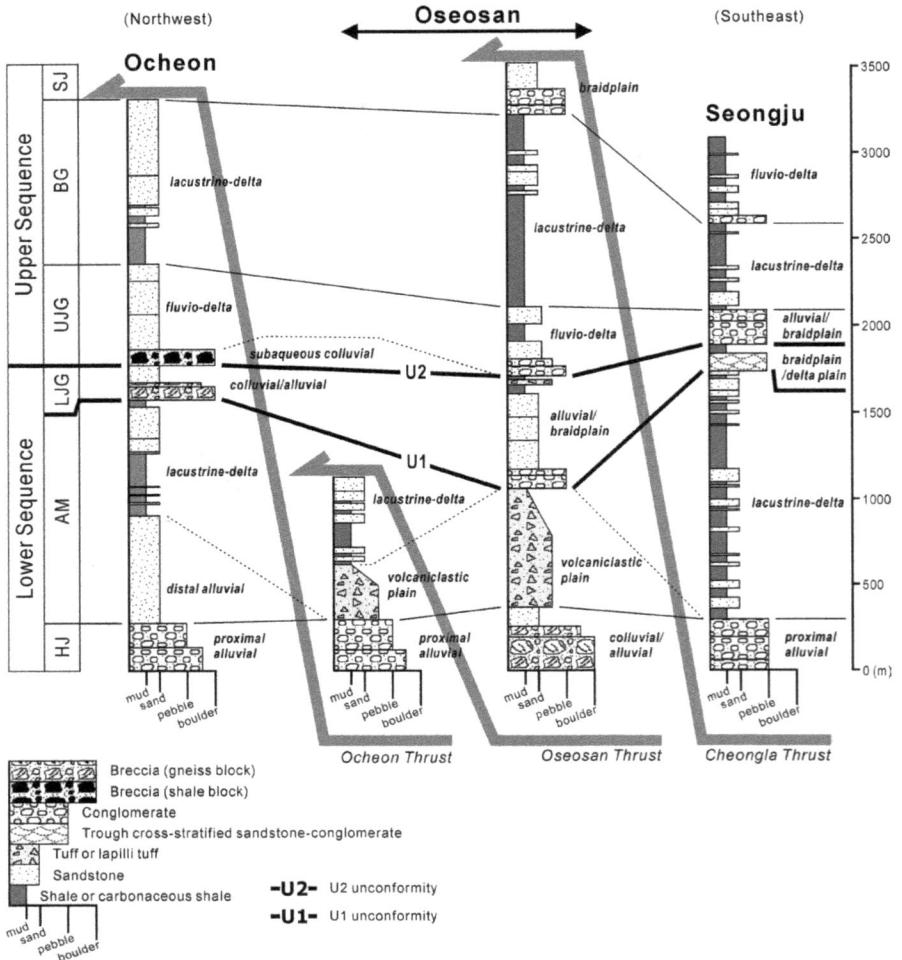

Figure 5. Strato-sedimentological interpretation of the Jurassic Nampo Group in the Ocheon, Oseosan and Seongju subbasins (modified after Egawa and Lee [7]). HJ, Hajo Formation; AM, Amisan Formation; BG, Baegunsa Formation; LJG, Lower Jogyeri Formation; SJ, Seongjuri Formation; UJG, Upper Jogyeri Formation.

3.2. Postdepositional thermal events

In the late stage of the Daebo orogeny (late Jurassic to earliest Cretaceous time), the orogenic activity was further accelerated by the oblique subduction of the paleo-Pacific plates with strike-slip motion [29, 78, 79, 80], leading to significant crustal shortening and thickening represented by thrust-imbricate stacking [4, 30, 31]. Most of the Daebo granites were synde-

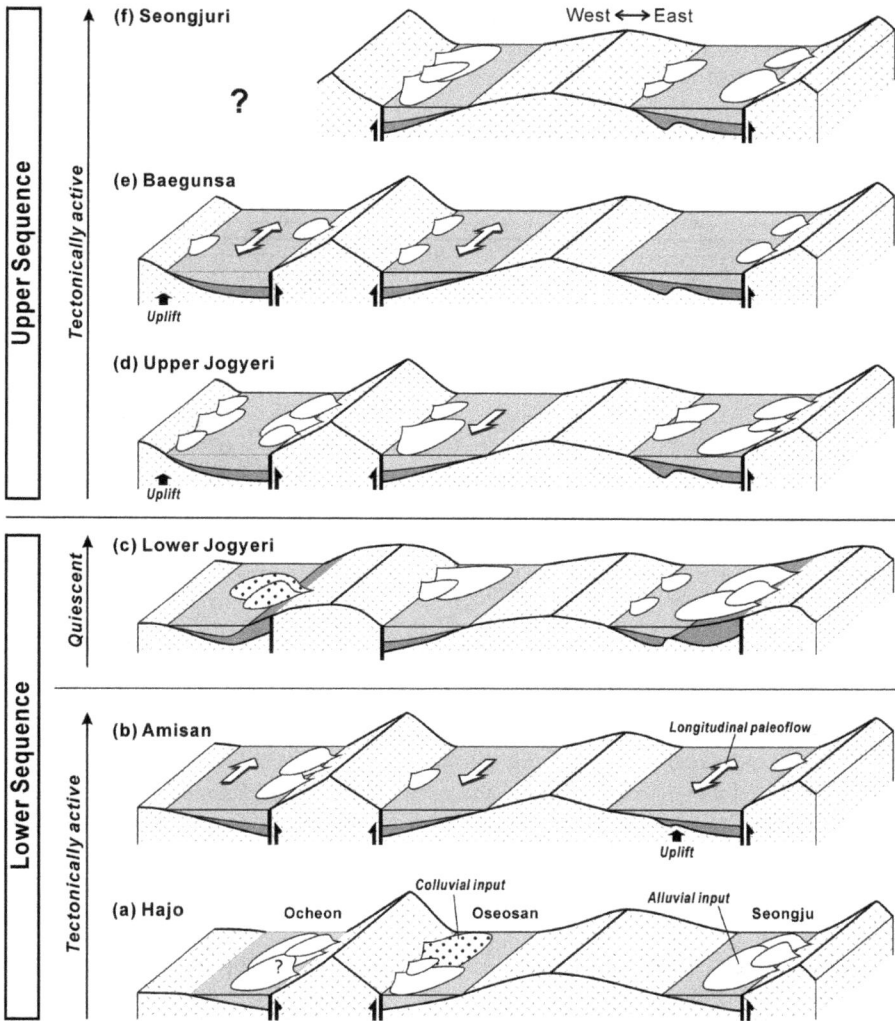

Figure 6. Schematic syntectonic evolution of the Chungnam Basin in the depositional stages of (a) the Hajo, (b) Amisan, (c) Lower Jogyeri, (d) Upper Jogyeri, (e) Baegunsa, and (f) Seongjuri formations (modified after Egawa and Lee [7]) (not to scale).

positionally intruded in the early stage of the orogeny, followed by a quiescent phase of magmatic activity of ca. 60 m.y. before the initiation of the Bulguksa orogeny (Fig. 7) [2, 5]. Such a magmatic hiatus is likely to have resulted from the existence of oceanic plateaus or ridges subducting underneath the East Asian continental crusts [81, 82]. South China, however, shows no interval of quiescent magmatism between the Indosinian and Yanshanian orogenies, and this is probably related to the slab delamination and rollback that occurred immediately after the flat subduction [6, 45].

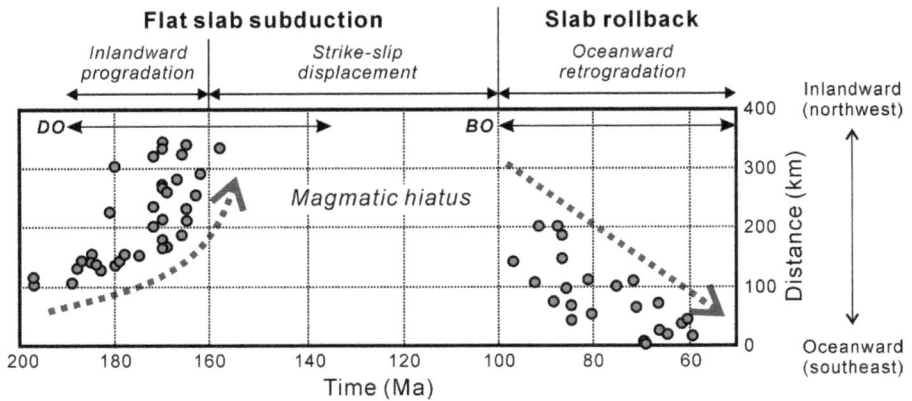

Figure 7. Inlandward and oceanward migration of the Daebo and Bulguksa granites in South Korea, respectively (modified after Kim [4], Sagong et al. [5], and Park [17]). BO, Bulguksa orogeny; DO, Daebo orogeny.

The Nampo Group has experienced high-grade diagenesis or low-grade metamorphism. This is evidenced by the presence of very high-rank coals (anthracite to meta-anthracite) and by the very high vitrinite reflectance values (5 to 6%) which occur entirely in the Seongju Subbasin [70, 71], as well as the high illitization occurring within the three subbasins which ranges in the thermal grade of anchizone to epizone [40]. Both coal and illite in sediments are commonly used as an indicator of paleotemperature, and Egawa and Lee [40] classified this postdepositional thermal event into early and late histories: tectonic burial metamorphism and hydrothermal alteration, respectively.

3.2.1. Tectonic burial metamorphism

The early tectonic burial resulted from crustal loading induced by the postdepositional basement overthrusting on the Nampo Group (Fig. 8). The grade of mechanical compaction textures in sandstones tends to increase down the sequence (Fig. 9), and the lowermost strata (Hajo Formation) appear to have been deformed in a ductile manner [40, 41, 83]. Similarly, the illite in sandstones shows a down sequence increase in its crystallinity, from anchizone to epizone (Fig. 9). Based on the equations proposed by Underwood et al. [84] and Kosakowski et al. [85], the measured illite crystallinity approximates the possible maximum paleotemperature and total burial depth of the Nampo Group in the Ocheon Subbasin as being 340 °C and 9700 m, respectively, although the total depositional thickness is 3300 m. This estimation is in good agreement with the observations of ductile deformation, epizonal metamorphism, and basement overthrusting.

(a) Middle Jurassic
Initiation of the basement overthrusting

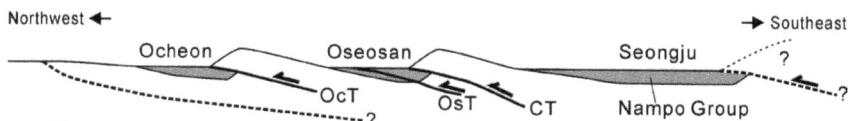

Northwest ←   → Southeast

Ocheon   Oseosan   Seongju   ?

OcT   OsT   CT   Nampo Group

?

(b) Late Jurassic to earliest Cretaceous
Tectonic burial and crustal loading

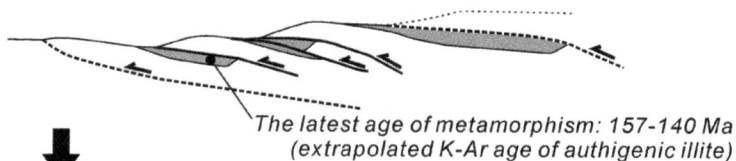

The latest age of metamorphism: 157-140 Ma
(extrapolated K-Ar age of authigenic illite)

(c) Cretaceous
Vertical rotation and hydrothermal alteration

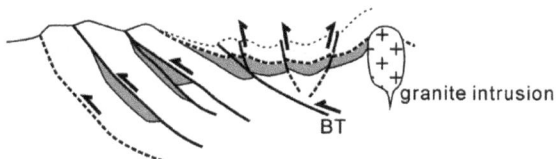

granite intrusion

BT

(d) Present

Ocheon   Oseosan   Seongju

OcT   OsT   CT   BT

?

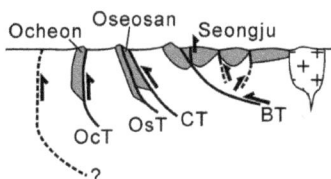

Figure 8. Conceptual structural models showing the postdepositional crustal shortening and thickening in the Chung-nam region (modified after Egawa and Lee [40, 42]) (not to scale). BT, Baegunsa Thrust; CT, Cheongla Thrust; OcT, Ocheon Thrust; OsT, Oseosan Thrust.

Radiometric dating of illite in sediments is helpful in constraining the latest diagenetic and low-grade metamorphic ages [86, 87], and is used to interpret the timing of regional over-

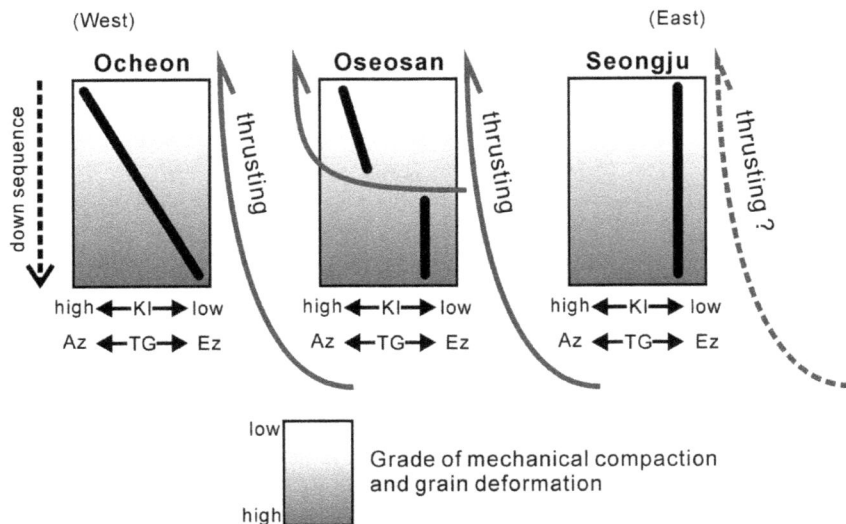

Figure 9. Simplified diagram showing the structural and diagenetic characteristics in the Ocheon, Oseosan and Seongju subbasins (modified after Egawa [43]). Az, anchizone; Ez, epizone; KI, Kübler Index; TG, thermal grade.

thrusting [88]. A mixture of authigenic ($1M_d$) and detrital ($2M_1$) components of illite is common in argillaceous sediments. Based on this knowledge, Egawa and Lee [42] measured the K–Ar ages of different-size clay fractions from the Amisan shale in the Ocheon Subbasin, and estimated the latest age of authigenic illite to be 157–140 Ma (Fig. 8), by using a linear regression model defined by the detrital amount and the K–Ar age of different size fractions [89, 90, 91]. The estimated age, therefore, is younger than the depositional age of the Nampo Group (~170 Ma) [30] and ranges within the duration of the Daebo orogeny (190–135 Ma) [4], which suggests that the tectonic burial metamorphism of the Nampo Group occurred in the late stage of the Daebo orogeny.

3.2.2. Hydrothermal alternation

The subsequent hydrothermal alternation was much affected by a magmatic intrusion and hot-fluid migration, probably during the Bulguksa orogeny [40]. The coal rank and illite crystallinity of the Seongju sediments plot into a very high thermal grade, with little stratigraphic variation (Fig. 9) [40, 70, 71]. When fluids warmed by pluton migrate along faults and fractures in the basin, they can transfer heat to the basin fills and lead to thermal alteration even at a relatively shallow depth of burial [85, 92, 93]. The Nampo Group in the Seongju Subbasin is highly faulted and folded in places, and there are granite intrusions into the southeastern subbasin (Fig. 4). These structures and intrusions probably enhanced the illitization and anthracitization after tectonic burial.

4. Conclusions

Mesozoic tectonism, magmatism, and sedimentation in East Asia were fundamentally controlled by a series of flat slab subduction and subsequent slab rollback of the northwestern paleo-Pacific plates, which allowed the evolution of an Andean-type continental arc several thousand kilometers-long. Paleo-Pacific oceanic crusts with buoyant materials (such as oceanic plateaus and ridges) had subducted and migrated inlandward underneath the Asian continent, leading to a significant magmatic progradation and crustal shortening and thickening. The subsequent delamination and rollback of the inland subducted slab resulted in the retrogradation of the magmatic front, together with crustal stretching and thinning. These dynamic events are closely associated with the evolution of major orogenies in Korea and South China: the flat slab subduction caused the Daebo and Indosinian orogenies, and the slab rollback produced the Bulguksa and Yanshanian orogenies. There is a clear time lag between the flat subduction- and rollback-induced orogenies in Korea and those in South China, which were initiated 60 m.y. and 80 m.y. later in Korea, respectively, probably due to the effect of the Chinese final amalgamation.

The Chungnam Basin in central western Korea was filled with a Lower to Middle Jurassic nonmarine succession, known as the Nampo Group, the deposition and structural development of which occurred simultaneously with the evolution of the flat subduction-induced continental-magmatic arc during the Daebo orogeny. An integrated stratigraphic, sedimentologic, diagenetic, and geochronologic analysis has demonstrated that the basin-filling processes and subsequent structural and thermal evolution of the Nampo Group were fundamentally controlled by subduction tectonics. The Nampo Group is composed of the two repeated, fining- to coarsening-upward alluvio-lacustrine sequences, separated by an interval of thick breccia–gravel progradation deposits and relative strong proximal unconformities. The observed relationships of the succession provide a record of sedimentation that was most likely controlled by the temporal variations of tectonism during the early stage of the Daebo orogeny. The postdepositional basement thrusting over the Nampo Group then led to a tectonic burial, resulting in low-grade metamorphism. Burial heating is strongly suggested by the down-sequence increase in illitization from anchizone to epizone, and in the degrees of mechanical grain compaction and ductile deformation. The maximum paleotemperature and burial depth of the Nampo Group are estimated to be 340°C and 10 km, respectively, and the extrapolated K–Ar illite dating of 157–140 Ma indicates that the tectonic burial metamorphism was completed at the end of the Daebo orogeny. A subsequent granite intrusion and hydrothermal alteration, probably occurring during the Bulguksa orogeny, have enhanced the illitization and anthracitization, regardless of the stratigraphy.

Acknowledgements

I am grateful to editor Yasuto Itoh and Ana Pantar for their contributions in improving the clarity of this publication. I would also like to thank Yong Il Lee, Daekyo Cheong, and Shigeru Otoh for their academic supports during my postgraduate study in Korea.

Author details

Kosuke Egawa[1,2,3*]

Address all correspondence to: egawa.k@aist.go.jp

1 School of Earth and Environmental Sciences, Seoul National University, Seoul, Republic of Korea

2 Institute for Geo-Resources and Environment, National Institute of Advanced Industrial Science and Technology, Tsukuba, Japan

3 Methane Hydrate Research Center, National Institute of Advanced Industrial Science and Technology, Sapporo, Japan

References

[1] Isozaki Y. Anatomy and genesis of a subduction-related orogen: A new view of geo-tectonic subdivision and evolution of the Japanese Islands. Island Arc 1996;5(3) 289–320.

[2] Maruyama S, Isozaki Y, Kimura G, Terabayashi M. Paleogeographic maps of the Japanese Islands: plate tectonic synthesis from 750 Ma to the present. Island Arc 1997;6(1) 121–142.

[3] Westerman GEG., editor. The Jurassic of the Circum–Pacific. Cambridge: Cambridge University Press, 1992.

[4] Kim JH. Mesozoic tectonics in Korea. Journal of Southeast Asian Earth Sciences 1996;13(3–5) 251–265.

[5] Sagong H, Kwon ST, Ree JH. Mesozoic episodic magmatism in South Korea and its tectonic implication. Tectonics 2005;24(5) TC5002.

[6] Li ZX, Li XH. Formation of the 1300-km-wide intracontinental orogen and postorogenic magmatic province in Mesozoic South China: A flat-slab subduction model. Geology 2007;35(2) 179–182.

[7] Egawa K, Lee YI. Jurassic synorogenic basin filling in western Korea: sedimentary response to inception of the western Circum-Pacific orogeny. Basin Research 2009;21(4) 407–431.

[8] Choi T, Lee YI, Orihashi Y. Mesozoic detrital zircon U-Pb ages of modern river sediments in Korea: implications for migration of arc magmatism in the Mesozoic East Asian continental margin. Terra Nova 2012;24(2) 156–165.

[9] Faure M, Lin W, Chen Y. Is the Jurassic (Yanshanian) intraplate tectonics of North China due to westward indentation of the North China block? Terra Nova 2012;24(6) 456–466.

[10] Watson MP, Hayward AB, Parkinson DN, Zhang ZM. Plate tectonic history, basin development and petroleum source rock deposition onshore China. Marine and Petroleum Geology 1987;4(8) 679–702.

[11] Lee YI, Sheen DH. Detrital modes of the Pyeongan Supergroup (Late Carboniferous–Early Triassic) sandstones in the Samcheog coalfield, Korea: implications for provenance and tectonic setting. Sedimentary Geology 1998;119(3–4) 219–238.

[12] Chough SK, Kwon ST, Ree JH, Choi DK. Tectonic and sedimentary evolution of the Korean peninsula: A review and new view. Earth Science Reviews 2000;52(1–3), 175–235.

[13] Şengör AMC, Natal'in BA. Paleotectonics of Asia: fragments of a synthesis. In: Yin A, Harrison M. (eds.) The tectonic evolution of Asia. Cambridge: Cambridge University Press; 1996. p486–640.

[14] Golonka J. Late Triassic and Early Jurassic palaeogeography of the world. Palaeogeography, Palaeoclimatology, Palaeoecology 2007;244(1–4) 297–307.

[15] Oh CW. Systematic changes in metamorphic styles along the Dabie–Hongseong and Himalayan collision belts, and their tectonic implications. Journal of Asian Earth Sciences 2010;39(6) 635–644.

[16] Kim HS, Ree JH. Permo-Triassic changes in bulk crustal shortening direction during deformation and metamorphism of the Taebaeksan Basin, South Korea using foliation intersection/inflection axes: Implications for tectnic movement at the eastern margin of Eurasia during the Songrim (Indosinian) orogeny). Tectonophysics 2013;587, 133–145.

[17] Park TH. Chronological and Petrological Study of Cretaceous to Paleogen Granitic Rocks, South Korea. PhD Dissertation, the University of Tokyo; 2009.

[18] Kee WS, Kim SW, Jeong YJ, Kwon S. Characteristics of Jurassic continental arc magmatism in South Korea: tectonic implications. The Journal of Geology 2010;118(3) 305–323.

[19] Chang KH. Aspect of Geologic History of Korea. Journal of the Geological Society of Korea 1995;31(1) 72–90.

[20] Adachi M, Suzuki K. Were Precambrian gneiss clasts in the Kamiaso conglomerate derived from the eastern Korean Peninsula? Bulletin of the Nagoya University Furukawa Museum 1993;9, 25–45.

[21] Ohta T. Geochemistry of Jurassic to earliest Cretaceous deposits in the Nagato Basin, SW Japan: implication of factor analysis to sorting effects and provenance signatures: Sedimentary Geology 2004;171(1–4) 159–180.

[22] Nutman AP, Sano Y, Terada K, Hidaka H. 743±17 Ma granite clast from Jurassic conglomerate, Kamiaso, Mino Terrane, Japan: the case for South China Craton provenance (Korean Gyeonggi Block?). Journal of Asian Earth Sciences 2006;26(1) 99–104.

[23] Joo YJ, Lee YI, Hisada K. Provenance of Jurassic accretionary complex: Mino terrane, inner zone of south-west Japan – implications for palaeogeography of eastern Asia. Sedimentology 2007;54(3) 515–543.

[24] Kim Y, Lee YI, Hisada K. Provenance of quartzarenite clasts in the Tetori Group (Middle Jurassic to Early Cretaceous), Japan: Paleogeographic implications: Journal of Asian Earth Sciences 2007;29(1) 116–126

[25] Lee YI. Paleogeographic reconstructions of the East Asia continental margin during the middle to late Mesozoic. Island Arc 2008;17(4) 458–470.

[26] Lee YI, Lee JI. Paleozoic sedimentation and tectonics in Korea: A review. Island Arc 2003;12(2) 162–179.

[27] Chough SK. Geology and Sedimentology of the Korean Peninsula. Oxford: Elsevier; 2013.

[28] Yanai S, Park BS, Otoh S. The Honam shear zone (South Korea): deformation and tectonic implication in the Far East. Scientific Papers of College of Arts and Sciences. University of Tokyo: 1985; 35, p181–210.

[29] Otoh S, Yanai S. Mesozoic inversive wrench tectonics in far east Asia: examples from Korea and Japan. In: Yin A, Harrison M. (eds.) The tectonic evolution of Asia. Cambridge: Cambridge University Press; 1996. p401–419.

[30] Koh HJ. Tectonic implication of the Mungyeong–Jeongseon tectonic line, the Yeongweol Nappe and the Bansong Group in the Ogcheon belt. In: Kee WS. (ed.) Mesozoic crustal evolution of Northeast Asia. Daejeon: Korea Institute of Geoscience and Mineral Resources; 2006. Report OAA2004009–2006(3), p228–259.

[31] Lim C. Cho M. Two-phase contractional deformation of the Jurassic Daebo Orogeny, Chungnam Basin, Korea, and its correlation with the early Yanshanian movement of China. Tectonics 2012;31(1) TC1004.

[32] Cluzel D. Formation and tectonic evolution of early Mesozoic intermontane basins in the Ogcheon belt (South Korea): a reappraisal of the Jurassic "Daebo orogeny". Journal of Southeast Asian Earth Science 1992;7(4) 223–235.

[33] Han R, Ree JH, Cho DL, Kwon ST, Armstrong R. SHRIMP U–Pb zircon ages of pyroclastic rocks in the Bansong Group, Taebaeksan Basin, South Korea and their implication for the Mesozoic tectonics. Gondwana Research 2006;9(1–2) 106–117.

[34] Jeon H, Cho M, Kim H, Horie K, Hidaka H. Early Archean to Middle Jurassic evolution of the Korean Peninsula and its correlation with Chinese cratons: SHRIMP U–Pb zircon age constraints. The Journal of Geology 2007;115(5) 525–539.

[35] Seo J, Choi SG, Oh CW. Petrology, geochemistry, and geochronology of the post-collisional Triassic mangerite and syenite in the Gwangcheon area, Hongseong Belt, South Korea. Gondwana Research 2010;18(2–3) 479–496.

[36] Oh CW, Kim SW, Choi SG, Zhai M, Guo J, Sajaav K. First finding of eclogite facies metamorphic event in South Korea and its correlation with the Dabie–Sulu collision belt in China. The Journal of Geology 2005;113(2) 226–232.

[37] Kim SW, Oh CW, Williams IS, Rubatto D, Ryu IC, Rajesh VJ, Kim CB, Guo J, Zhai M. Phanerozoic high-pressure eclogite and intermediate-pressure granulite facies metamorphism in the Gyeonggi massif, South Korea: Implications for the eastward extension of the Dabie–Sulu continental collision zone. Lithos 2006;92(3–4) 357–377.

[38] Kim SW, Williams IS, Kwon S, Oh CW. SHRIMP zircon geochronology, and geochemical characteristics of metaplutonic rocks from the south-western Gyeonggi block, Korea: Implications for Paleoproterozoic to Mesozoic tectonic links between the Korean Peninsula and eastern China. Precambrian Research 2008;162(3–4) 475–497.

[39] Kwon S, Kim SW, Santosh M. Multiple generations of mafic-ultramafic rocks from the Hongseong suture zone, western South Korea: Implications for the geodynamic evolution of NE Asia. Lithos 2013;160–161, 68–83.

[40] Egawa K, Lee YI. Thermal maturity assessment of the Upper Triassic to Lower Jurassic Nampo Group, mid-west Korea: Reconstruction of thermal history. Island Arc 2008;17(1) 109–128.

[41] Egawa K, Lee YI. Stratigraphy of the Nampo Group in the Ocheon and Oseosan areas: significance of conglomerates of the Jogyeri Formation for unconformity. Journal of the Geological Society of Korea 2006;42(4) 635–643. (in Korean)

[42] Egawa K, Lee YI. K–Ar dating of illites for time constraint on tectonic burial metamorphism of the Jurassic Nampo Group (West Korea). Geosciences Journal 2011;15(2) 131–135.

[43] Egawa K. Stratigraphy, sedimentation, and basin evolution of the Jurassic Chungnam Basin, western Korea. PhD Dissertation, Seoul National University; 2009.

[44] Egawa K. Geologic records of Andean-type orogeny in Korea. Proceedings of the 21st Century Science and Human Symposium, Tokai University, Tokyo; 2010. (in Japanese with English abstract)

[45] Li XH, Li ZX, Li WX, Liu Y, Yuan C, Wei G, Qi C. U–Pb zircon, geochemical and Sr–Nd–Hf isotopic constraints on age and origin of Jurassic I- and A-type granites from

central Guangdong, SE China: A major igneous event in responce to founding of a
subducted flat-slab? Lithos 2007;96(1–2) 186–204.

[46] Kiminami K, Kishita S, Imaoka T. Marked change in sandstone composition during
the Middle Jurassic in Jurassic accretionary complexes of SW Japan, and geologic significance. Journal of the Geological Society of Japan 2009;115(11) 578–596. (in Japanese with English abstract)

[47] Gutscher MA, Olivet JL, Aslanian D, Eissen JP, Maury R. The "lost Inca Plateau":
cause of flat subduction beneath Peru? Earth and Planetary Science Letters
1999;171(3) 335–341.

[48] Gutscher MA, Maury R, Eissen JP, Bourdon E. Can slab melting be caused by flat
subduction? Geology 2000;28(6) 535–538.

[49] Gutscher MA, Spakman W, Bijwaard H, Engdahl ER. Geodynamics of flat subduction: Seismicity and tomographic constraints from the Andean margin. Tectonics
2000;19(5) 814–833.

[50] McQuarrie N, Horton BK, Zandt G, Beck S, DeCelles PG. Lithospheric evolution
ofthe Andean fold–thrust belt, Bolivia, and the origin of the central Andean plateau.
Tectonophysics 2005;399(1–4) 15–37.

[51] van Hunen J, van den Berg AP, Vlaar NJ. On the role of subducting oceanic plateaus
in the development of shallow flat subduction. Tectonophysics 2002;352(3–4) 317–
333.

[52] Wakita K, Metcalfe I. Ocean plate stratigraphy in East and Southeast Asia. Journal of
Asian Earth Sciences 2005;24(6) 679–702.

[53] Takami M, Itaya T. Episodic accretion and metamorphism of Jurassic accretionary
complex based on biostratigraphy and K–Ar geochronology in the western part of
the Mino–Tanba Belt, Southwest Japan. Island Arc 1996;5(3) 321–336.

[54] Koizumi K, Ishiwatari A. Oceanic plateau accretion inferred from Late Paleozoic
greenstones in the Jurassic Tamba accretionary complex, southwest Japan. Island Arc
2006;15(1) 58–83.

[55] Ichiyama Y, Ishiwatari A, Koizumi K. Petrogenesis of greenstones from the Mino–
Tamba belt, SW Japan: Evidence for an accreted Permian oceanic plateau. Lithos
2008;100(1–4) 127–146.

[56] Otofuji Y, Matsuda T, Nohda S. Opening mode of the Japan Sea inferred from the pa-
laeomagnetism of the Japan Arc. Nature 1985;317(6038) 603–604.

[57] Sugamori Y. Upper Permian Takatsuki Formation, Middle Triassic Shimamoto Formation and Triassic sedimentary complex in the Nishiyama area, Osaka and Kyoto
prefectures, SW Japan: Journal of the Geological Society of Japan 2006;112(6) 390–406.
(in Japanese with English abstract)

[58] Uchida E, Choi SG, Baba D, Wakisaka Y. Petrogenesis and solidification depth of the Jurassic Daebo and Cretaceous Bulguksa granitic rocks in South Korea. Resource Geology 2012;62(3) 281–295.

[59] Beate B, Monzier M, Spikings R, Cotten J, Silva J, Bourdon E, Eissen JP. Mio-Pliocene adakite generation related to flat subduction in southern Ecuador: the Quimsacocha volcanic center. Earth and Planetary Science Letters 2001;192(4) 561–570.

[60] Zhou X, Sun T, Shen W, Shu L, Niu Y. Petrogenesis of Mesozoic granitoids and volcanic rocks in South China: A response to tectonic evolution. Episodes 2006;29(1) 26–33.

[61] Kinoshita O. Migration of igneous activities related to ridge subduction in Southwest Japan and the East Asian continental margin from the Mesozoic to the Paleogene. Tectonophysics 1995;245(1–2) 25–35.

[62] Lister G, Forster M. Tectonic mode switches and the nature of orogenesis. Lithos 2009;113(1–2) 274 –291.

[63] Kim KH, Van der Voo R. Jurassic and Triassic paleomagnetism of South Korea. Tectonics 1990;9(4) 699–717.

[64] Kang SS, Kim JM, Jang BA. Paleostress fields from calcite twins in the Pyeongan Supergroup, South Korea. Island Arc 2005;14(2) 137–149.

[65] Seo HG, Kim DS, Lee CB, Bae DJ, Jo MJ. Geology of the Ungcheon and Misan areas, Chungnam Coalfield (II). Daejeon: Korea Institute of Energy Research; KIER researches of coal resources 1982;4. (in Korean with English abstract)

[66] Choi HI, Kim DS, Seo HG. Stratigraphy, depositional environment and basin evolution of the Daedong strata in the Chungnam Coalfield. Daejeon: Korea Institute of Energy Research; KIER research report (KR–87–B–3) 1987. (in Korean with English abstract)

[67] Paola C, Heller PL, Angevine CL. The large-scale dynamics of grain-size variation in alluvial basins, 1: Theory. Basin Research 1992;4(2) 73–90.

[68] Heller PL, Paola C. The large-scale dynamics of grain-size variation in alluvial basins, 2: Application to syntectonic conglomerate. Basin Research 1992;4(2) 91–102.

[69] Lee CH, Lee HK, Kim KW. Petrochemistry and environmental geochemistry of shale and coal from the Daedong Supergroup, Chungnam Coal Field, Korea. Economic and Environmental Geology 1997;30(5) 417–431. (in Korean with English abstract)

[70] Park SW, Park HS. Property of the Jurassic anthracite (Anthracite from the Seongju area of the Chungnam Coalfield). Journal of the Korean Institute of Mining Geology 1989;22(2) 129–139. (in Korean with English abstract)

[71] Choi SW, Kim JH. Organic metamorphism of the coals in the Chungnam Coalfield, Korea. Journal of the Korean Earth Science Society 1997;18(2) 146–154. (in Korean with English abstract)

[72] Blair TC. Tectonic and hydrologic controls on cyclic alluvial fan, fluvial, and lacustrine rift-basin sedimentation, Jurassic–lowermost Cretaceous Todos Santos Formation, Chiapas, Mexico. Journal of Sedimentary Petrology 1987;57(5) 845–862.

[73] Dorsey RJ, Umhoefer PJ, Falk P. Earthquake clustering inferred from Pliocene Gilbert-type fan deltas in the Loreto basin, Baja California Sur, Mexico. Geology 1997;25(8) 679–682.

[74] Ricci Lucchi F. The Oligocene to recent foreland basins of the northern Apennines. In: Allen PA, Homewood P. (eds.) Foreland Basins. London: Special Publication of the International Association of Sedimentologisits 8; 1986. p105–139.

[75] Lawton TF, Pollock SL, Robinson RAJ. Integrating sandstone petrology and nonmarine sequence stratigraphy: application to the late cretaceous fluvial systems of southwestern Utah, U.S.A. Journal of Sedimentary Research 2003;73(3) 389–406.

[76] Ori GG, Friend PF. Sedimentary basins formed and carried piggyback on active thrust sheets. Geology 1984;12(8) 475–478.

[77] Chiang CS, Yu HS, Chou YW. Characteristics of the wedge-top depozone of the southern Taiwan foreland basin system. Basin Research 2004;16(1) 65–78.

[78] Cheong CS, Kee WS, Jeong YJ, Jeong GY. Multiple deformations along the Honam shear zone in southwestern Korea constrained by Rb-Sr dating of synkinematic fabrics: Implications for the Mesozoic tectonic evolution of northeastern Asia. Lithos 2006;87(3–4) 289–299.

[79] Otoh S, Tsukada K, Sano K, Nomura R, Jwa YJ, Yanai S. Triassic to Jurassic dextral ductile shearing along the eastern margin of Asia: A synthesis. In: Metcalfe I, Jishun R, Charvet J, Hada S. (eds.) Gondwana dispersion and Asian accretion. IGCP 321 final results volume. Rotterdam: Brookfield, A.A.Balkema; 1999. p89–113.

[80] Niwa M. The structure and kinematics of an imbricate stack of oceanic rocks in the Jurassic accretionary complex of Central Japan: an oblique subduction model. Journal of Structural Geology 2006;28(9) 1670–1684.

[81] McGeary S, Nur A, Ben-Avraham Z. Spatial gaps in arc volcanism: The effect of collision or subduction of oceanic plateaus. Tectonophysics 1985;119(1–4) 195–221.

[82] Rosenbaum G, Giles D, Saxon M, Betts PG, Weinberg RF, Duboz C. Subduction of the Nazca Ridge and the Inca Plateau: Insights into the formation of ore deposits in Peru. Earth and Planetary Science Letters 2005;239(1–2) 18–32

[83] Seo HG, Kim DS, Lee CB, Paik DJ. Geology of the Hongsan and Naesan areas, Chungnam Coalfield (III). Daejeon: Korea Institute of Energy Research; KIER researches of coal resources 1983;5. (in Korean with English abstract)

[84] Underwood MB, Laughland MM, Kang SM. A comparison among organic and inorganic indicators of diagenesis and low-temperature metamorphism, Tertiary Shimanto Belt, Shikoku, Japan. Geological Society of America Special Paper 1993;273, 45–61.

[85] Kosakowski G, Kunert V, Clauser C, Franke W, Neugebauer HJ. Hydrothermal transients in Variscan crust: Paleo-temperature mapping and hydrothermal models. Tectonophysics 1999;306(3–4) 325–344.

[86] Clauer N, Chaudhuri S, Kralik M, Bonnot-Courtois C. Effect of experimental leaching on Rb–Sr and K–Ar isotopic systems and REE contents of authigenic illite. Chemical Geology 1993;103(1–4) 1–16.

[87] Dong H, Hall CM, Peacor DR, Halliday AN. Mechanisms of argon retention in clays revealed by laser ^{40}Ar–^{39}Ar dating. Science 1995;267(5196) 355–359.

[88] Hoffman J, Hower J, Aronson JL. Radiometric dating of time of thrusting in the distributed belt of Montana. Geology 1976;4(1) 16–20.

[89] Pevear DR. Illite age analysis, a new tool for basin thermal history analsis. In: Kharaka YK, Maest AS. (eds.) Water-Rock Interaction. Rotterdam: A.A. Balkema; 1992. p1251–1254.

[90] Grathoff GH, Moore DM. Illite polytype quantification using WILDFIRE calculated X-ray diffraction patterns. Clays and Clay Minerals 1996;44(6) 835–842.

[91] Haines SH, van der Pluijm BA. Clay quantification and Ar–Ar dating of synthetic and natural gouge: Application to the Miocene Sierra Mazatán detachment fault, Sonora, Mexico. Journal of Structural Geology 2008;30(4) 525–538.

[92] Uysal IT, Glikson M, Golding SD, Audsley F. The thermal history of the Bowen Basin, Queenland, Australia: Vitrinite reflectance and clay mineralogy of Late Permian coal measures. Tectonophysics 2000;323(1–2) 105–129.

[93] Harrison MJ, Marshak S, Onasch CM. Stratigraphic control of hot fluids on anthracitization, Lackawanna synclinorium, Pennsylvania. Tectonophysics 2004;378(1–2) 85–103.

Chapter 5

Tectonic Process of the Sedimentary Basin Formation and Evolution in the Late Cenozoic Arc-Arc Collision Zone, Central Japan

Akira Takeuchi

Additional information is available at the end of the chapter

1. Introduction

The Japanese archipelagoes in the northwest Pacific consist of southwest Japan arc, Izu - Ogasawara arc, Ryukyu arc and Northeast Japan arc and Krile arc, and the Japan Sea is a marginal sea of northeast Asian continent separated from the northwest Pacific by three islands of Sakhalin, Hokkaido, and Honshu. The thrust-and-fold zones of the Neogene and Quaternary systems in the coastal tectonic belt along the eastern margin of Japan Sea continue their activity until now. Particularly, the seismogenic zone continued from Sakhalin is arrested with the convergent boundary between Amur Plate and the Okhotsk Plate because reverse-fault type of earthquakes with magnitudes larger than 7.4 in Richter's scale have been occurred along this heteromorphic belt every several hundred years. From this point of view to the plate tectonic condition around Japan, the central Japan acts as a multiple junction area unique in the earth, where four pieces of plates, such as the Amur, the Okhotsk, the Philippine Sea, and the Pacific plates, gather and converge together in and around the Japanese archipelagoes (Figure 1). Although the GPS geodetic observations confirmed the presence of the micro-plates [3-5], the structural features of the incipient boundary between the Okhotsk and Amur Plates seemed still immature but recognized as strain accumulated zone along the eastern margin of Japan Sea [6-11].

The Pacific plate have subducted beneath the Okhotsk, the Philippine Sea, and Amur plates which are converging together. In particular, central Japan is the place where the Southwest Japan, the Northeast Japan, the Izu-Ogasawara arcs mutually collide, and therefore the process and mechanism of the development of geologic structures here is quite complicated and hard to be interpreted. Lately, in such situation, the damaging earthquakes of the middle scale have

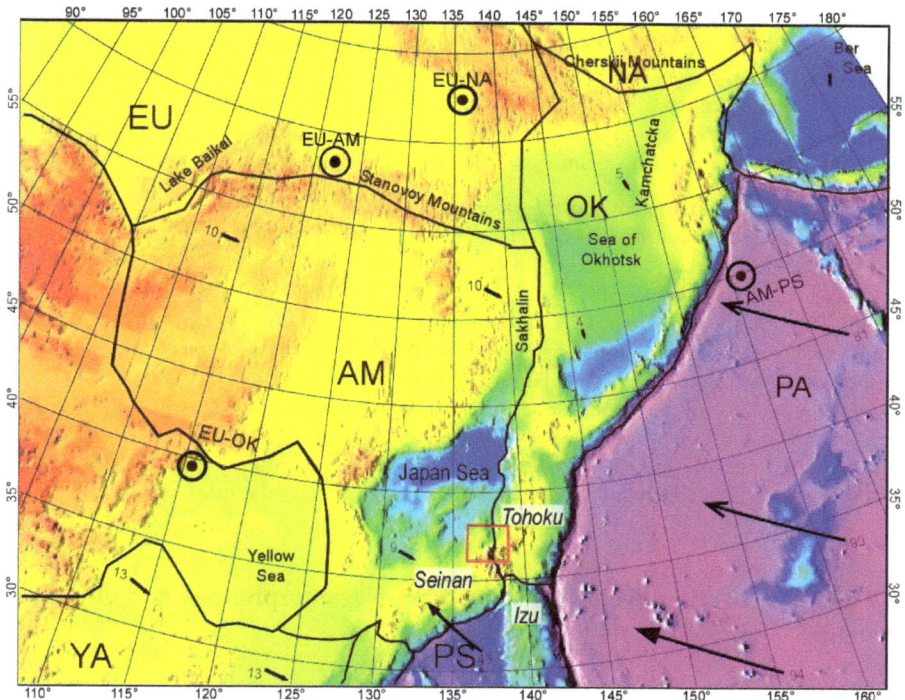

Figure 1. Index Map of Plate Framework in the Northeast Asia. Study area is depicted by open red box. Boundaries of the Okhotsk (OK) and Amur (AM) plates are shown. Surrounding plates include Eurasia (EU), North America (NA), Pacific (PA), Philippine Sea (PS), and Yangtze (YA). Black vectors give model velocities (with numbers in mm/a) relative to plate whose identifier is underlined. Black circles are locations of Euler poles. Simplified from [1] with an addition of Euler pole EU-AM [2].

occurred in the Japan Sea side of the central Japan in succession. In 2004, the Chuetsu earthquake ($M_W6.6$) occured in the inland Chuetsu area, Niigata Prefecture. Two years and nine months later, in 2007, the Noto Peninsula earthquake ($M_W6.7$) occurred at a 11km deep hypocenter beneath the west coast of northern Noto Peninsula. Then, only 3.7 months later, the Niigata Prefecture Chuetsu-oki earthquake ($M_W6.7$) occurred approximately 15km in depth, 32km distant from the epicenter of the 2004 earthquake, attracting an attention to the relations of three earthquakes from a time-space point of view. Furthermore, the Naganoken Hokubu earthquake (M_W 6.35) was generated by the hypocenter 8km in depth at a moment 13 hours 13 minutes after the 2011 off the Pacific coast of Tohoku Earthquake (M_W 9.0) occurred in the Japan Trench on March 11, 2011.

As for the generation style of middle scale and larger earthquakes occurring along the eastern margin of Japan Sea, including the above mentioned earthquakes in central Japan, major listric faults contributed with back arc spreading during the Miocene has been explained as inversion tectonics which inverted the sense of faulting from normal dip-slip into the reverse slip since the Pliocene. However, the western Noto Peninsula earthquake and the Naganoken Hokubu earthquake were the focal mechanisms that they could not be explained by means of simple inversion tectonics concerning the geologic structure of shallower layers so far. Main shocks of these earthquakes are commonly occuured around 15km in depth, which is almost equivalent to the basal depth of the seismogenic layer.

Since the thick sedimentary cover of the seismogenic layer has been remarkably deformed into faulted and folded structures, the character of tectonics in progress at the Present is questionable whether thin-skinned tectonics or basement-involved tectonics, being concerned with the presence of the detachment surface between the seismogenic layer and the sedimentary cover [12-14]. Thus, this paper discusses the formation process of sedimentary basins in the collision zone between island arcs from the viewpoint of earthquake tectonics about the recent crustal earthquakes. For the purpose of elucidating the active tectonics (crustal movement in progress) along the Japan Sea coast in central Japan, this paper focuses on the specificity of the geologic structure and geomorphology of Eastern Hokuriku district and Fossa Magna - Toyama Trough region, and traces the history of geomorphologic and geologic development in order to propose a comprehensive relation with earthquake occurrence and crustal movement.

2. The features and development process of geomorphology and geologic structure

2.1. Target area

The target area for this paper, the Hokuriku-Shin'etsu district, is composed of three adjacent Neogene sedimentary basins located in the Japan Sea side of central Japan including the seabed area (Toyama Trough) between Noto Peninsula and Sado Island (Figure 2). The trough is administratively enclosed by Ishikawa, Toyama, and Niigata Prefectures. During the early to middle Miocene periods the Hokuriku, Shin'etsu, and Niigata basins had developed obliquely upon the basement geologic zones geotectonically belonging to the inner belt of pre-Cenozoic Southwest Japan. Although Shin'etsu, and Niigata basins tends to be treated as a single sedimentary basin, herein, the most part of Niigata basin is excluded from the Fossa Magna area as long as geomorphology and geologic history are concerned [17, 18]. In addition, the north-south trending, narrow basin in the central part of continental slope offing the Japan Sea side of Honshu, the Toyama Trough, borders Northeast Japan and Southwest Japan in the seabed area.

2.2. Tectonic provinces of target area

When a zonal division is available on the basis of regional characteristics of fault distribution such as fault length and orientation, inclination, type of displacement sense (normal, reverse, or strike-slip), and the density of fault distribution in a certain geological age, a tectonic unit in this paper is defined as a fault province. A fault province composed of active faults is called

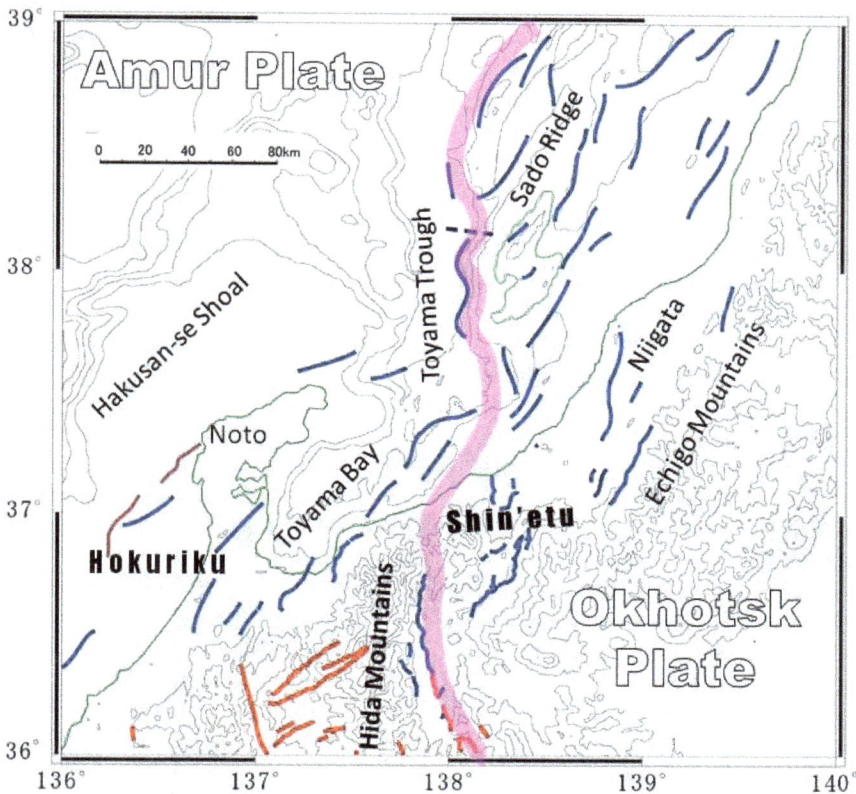

Figure 2. Index Map of Active Faults for the central Japan. Blue line denotes reverse fault, red does strike-slip fault. Thick pink line indicates plate boundary between Amur and Okhotsk plates. Place names are also indicated. Simplified and compiled from [15, 16].

an active fault province which reflects regional characteristics of the seismogenic stress field. From this point of view to the recent crustal movement regionally in a geodetic to geological time scale, a strain concentration zone denotes a geodetic zone where a pattern of displacement field demonstrates a belt of larger strain rate, and geologically it corresponds to a zone where deformation structures such as faults and/or folds develop intensively [6, 9, 19, 20].

Including Mizuho-Fossa Magna folded belt [21], concentrated deformation belts were known in many places in the Cenozoic Japan, but existence of the Niigata - Kobe tectonic belt [19] becomes recognized by the GPS precise geodetic observation network of Geospacial Information Authority of Japan (GSI) having been maintained in and after 1995.

Figure 2 shows active fault distribution [16] and the active fault province [15, 22] of the inner Chubu District. The reverse fault province occupies the inner Tohoku arc and the strike-slip fault province is located inland area adjacent to the Hida mountain range. Concerning generation of earthquakes in the inland crust, the strong shortening in the Niigata - Kobe tectonic belt lately has attracted attention by intersecting the active fault provinces and the plate boundary between the Amur and Okhotsk plates, Itoigawa - Shizuoka tectonic line as is recognized from Figure 3.

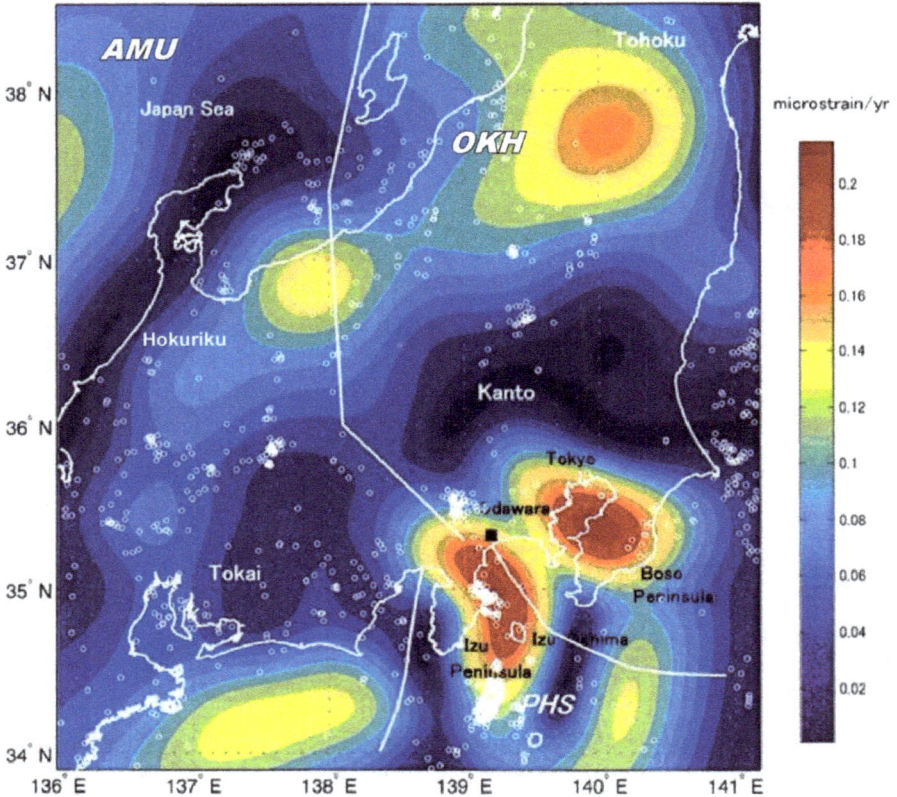

Figure 3. Maximum shear strain rates in central Japan. Estimated from the two-year improved time series data from April 1996 to March 1998 [25]. White circles indicate epicenters of the earthquakes with depths shallower than 30 km and magnitudes greater than 3.0 during the period from January 1996 to March 1998. Note that the strain distribution belt intersects the Itoigawa-Shizuoka tectonic line bounding Amur and Okhotsk plates. This belt corresponds to the Shinanogawa seismic zone and Atotsugawa fault zone in the Niigata Kobe tectonic zone [19].

2.3. Plate tectonic framework

From the plate tectonic point of view, the central Japan acts as a multiple junction area unique in the earth where four pieces of plates, such as the Amur, the Okhotsk, the Philippine Sea, and the Pacific plates, gather and converge together in and around the Japanese archipelagoes. The border of Amur plate and the Okhotsk plate has just jumped from the west margin of the Hidaka Mountain Range into the eastern margin of Japan Sea at about 0.5Ma. The former plate boundary between the North American plate and the Eurasian plate had been situated in the central Hokkaido where another collision between the Kurile and the Tohoku arcs had performed. As for the Seinan fore-arc, the commencement of subduction with the northing of Philippine Sea plate was represented by the 15Ma intrusions of outer-zone granite and the bended structure of the earlier Nankai trough caused by the paleo-Izu indentation at 15-14Ma. This remarkable transition might have affected the convergent boundary between the Eurasia plate and the North America plate and the both continental plates would be put together in the collision state. Contrastingly, the Pacific plate has continued almost steady subduction along the Japan Trench for the past 40 million years without significant change in the north-westward motion, despite tectonic episodes of back-arc spreading in Japan Sea, Okhotsk Sea and Shikoku Basin.

In the eastern margin of Japan Sea and the Fossa Magna region, the environment of the crustal movement switched totally from the calm period in the late half of Miocene to the Pliocene contraction tectonics. The start of folding in the northern Fossa Magna region dates up by evidence of the paleomagnetism in at least 4Ma [24]. However, the start of folding was much older because of the sedimentological fact that turbidite flowed down the trough-like basins of syncline and the stratigraphic fact that the base of Pliocene andesites (5.4Ma) covered obliquely the anticline which has already begun growth [25-27].

By the way, due to the migration of trench triple junction, the moving direction of the Philippine Sea plate switched at 3Ma from the north direction to northwest [28], and, therefore, the colliding force against the border area between east and west Japan as well as the southern Fossa Magna should have weakened in comparison with the past. The contraction tectonics in the Japan Sea side could be attributed to starting of eastward motion of the Amur plate, because the start of the contractinal tectonics in the eastern margin of Japan Sea was significantly older than 3Ma.

2.4. Time scale setting

As for the upper Cenozoic system distributed over the Hokuriku and Shin'etsu areas, the biochronological stratigraphy was almost established in the 1980s [29-31]. A complicated stratigraphy on terrestrial sediments of the lower Cenozoic system widely distributed in Noto Peninsula has become elucidated based on age-determination data of volcanic rocks [31].

In addition, in late years for the purpose of analysis of the marine paleoenvironment, high precision chronostratigraphy is performed by means of age-marker for the period after the Pleistocene in particular.

Based on the recent advance in the Pliocene stratigraphic correlation and age determination of tephra distributed widely over the central Japan [32, 33], there was large progress for historical studies on the fault activities and upheaval of Hida Mountain ranges [34-36]. In conformity with these results, this paper also obeys a new definition of the Quaternary period recently revised by the International Stratigraphic Committee (http://quaternary.stratigraphy.org/definitions/).

3. Formation and development of sedimentary basins

The Present Hida Plateau and Noto Peninsula are upheaval zones which expose the basement rocks of pre-Cenozoic system, and are different from the Present coastal plains and near shore waters which comprise the thick sedimentary layers. As for the approximately 5 million years period previous than 1 million years ago, it is thought that the area of Noto Peninsula is a large terriginous flat or is a very shallow archipelago [30], and that this area formed a peninsula after 0.5 Ma [38]. This paper considers the geomorphplogical development of the seabed and coastal places of Japan Sea mainly for block structures of Honshu Island since the Oligocene, with paying attention to the following five stages of crustal movement concerned with a geological development of Southwest Japan west of the Itoigawa-Shizuoka tectonic line. Besides, Northeast Japan saying in this paper includes the northern Fossa Magna region for convenience and excludes the southern Fossa Magna region.

3.1. Rifting phase [32 Ma — 28 Ma]

The drillings into Japan Basin and Yamato Basin conducted in 1989 by International Ocean Drilling Program (ODP) provided an important data related to the timing of formation of the Japan Sea area. According to [39], the formation of the Japan Basin began by thinning of the continental crust in the early Oligocene (32 Ma), and such a tectonic style changed into expanding of the sea floor in the late Oligocene (28 Ma).

3.2. Sea floor spreading phase [28Ma — 18Ma]

The tectonic domain of the sea floor spreading in the Japan Sea area had moved from the widened Japan Basin area to the southwest, and formed both Yamato Basin and Tsushima Basin by crustal expansion, but it ceased in 18 million years ago [39-42]. The rifted structures with trends of north-south direction or northwest-southeast were formed in Toyama Trough and the Hokuriku and Niigata areas in the period from the end of Oligocene to the early Miocene [43-45]. In the Hokuriku district in the middle Early Miocene (20Ma-18Ma), submarine volcanic activities occurred and tearing of the basement, i.e. intra-arc rifting, formed graben-like depressions. According to[46], tectonically distinct boundary between Tohoku and Seinan Honshu arcs had been formed or activated at the end of this phase. Toyama Bay was originally an embayment that branched off the Toyama Trough into the Hokuriku area, and the sea-bottom faults along the coastal line were activated for the period of opening of Japan Sea [9].

3.3. Sedimentary-basin forming stage [18Ma — 15Ma]

Marine sedimentary basins were formed by sudden subsidence to reach 3,000m during this period in the Hokuriku district [30, 47, 48]. These sedimentary basins are located in the almost same places of the Present coastal alluvial plains including Kaga and Toyama Plains.

Intense volcanic activities occurred in the inner belt of Honshu arc on the Japan Sea side of Tohoku region, and submarine volcaniclastics known as 'green tuff' deposited in the period from 24Ma to 14Ma. In detail, around the time of 17Ma, southerly warm current water called 'paleo-Kuroshio' had become to emerge the back arc area, and then in substitution relatively calm subsidence commenced at 16Ma. During the period from 15Ma to 14Ma, marine transgression enlarged the entire intra-arc area, and the bathypelagic black mudstones (Nanatani Formation and its correlatives) were deposited in seabed area where the previous morphologic ups and downs would be buried.

In the Hokuriku district, the depth of the bedrock is at least 2,000m - 3,000m for last Neogene of Kaga - Toyama plains sandwiched between Noto Peninsula and the Hida Highlands [50]. In addition, the zone of relatively high-density rocks such as andesite and basalt lavas occupying the graben-like depressions of the basement is expressed as the narrow zone of highly positive features of Bouguer gravity anomaly [51-54, 35]. These kinds of volcanism are not product of the pervious syn-rift phase of back-arc spreading but of the intra-arc rifting due to commencement of arc volcanism of the Honshu arc [29, 31, 35].

3.4. Basin differentiation phase [15Ma — 5Ma]

Lateral variations in thickness of the middle to upper Miocene strata among the sedimentary basins became remarkable in this period. Spatial variety in sedimentary thickness of individual deposition centers were well documented in the Shin'etsu sedimentary basin, suggesting a syn-sedimentary fault-block movement [25]. In the Hokuriku district, however, the Present-day hilly and mountainous countries including Iouzen-Hodatsu Hill, Imizu Hill, and Yatsuo area and Noto Peninsula bordered the sedimentation basins, where the rates of sedimentation reduced during the time from 15 Ma until 13 Ma[55].

As for the Shimane Peninsula in the San-in district of the western Seinan arc, [56] mentioned that the sedimentary basin had begun its inversion tectonics under the crustal stress field of the north-south compression in 14 million years ago, and the formation of the Shinji folded zone was completed in 6 million years ago [57-59]. The expanse of such the north-south compression field became broader, and the concentrated zone of east-west trending reverse faults and related folds parallel to the Southwest Japan arc developed from San-in to Hokuriku districts in the Japan Sea side from 8 million years ago [60]. The late-Miocene east-westerly deformation zone in the Hokuriku district includes Houdatsu-san Kita fault zone in the southern part of Noto peninsula and Wakayama-gawa fault zone and in northern Noto Peninsula. Landfill underwent ahead through the Hokuriku sedimentary basin from the side of Hida area towards the former Toyama Bay. In the Noto Peninsula, however, nanofossil chronostratigraphy detected several times of hiatuses when glauconites produced on the seabed after the Middle Miocene [55, 61]. Therefore the Present Toyama Bay is the remains of the Miocene graben half of which had been filled with the sediment [9].

3.5. Evolving stage [5Ma — Present]

Upheaval and subsidence (i.e. undulation of the basement) with axes striking north-south began in the later Pliocene in the eastern part of Southwest Japan, but another tectonic regime of northwest-southeast compression has superposed by the collision with the Izu arc, and the structural trend in the northeast-southwest direction reaches the expression of remarkable current active structure [22, 59, 62, 63]. The fault block movement of this stage is a process of modification where the existing geologic structure change into new one, and is deeply participated in the geomorphology development such as Hida Mountain Ranges, Noto Peninsula, and Toyama Bay in the Present period. This process did not begin at the same time in all the areas of Southwest Japan, but a tendency to migrate from the southeast (the Tokai district) to the northwest (the Hokuriku district and the Kinki district) and to the northeast (the northern Fossa Magna) is recognized [64, 65]. In addition, the inversion process included locally the one of fault-slip sense where former normal faults trending north-south to north-east-southwest directions became re-activated as reverse faults (e.g., Kureha-yama fault: [66]). In the Hokuriku district of the Neogene sedimentary basins, however, alluvial plains continue their sedimentation without performing "basin inversion" like the Miocene Shin'etsu sedimentary basin, where the whole area of subsidence with thick sedimentary layers had changed reversely into the upheaval zone [27-29].

4. Tectonic inversion of sedimentary basins and related faults

The style and degree of basin inversion varies from the Tohoku arc to the Seinan arc. Although any collisional inversion of continental margin rift complexes did not occur in the Japan margin of Amur Plate, the whole basin uplift and major structural inversions with substantial thrust reactivation of earlier extensional structures have performed along the coastal belt of Japan Sea, such as the Shin'etsu basin and Sado ridge on the Tohoku side and Noto and San'in districts on the Seinan side. A gentle inversion of intra-arc rifts has occurred in the Niigata basin. While, in the Hokuriku district, the Miocene sedimentary basins remains as coastal plains or relatively low-lying area where remobilization of earlier master faults is not clear.

Some inversion mechanisms are intrinsic to the existence and lithospheric structure of the basin, and the likelihood of fault reactivation depends on the attitude of the existing fault plane such as the dip and strike to the principal stress axes [67]. If the existing fault were too steep, antithetic accessary faults might develop as new reverse faults in the footwall of the earlier extensional fault.

Figure 4 illustrates the spatial variation of inversion tectonics. The section is obtained by a tomographic inversion method in the analytical line from the western Fukushima through Echigo Plain, Sado Island and Toyama Trough to Yamato Trough [68]. The earlier normal faults are distinctively distributed in the lower part of Toyama Trough to the northwest on the

Hakusan-se Shoal and in the Yamato Trough, while the later inverted reverse faults developed on Sado Island and Niigata sedimentary basin. The latter corresponds to the strain concentration zone along the eastern margin of Japan Sea [9].

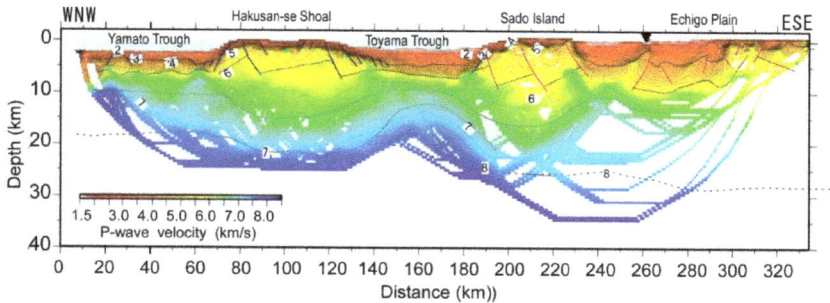

Figure 4. P wave velocity image for the crustal structure. The section is obtained by a tomographic inversion method in the analytical line from the western Fukushima through Echigo Plain, Sado Island and Toyama Trough to Yamato Trough. Faults are distinguished into earlier normal faults (blue) and later inverted reverse faults (red). Note their distinctive distribution. Compiled from [63].

As for the Tohoku arc, the start of eastward motion of Amur Plate at around 5 Ma [69] might have resulted in a new plate boundary along the strain concentration zone since 0.5 Ma. Moreover, [36] examined the U-Pb age data of Kurobegawa granite in the Hida mountain range and concluded that the granites were emplaced incrementally through the amalgamation of many intrusions since the late Miocene up to the latest intrusion event at 0.8 Ma, and that such magmatic intrusions caused rapid uplift and erosion of the Hida mountain range in the Quaternary.

As mentioned already, Japanese archipelagoes forming marginal seas between the northeastern Eurasian Continent and the northwestern Pacific Ocean comprise five island arcs (Kurile, Northeast Japan, Izu-Ogasawara, Southwest Japan, and Ryukyu arcs) which perform collisions each other in their adjacent terminations. Especially in central Japan three arcs (Northeast Japan, Izu-Ogasawara, and southwest Japan arcs) are mutually colliding, where deformed structures and active faults associated with inland crustal earthquakes are concentrated along the fringing zone east and south of Japan Sea. The mobile belt along the Japan margin of Amur Plate runs from Sakhalin - Hokkaido on the Okhotsk plate side, through the volcanic inner zones of the Northeast Japan arc, to the Southwest Japan arc on the Amur plate side [1, 2, 70, 71]. In detail, this belt includes the tectonic zone along eastern margin of Japan Sea, the Noto – San'in tectonic zone, and the Niigata - Kobe tectonic zone. Therefore, such a tectonic phenomenon could not be attributed to back arc compression of a single island-arc due simply to subduction of the oceanic plate on the Pacific side. The belt is situated in a circumference equivalent to the outer margin of the domain of back arc spreading of the Honshu arc.

Such characteristics of deformations and active faults in the inland crust as remarkable along the Japan Sea east margin is not seen at the epicentral and adjacent areas of Mw9.0 class trench-type earthquakes such as the Chile earthquake (on May 22, 1960 Mw 9.5), the Alaska earthquake (on March 28, 1964 Mw9.2), the Sumatra earthquake (on December 26, 2004 Mw9.0) and others. Conformably, the inland crustal strain accumulation and deformation directly by northwest motion of the Philippine Sea plate is not admitted conspicuous in the inner zone of southwest Japan either. After all the above-mentioned thing is the reason that cannot let the cause of inland East-West compression and the crustal earthquake generation in northeast Japan belong to northwest motion of the Pacific plate directly.

A feature of the tectonic stress field to produce crustal earthquakes can explain this most clearly. As shown in Figure 5, it is significant that a uniform compressional stress field wide spreads over the Japan Sea side of Hokkaido and Honshu, whereas variety of regionality is remarkable as for the Pacific coasts from Hokkaido to Kyushu [72].

5. Summary

5.1. Basin formation

Based on chronostratigraphy of Hokuriku established in late years, geomorphology, geologic structures and history of Hokuriku-Shin'etsu area were briefly summarized as follows.

After the marginal sea, i.e. Japan Sea, had been formed in the back arc area of the Honshu arc during the period from the Oligocene to Miocene time, there occurred broad transgression associated with calming of magma activity followed by cooling in central Japan. As the northward motion of the Philippine Sea plate commenced at around 15Ma, the western half of Honshu arc rotated clockwise with a decrease in area of the Japan Sea, while a buoyant subduction of Izu Arc into Honshu Arc had started.

Consequently, the mega-chasm from Fossa Magna to Toyama Trough was formed above the subducted paleo-Izu arc and the northern extension, and then the single Honshu arc differentiated into the Seinan arc and the Tohoku arc. During this process until 13Ma, the Hokuriku sedimentary basin in the Seinan arc and Shin'etsu and Niigata sedimentary basins in the Tohoku arc were developed in the Japan Sea side in the short term.

5.2. First tectonic inversion

According to [67], it is possible that the tectonic inversion was attributed to temporal variations in stress patterns within plates, resulting from forces caused by changes in plate boundary configuration. The sedimentary basins mentioned above had evolved individually in the period from the late half of middle Miocene to the beginning of the Pliocene. Namely, across-arc contraction tectonics with the E-W trending reverse faults and folds proceeded in the inner zone of Seinan arc, while along-arc subsiding piled up the thick sedimentary sequence in the inner zone of Tohoku arc. The start of buoyant subduction or collision of the Izu arc against the Seinan arc would have changed the configuration and relative motion at nearby plate boundaries as shown in Figure 6.

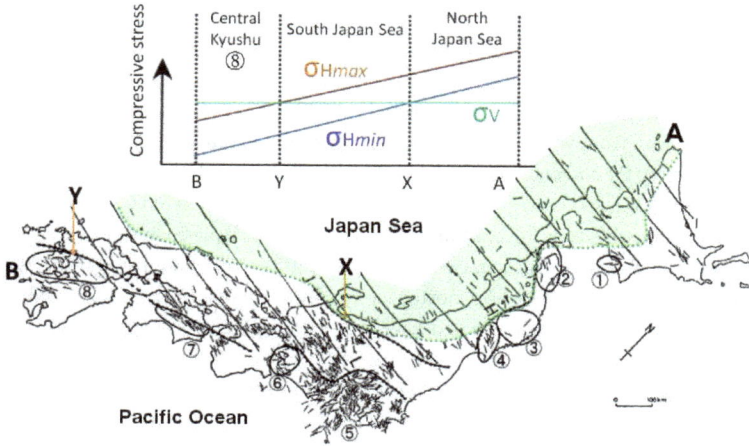

Figure 5. Regional seismogenic stress provinces in Japanese Islands. Inset is a simplified model for variation of fault types due to the along-arc stress gradient of horizontal compression. After [72] with a slight addition.

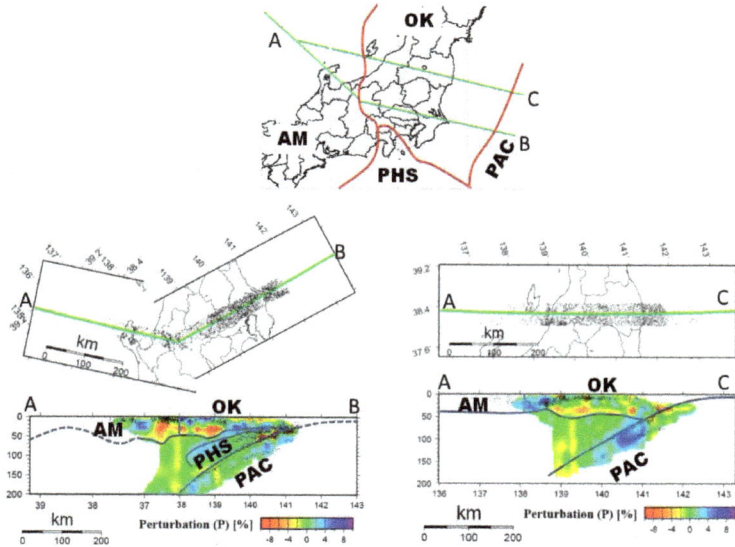

Figure 6. P-wave perturbation structures beneath the central Japan. This section image obtained by high-density seismic stations by seismic tomography using a viewer software developed by NIED to estimate the 3D seismic velocity structure typical of under Japan. Data are quoted from [73]. Tohoku and Seinan arcs are distinctive in rheology structure due to difference in plate configuration.

5.3. Superposition of second inversion

An eastward motion of the Amur plate happened to commence in association with the structural development of Himalaya - Tibetan plateau since the end of Miocene at around 7 Ma. Then the Tohoku, Izu and Seinan arcs started collisions mutually in central Japan, forming a core site of orogeny from where compressional tectonics has gradually spread to *far from the central Fossa Magna to the Niigata*. Therefore, sedimentary basins in the Hokuriku-Shin'etsu area developed deformation structures with reducing the deposition area of clastic sediments and being restricted into the present coastal plains and inland fault-basins. In the Hokuriku-Shin'etsu area after the late Miocene, overlapping of faults and folds in the three structural trends of north-south, east-west, and northeast-southwest is well recognizable and such a superposed structure has been illustrated also by current geomorphology. This intersecting feature has a broad expanse throughout the eastern margin of Japan Sea and is displayed in the seafloor topography conspicuously in particular along the continental slope of Okushiri and the Sado ridges. In addition, Present crustal earthquakes of moderate magnitudes occur by reverse faulting with a sense to promote the geomorphology development of the northeast - southwest direction of hills and mountains surrounding coastal plains evolved from the Miocene sedimentary basin.

5.4. Lateral variation in modes of active tectonics

This paper also noticed the present-day deformed structures and the spatial variety about the existence of basin inversion is also recognized. Namely, in the reverse fault province of Hokuriku, the inversion structure by earlier normal fault which formed the Miocene sedimentary basin is not seen, but typical basin inversion structures are seen in the reverse fault province such as the faults along the western margin of Nagano basin and the middle and northern segment of ISTL in the northern Fossa Magna area.

This cause can be considered as area characteristics of the principal stress axes arrangement by the stress gradient in the seismogenic upper crust. In the former province, stress field is in a state of strong horizontal compression ($\sigma2>>\sigma3=\sigma V$) and the latter state is somehow neutral ($\sigma2 \approx \sigma 3$) where a strike-slip faulting is easily exchanged into a reverse faulting ([72]; see Figure 5). Moreover, [35, 48, 74, 75] presented a possible model for the deeper geologic structure, where high-angled block faults among tectonic provinces originated as transform faults and rooted in vertical weak zones in the lower crust beneath the basement of the sedimentary basin.

Based on the sedimentary basin evolution discussed in this paper and in accordance to the results of GPS geodesy and related studies [3-5, 76], the hypothesis of tectonic belt along the eastern margin of Amur Plate [70] is promising for the origin of strain concentration belt running oblique through the zone. This hypothesis includes an eastward motion of the Amur plate with convergence along the east Japan margin and transpression along the west Japan margin as well as its collision in central Japan [1, 2, 71].

6. Conclusion

The development of the thrust/fold belt is attributed not only to horizontal compression but also to vertical block movements as a basement-involved tectonics. In response to the Pliocene and later compression regime, not only master fault but also secondary antithetic faults of the earlier fault-block boundaries are reactivated, and continued differential block movement such as subsiding of the sedimentary basin and uplifting of the igneous provinces.

The Neogene thrust-fault and folded belts in the Tohoku arc comprises the present-day tectonic zone of strain concentration in the sedimentary cover along the eastern margin of Japan Sea and Fossa Magna, while the stress regime of strike-slip faulting occupies the basement as inferred from focal mechanism solutions for small events. In order to account for the tectonic environment, the existence of subducted slab of the Philippine Sea plate, i.e. the paleo-forearc sliver of Izu arc, and related mechanism of rheological accommodation are possibly appreciated to have been worked in the asthenosphere mantle of the late Cenozoic arc-arc collision zone.

In the present study we conclude that an understanding of the tectonics of central Japan arc system provides useful insight into basin formation and evolution in general. The arc-to-arc colliding system in central Japan thus provides one of typical example for understanding how the development of a sedimentary basin is related to plate tectonics, because the GPS geodesy, seismicity, and active fault distribution are constraining the present process better than elsewhere.

Acknowledgements

The author is deeply grateful to Dr. Yasuto Itoh and Dr. Shigekazu Kusumoto for their encouraging comments and suggestions.

Author details

Akira Takeuchi

Graduate school of Science and Engineering for Research, University of Toyama, Japan

References

[1] Bird P. An updated digital model of plate boundaries. *Geochemistry Geophysics Geosystems* 2003; 1027 doi: 10.1029/2001GC000252.

[2]   Ashurkova S, San'kova V, Miroshnichenkoa A, Lukhneva A, Sorokinb A, Serovb M, Byzov L. *Russian Geology and Geophysics* 2011; 52(2) 239–249. doi:10.1016/j.rgg. 2010.12.017.

[3]   Heki K. Vertical and horizontal crustal movements from three dimensional VLBI kinematic reference frame: implication for the geomagnetic reversal timescale revision. *Journal of Geophysical Research.* 1996;101(B2) 3187-3198.

[4]   Heki K, Miyazaki S, Takahashi H, Kasahara M, Kimata F, Miura S, Vasilenco N, Ivashchenco A, An K. The Amurian plate motion and current plate kinematics in Eastern Asia. *Journal of Geophysical Research.* 1999;104(B12) 29147-29155.

[5]   Heki K, Miyazaki S. Plate convergence and long-term crustal deformation in Central Japan. *Geophysical Research Letters* 2001;28 2313-2316.

[6]   Okamura Y, Watanabe M, Morijiri R, Satoh M. Rifting and basin inversion in the eastern margin of the Japan Sea. *The Island Arc* 1995;4(3) 166–81.

[7]   Takeuchi A, Shipboard Scientific Party of R/V Yokosuka Japan Sea Cruise. Submersible observations in the epicenter area of the 1993 earthquake off southwestern Hokkaido Sea of Japan. *Journal of Geophysical Research* 1998;103(B10) 24109–24125 doi: 10.1029/98JB00572.

[8]   Okamura Y. Inversion tectonics along the eastern margin of the Japan Sea. *Journal of the Japanese Association for Petroleum Technology* 2000;65(1) 40-47. (In Japanese with English abstract)

[9]   Okamura Y. The Neogene and later strain concentration zone. In: Otake M, Ota Y, Taira A. (ed). *Earthquake tectonics of the active faults along the eastern margin of Japan Sea.* Tokyo: University of Tokyo Press; 2002 Chapter 7 p111-121. (In Japanese)

[10] Okamura Y. Fault-related folds and an imbricate thrust system on the northwestern margin of the northern Fossa Magna region central Japan. *The Island Arc* 2003;12(1) 61–73 doi: 10.1046/j.1440-1738. 2003.00379.x.

[11] Taira A. Tectonic Evolution of the Japanese Island Arc System. *Annual Review of Earth and Planetary Sciences* 2001;29 109–34 doi: 10.1146/annurev.earth.29.1.109.

[12] Coward MP. Thrust tectonics, thin skinned or thick skinned, and the continuation of thrusts to deep in the crust. *Journal of Strutural Geology* 1983;5 113–123.

[13] Tozer R, Butler R, Corrado S. Comparing thin- and thick-skinned thrust tectonic models of the Central Apennines, Italy. *EGU Stephan Mueller Special Publication Series* 2002;1 181–194.

[14] Lacombe O, Mouthereau F. Basement-involved shortening and deep detachment tectonics in forelands of orogens: Insights from recent collision belts (Taiwan, Western Alps, Pyrenees). *Tectonics* 2002;21(4), 1030, doi:10.1029/2001TC901018.

[15] Research Group for Active Faults in Japan. Active Faults in Japan Sheet Maps and Inventories. Tokyo: University of Tokyo Press; 1991. (In Japanese)

[16] Nakata T, Imaizumi T. (eds) [DVD-ROM] *Digital active fault map of Japan* DVD-ROM2. Tokyo: University of Tokyo Press; 2002. (In Japanese)

[17] Uemura T, Yamada T. (eds.) Regional Geology of Japan Part 4 (Chubu I). Tokyo: Kyoritsu Publishing; 1988. (In Japanese)

[18] Uemura T. Late Cenozoic Folding Mechanism and Crustal Dynamics in the Southern Back Arc Area of Northeast Honshu, Japan. In: *Development of Sedimentary Basins and Its Relation to Folding.* Memoirs of the Geological Society of Japan 34 p199-209. Geological Society of Japan; 1990.

[19] Sagiya T, Miyazaki S, Tada T. Continuous GPS Array and Present-day Crustal Deformation of Japan. *Pure and Applied Geophysics* 2000;157(11-12) 2303-2322.

[20] Okamura Y. Ishiyama T. and Yanagisawa Y. Fault-related folds above the source fault of the 2004 mid-Niigata Prefecture earthquake in a foled-and-thrust belt caused by basin inversion along the eastern margin of Japan Sea. *Journal of Geophysical Research* 2007;112(B3) B03S08 doi:10.1029/2006JB004320.

[21] Otuka,Y. Active folded structures. *Jishin (Series 1)* 1942;14, 46-63. (In Japanese)

[22] Takeuchi A. *Vicissitude of the stress field and tectonic evolution in the Hokushin'etsu area since the Pliocene.* The Earth Monthly 1999;21, 583-588. (In Japanese)

[23] Li J, Miyashita K, Kato T, Miyazaki S. GPS time series modeling by autoregressive moving average method: Application to the crustal deformation in central Japan. *Earth Planets Space* 2000;52 155–162.

[24] Niitsuma S, Niitsuma N, Saito K. Evolution of the Komiji Syncline in the North Fossa Magna central Japan: Paleomagnetic and K-Ar age insights. *The Island Arc* 2003;12(3) 310-323.

[25] Takeuchi A. Temporal changes of regional stress field and tectonics of sedimentary basin. *Journal of the Geological Society of Japan* 1981;87(11) 737-751.

[26] Takano O. Changes in depositional systems and sequences in response to basin evolution in a rifted and inverted basin: an example from the Neogene Niigata–Shin'etsu basin Northern Fossa Magna central Japan. *Sedimentary Geology* 2002;152(1-2) 79–97 doi: 10.1016/S0037-0738(01)00286-X

[27] Takano O, Tateishi M, Endo M. Tectonic controls of a backarc trough-fill turbidite system: The Pliocene Tamugigawa Formation in the Niigata–Shin'etsu inverted rift basin Northern Fossa Magna central Japan. *Sedimentary Geology* 2005;176(3-4) 247–279 doi: 10.1016/j.sedgeo.2005.01.004

[28] Takahashi M. Tectonic Development of the Japanese Islands Controlled by Philippine Sea Plate Motion. *Journal of Geography* 2006;115(1) 116-123. (In Japanese with English abstract)

[29] Kano K, Kato H, Yanagisawa Y, Yoshida F. (ed.) Stratigraphy and Geologic History of the Cenozoic of Japan. *Report of Geological Survey of Japan* no.274 Tsukuba: Geological Survey of Japan; 1991. (In Japanese with English abstract)

[30] Fujii S, Kaseno Y, Nakagawa T. Neogene paleogeography in the Hokuriku region, Central Japan, based on the revised stratigraphic correlation. *Memoirs of Geological Society of Japan* 1992;(37) 85-95. (In Japanese with English abstract)

[31] Geological Society of Japan (ed.) *Regional Geology of Japan 4-Chubu*. Tokyo: Asakura Publishing; 2006. (In Japanese)

[32] Tamura I, Yamazaki H, Nakamura Y. The wide-spread tephra in the Hokuriku Group and the Quaternary tectonics of the Toyama Basin, Japan. Journal of the Geological Society of Japan 2010;116 (Sup.) S1-S20. doi: 10.5575/geosoc.116.S1

[33] Kurokawa K. Studies of subaqueous tephra beds stratigraphy sedimentology volcanology -stratigraphy, sedimentology and volcanology-. *Chikyu Kagaku* 2005;59(1) 62-67.

[34] Harayama S, Wada H, Yamaguchi Y. Quaternary and Pliocene granites in the Northern Japan Alps. In: *Hutton Symposium V field Guidebook for TripA1*. Interim-Report 28 p3-21. Tsukuba: Geological Survey of Japan; 2003.

[35] Takeuchi A. Duplex Stress Regime in the North Fossa Magna, Central Japan. *Bulletin of the Earthquake Research Institute* 2008;83 1-8. Tokyo: University of Tokyo; 2008.

[36] Ito H, Yamada R, Tamura A, Arai S, Horie K, Hokada T. Earth's youngest exposed granite and its tectonic implications: the 10–0.8 Ma Kurobegawa Granite. *Scientific Reports* 2013;3 doi:10.1038/srep01306.

[37] International Stratigraphic Committee. ISC: Formal stratigraphical definitions. http:// quaternary.stratigraphy.org/ definitions/ (accessed 28 June 2013)

[38] Ota Y, Hirakawa K. Marine Terraces and their Deformation in Noto Peninsula, Japan Sea Side of Central Japan. *Geographical Review of Japan* 1979;52(4) 169-189. (In Japanese with English abstract)

[39] Tamaki K, Suyehiro K, Allan J, Ingle J, Pisciotto K. Tectonic synthesis and implications of Japan Sea ODP drilling program. *Proceedings of the Ocean Drilling Program, Scientific Results,* 1992; 127/128(Part 2) p1333–1348.

[40] Jolivet L, Tamaki K. Neogene kinematics in the Japan Sea region and the volcanic activity of the Northeast Japan arc. In: *Proceedings of the Ocean Drilling Program, Scientific Results.* 1992;127/128(Part 2) 1311–1331.

[41] Jolivet L, Tamaki K, Fournier M. Japan Sea opening history and mechanism: A synthesis. *Journal of Geophysical Research* 1994; 99(B11) 22237-22259.

[42] Itoh Y, Uno K, Arato H. Seismic evidence of divergent rifting and subsequent deformation in the southern Japan Sea and a Cenozoic tectonic synthesis of the eastern Eurasian margin. *Journal of Asian Earth Sciences* 2006;27 933–942.

[43] Shiki T, Tateishi M. On the hypothesis of aulacogen for the Fossa Magna. In: Proceedings Memorial retired Professor Seiya Ueda -Active Margin-. *The Earth Monthly* 1991;*extra* 3 106-112. (In Japanese)

[44] Ishida H. Characteristics of the basement structure and formation of oil folding: examples of Toyama Bay through Off-Niigata-Kubiki. Annual Report of Technology and Research Center (TRC)'s Activities for the Year 1995. p17–24. Chiba: Japan Oil, Gas, Metal National Corporation (JOGMEC); 1995.

[45] Kobayashi I. (ed) Chubu District I In: Editorial Committee for the Supplement Edition of Geology of Japan (ed.) Geology of Japan – Supplement. Tokyo: Kyoritsu Shuppan; 2005 p129-166.

[46] Sakamoto T. Cenozoic Strata and Structural Development in the Southern Half of the Toyama Basin, Central Japan. *Report of Geological Survey of Japan* 1966; no.213, 1-28. (In Japanese with English abstract)

[47] Kaseno Y. *Geology of Ishikawa Prefecture –Supplements*. Kanazawa: Hokuriku Geological Institute; 2001. (In Japanese)

[48] Takeuchi A. Basement-involved tectonics in the North Fossa Magna central Japan: The significance of the northern Itoigawa-Shizuoka Tectonic Line. *Earth Planets Space* 2004;56 1261-1269.

[49] Takahashi M. Tectonic Boundary between Northeast and Southwest Japan Arcs during Japan Sea Opening. *Journal of the Geological Society of Japan* 2006;112(1) 14-32.

[50] Furuse N, Kono Y. A Three Dimensional Gravity Inversion Method for Layered Structures with Lateral Density Variation. An Application to the Hokuriku District, Central Japan. *Jishin 2*, 1990;43, 1-11. (In Japanese with English abstract)

[51] Senna S, Fujiwara H, Kawai S, Aoi S, Kunugi T, Ishii T, Hayakawa M, Morikawa N, Honda R, Kobayashi K, Oi M, Yasojima Y, Okumura N. A Study on Strong-Motion Maps for Scenario Earthquakes in Morimoto Togashi Fault Zone. *Technical Note* no. 255. Tsukuba: Institute of Disaster and Earth Sciences (NIED); 2004. (In Japanese)

[52] Senna S, Fujiwara H, Kawai S, Aoi S, Kunugi T, Ishii T, Hayakawa M, Morikawa N, Honda R, Kobayashi K, Oi M, Yasojima Y, Okumura N. A study on strong-motion maps for scenario earthquakes in Tonami plain fault zone. [CD-ROM] Tsukuba: *Technical Note* no.263. Tsukuba: Institute of Disaster and Earth Sciences (NIED); 2005. (In Japanese)

[53] Senna S, Fujiwara H, Kawai S, Aoi S, Kunugi T, Ishii T, Hayakawa M,   Morikawa N, Honda R, Kobayashi K, Oi M, Yasojima Y,   Okumura N. A Study on Strong-Motion Maps for Scenario Earthquakes in Takayama-Oppara Fault Zone. [CD-ROM] Tsukuba: *Technical Note* no.282. Tsukuba: Institute of Disaster and Earth Sciences (NIED); 2005. (In Japanese)

[54] Ishiwatari A. Diverse lava and pyroclastic rocks of continental arc: the Oligocene-Miocene volcanic rocks in Noto. In: Geological Society of Japan (ed.) *Regional Geology of Japan 4-Chubu.* Tokyo: Asakura Publishing; 2006 p336-337. (In Japanese)

[55] Yanagisawa Y, Yoshikawa T. Diatomaceous mudstone and Glauconite sandstone in the Suzu district. In: Geological Society of Japan (ed.) *Regional Geology of Japan 4-Chubu.* Tokyo: Asakura Publishing; 2006 p346-347. (In Japanese)

[56] Kano K, Yamauchi Y, Miyake Y. Miocene subaqueous lavas and volcaniclastic rocks in the Shimane Peninsula, SW Japan. In: *Field excursion guidebook for 107th annual meeting of Geological Society of Japan.* Tokyo: Geological Society of Japan; 2000 p23-34. (In Japanese)

[57] Tai Y. On the "Shinji Folded Zone". *Memoir of Geological Society of Japan* 1973;no.9 137-146. (In Japanese with English abstract)

[58] Kano K, Yoshida F. Geology of the Sakaiminato District. 1:50000 Quadrangle Series 12-Okayama-7-57. Tokyo: Geological Society of Japan; 1985.

[59] Nomura R. Geology of the central part in the Shiimane Peninsula - Part 1 Miocene stratigraphy -. *Journal of Geological Society of Japan* 1986;92(6) 405-420. (In Japanese with English abstract)

[60] Tokuyama H, Honza E, Kimura M, Kuramoto S, Ashi J, Okamura Y, Arato H, Ito Y, Soh W, Hino R, Nohara T, Abe H, Sakai S, Mukaiyama K. Tectonic development in the regions around Japan since latest Miocene. Marine Surveys and Technology 13(1). [CD-ROM] Japan Society for Marine Surveys and Technology; 2001.

[61] Watanabe M. Glauconite sandstone along the boundary-layer between Yabuta and Sugata formations. In: Geological Society of Japan (ed.) *Regional Geology of Japan 4-Chubu.* Tokyo: Asakura Publishing; 2006 p348-349. (In Japanese)

[62] Kaizuka S. Quaternary morphogenesis and tectogenesis of Japan. *Zeitschrift für Geomorphologie* 1987;(Supplement Bd63) 61-73.

[63] Takeuchi A. Recent crustal movements and strains along the eastern margin of Japan Sea Floor. In: Isezaki N (ed). *Geology and Geophysics of the Japan Sea.* Volume 1 of *Japan-USSR Monograph.* Tokyo: Terra Scientific Pub Co; 1996. p385-398.

[64] Takeuchi A, Okada A. Crustal movement. In: Chapter 7.1 of *Geology of Japan 5 – Chubu Area.* Tokyo: Kyoritsu Publishing; 1988 p192-194. (in Japanease)

[65] Takeuchi A. Geologic structure and late Cenozoic evolution of the Hokuriku and Shin'etsu regions, north-central Japan. *Journal of the Geological Society of Japan* 2010;116(11) 624-635.

[66] Toyama Prefecture. Survey on the Kurehayama Fault. In: *Result Report FY 1995 grant for Earthquake Research.* Science and Technology Agency of Japan; 1996 p235. (In Japanese)

[67] Dewey JF. *Kinematics and dynamics of basin inversion.* Geological Society London Special Publications. 1989;44(1) 352.

[68] Japan Agency for Marine-Earth Science and Technology (JAMSTEC). Multichannel Seismic survey of strain concentration zone along the eastern margin of Japan Sea. In: *FY2009 Report of Observation and survey research project focused on strain concentration zone.* Tsukuba: National Research Institute for Earth Science and Disaster Prevention (NIED);2010. p236-250.

[69] Zonenshain LP, Savostin LA. Geodynamics of Baikal rift zone and plate tectonics of Asia, *Tectonophysics* 1981;76 1–45, 1981 doi:10.1016/0040-1951(81)90251-1

[70] Ishibashi K. The 1995 Kobe Japan earthquake (M7.1) in the "Mobile Belt along the Eastern Margin of Amur Plate" and its implication to the regional seismic activity (preliminary report). *Chishitsu Nyusu* 1995;490 14–21 Geological Survey of Japan (In Japanese).

[71] Wei D, Seno T. Determination of the Amurian plate motion. In: Flower M, Chung SL, Lo CH, Lee TY (ed.) *Mantle dynamics and plate interactions in East Asia.* Geodynamics Series 27 AGU;1998 p337-346 doi: 10.1029/GD027p0337.

[72] Tsukahara H. Regional seismogenic stress provinces in Japanese Islands. *The Earth Monthly* 2003;25 302-309. (In Japanese)

[73] Matsubara M, Obara K. The 2011 Off the Pacific Coast of Tohoku earthquake related to a strong velocity gradient with the Pacific plate. *Earth Planets Space* 2011;63 663-667.

[74] Iio Y, Sagiya T, Kobayashi Y, Shiozaki I. Water-weakened lower crust and its role in the concentrated deformation in the Japanese Islands. *Earth Planetary Science Letters* 2002;203(1) 25-253.

[75] Shibazaki B, Garatani K, Iwasaki T, Tanaka A, Iio Y. Faulting processes controlled by the nonlinear flow in the deeper crust and upper mantle beneath the northeastern Japanese island arc. *Journal of Geophysical Research* 2008;113(B8) B08415 doi: 10.1029/2007JB005361.

[76] Hyodo M. and K. Hirahara. A viscoelastic model of interseismic strain concentration in Niigata-Kobe Tectonic Zone of central Japan. *Earth Planets Space* 2003;55 667–675.

Intra-Arc & Foreland Basins Under the Influence of Fluctuation of Stress Regimes

Late Cenozoic Tectonic Events and Intra-Arc Basin Development in Northeast Japan

Takeshi Nakajima

Additional information is available at the end of the chapter

1. Introduction

On 11 March 2011, a catastrophic earthquake (Mw 9.0) took place along the Japan Trench off the Pacific coast of northeast Japan and about 20 thousands people were lost mainly by Tsunamis. The 2011 great earthquake occurred along a subduction zone where the Pacific Plate subducts perpendicularly toward the North American Plate at a rate of 8~9 cm/year (Figure. 1). Coseismic slip caused by the 2011 earthquake exceeded 50 m around the epicenter [1]. There has been a discrepancy between short-term (geodetic) and long-term (geologic) strain rates in both horizontal and vertical directions over the northeast Japan arc [2-3]. Geodetic observations [1, 3] have revealed horizontal shortening rate at around several tens mm/yr across northeast Japan arc, which is almost half of the subduction rate, whereas geological observations [4] have revealed horizontal shortening rate at around 3~5 mm/yr, which is one order of magnitude slower than geodetic observations [3]. Only a fraction (~10%) of plate convergence is, therefore, accommodated within the northeast Japan arc as long-term deformation. The 2011 great earthquake was one of the decoupling events that effectively released the accumulated elastic strain due to a plate coupling[3]. Study on long-term (geologic) deformation within the northeast Japan arc was thus proved to be crucial for assessing such extraordinary large decoupling events because they are too rare events (~one per kyr) to be detected by short-term (geodetic) observations.

The Japanese Islands are divided into northeast and southwest Japan by an inter-arc rift system called "Fossa Magna" bounded by the Itoigawa-Shizuoka tectonic line (ISTL; Figure. 1) and the Tonegawa tectonic line (TTL; Figure. 1) that divides northeast Japan from southwest Japan [5]. This chapter focuses on the Northeast Honshu where three lines of ranges, the Kitakami and Abukuma mountains, the Ou backbone Range, and the Dewa Hills from east to west, run parallel to the Japan Trench (Figure. 1). This chapter aims to clarify Late Cenozoic tectonic events in Northeast Japan based on reviews of recent high-resolution studies on modes and development patterns of intra-arc basins.

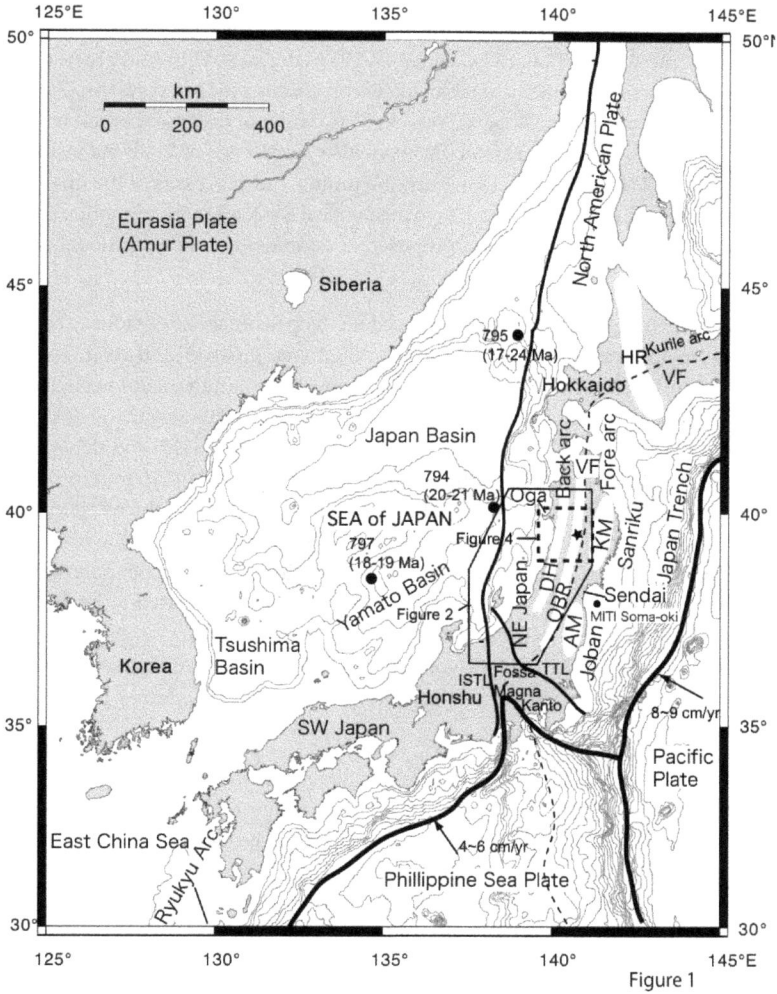

Figure 1

Figure 1. Index map showing the tectonic setting of Northeast Japan. OBR: Ou Backbone Range, DH: Dewa Hills, KM: Kitakami Mountains, AM: Abukuma Mountains, HR: Hidaka Range, VF: Quaternary volcanic front (dotted line). ISTL: Itoigawa-Shizuoka Tectonic Line, TTL: Tonegawa Tectonic Line. Solid circle represents ODP sites with ages of basement rocks and MITI soma-oki well. Star represents location of the Yuda Basin. A hexagone and a dotted square denote the areas shown in Figure 2 and Figure 4, respectively.

2. Opening of the Sea of Japan

The Sea of Japan is a large marginal sea formed in the backarc side of the Southwest and Northeast Japan and comprises component backarc basins, the Japan, Yamato and Tsushima basins (Figure. 1). The Japan Basin is underlain by an oceanic-type crust 11-12 km thick, whereas the Yamato Basin is floored by a crust 17-19 km thick and is unlikely to be oceanic [6]. The component structure of the Sea of Japan likely constitutes a multi rift system [7]. The Japan Basin was dated around 24-17 Ma based on ^{40}Ar-^{39}Ar dating of basement basalt in Site 795 of ODP Leg. 127 (Figure. 1)[8]. The Yamato Basin yields younger age of 21-18 Ma in Site 794 and 797 of ODP Leg. 127 (Figure. 1)[8]. However, the timing and processes of the opening of the Sea of Japan has been a matter of debate. Two mechanical models for the formation of the Japan arc – the Sea of Japan system has been proposed; a double-door opening model and a pull-apart basin model.

A double-door opening model is primarily based on the paleomagnetic evidence on the Japan arc. The model includes a clockwise rotation of the southwest Japan [9] and a counter-clockwise rotation of the northeast Japan [10-11]. This model implies a simultaneous occurrence of the rotation of the Japan arc and the opening of the Sea of Japan. Paleomagnetic data suggest a rapid 50° clockwise rotation of the Southwest Japan at around 15 Ma [12] and rather prolonged counter-clockwise rotation of the Northeast Japan from 17 to 14 Ma [13]. In terms of discrepancy between the timing of the opening of the Sea of Japan and that of the rotation of SW and NE Japan, Nohda [14] reassessed the ^{40}Ar-^{39}Ar age data of basement basalt with reference to Nd-Sr isotopic data in Site 797 of ODP Leg. 127. The basalts from the upper part of basements in Site 797 have not been dated and were overlain by a felsic tuff dated to be 14.86 Ma. He concluded that the upper basalts at Site 797 may be inferred to be younger than the lower basalts and that the inferred timing of volcanic activity in the Sea of Japan region (ca. 21-15 Ma) is consistent with the timing of rotational crustal movements inferred from palaeomagnetic studies in the Japanese arcs.

A pull-apart basin model is primarily based on the structural studies. In this model, the opening of the Sea of Japan was attributed to pull-apart basins formed by lateral displacement of the Japan arc associated with dextral transcurrent fault systems [15-17]. Jolivet et al. [18] revised their previous model and incorporated paleomagnetic evidence of rotation into their model; strike-slip displacement of the Japan arcs was associated by clockwise and counter-clockwise rotations of numerous blocks in southwest and northeast Japan, respectively.

There has been still no general agreement on the timing and processes of the opening of the Sea of Japan. However, Takahashi [5] proposed that the Early Miocene volcanic front of the northeast Japan was displaced more than 200 km toward east from that of southwest Japan because of the right lateral strike-slip motion of the Tonegawa tectonic line (TTL; Figure. 1). This suggests a differential rotation between southwest and northeast Japan accompanied by southward displacement of northeast Japan.

3. Intra-arc rifting

3.1. Incipient rifting

In response to opening of the Sea of Japan, both Northeast and Southwest Japan Arcs were subjected to intra-arc rifting. Recent studies demonstrated that the two-stage intra-arc rifting occurred along the eastern margin of the Sea of Japan [19-21]. Kano et al. [21] reviewed that Late Eocene and Oligocene (ca. 35-23 Ma) marine sediments distribute sparsely along the eastern margin of the Sea of Japan from Sakhalin to Kyushu through Honshu. It is suggested that rifting started and incipient rift system was formed by the Oligocene time on the back-arc side prior to the opening of the Japan and Yamato basins. During Oligocene-Early Miocene (34-21 Ma) volcanic rocks accumulated in southwest Hokkaido and Oga Peninsula with petrological and geochemical features similar to those of the volcanic rocks from continental rifts, suggesting that the former volcanic rocks were formed under rifting in the Eurasian continental arc during the pre-opening stage of the Sea of Japan [21, 23]. The early phase of incipient rifting during Late Eocene to Oligocene was characterized by slow subsidence (< 800m/m.y.) in the rifted zones [21]. The incipient rifting was interrupted by a regional unconformity prior to the succeeding rapid rifting [21].

Figure 2. Index map showing the Eastern Japan Sea Rift System [24].

3.2. Rapid rifting

Besides large backarc basins such as the Japan and Yamato basins in the Sea of Japan, the Eastern Japan Sea Rift System [24] was generated along the Sea of Japan coast of the northeast Honshu (Figure. 2) as a series of the NE-SW trending rift basins at around 16 Ma [25], which corresponded to the final phase of the backarc opening. The Eastern Japan Sea Rift System consists of several composite basins, such as the Niigata Basin in the south and the Akita Basin in the north (Figure. 2). In the Uetsu district between the Niigata and Akita basins (Figure. 2), many half grabens trending NNE-SSW to NE developed [26]. Subsidence analysis for syn-rift basins showed that rapid rifting started around 18 Ma and ceased around 15 Ma [26]. The maximum subsidence rate exceeded 1 km/m.y., much faster than that in major continental rifts [26]. Intra-arc rifting in outer arc and in most of inner arc of northeast Honshu ended at around 15 Ma [25-26].

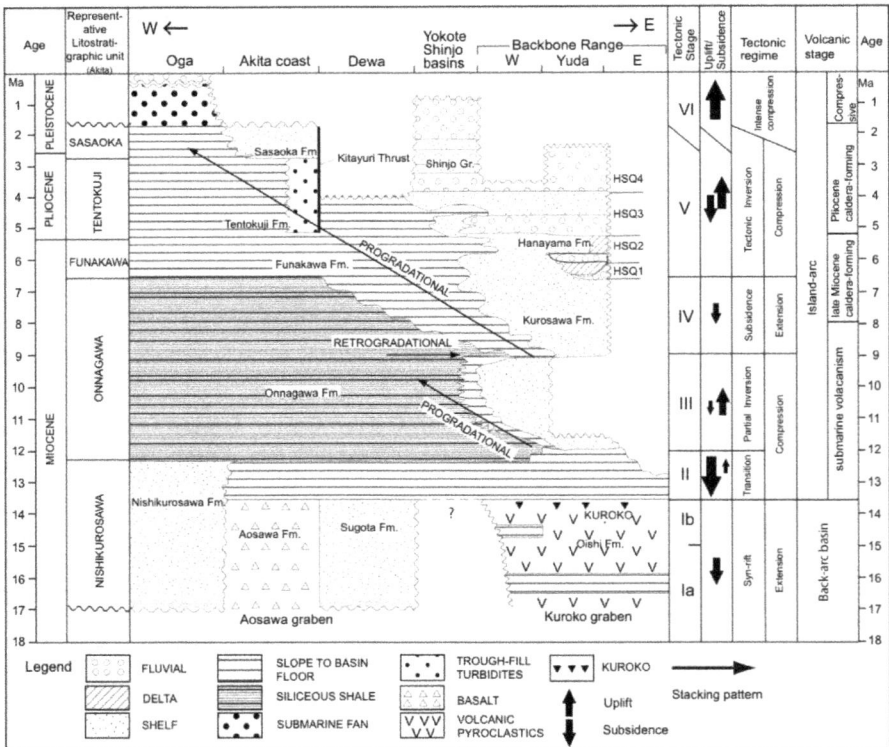

Figure 3. Tectonostratigraphic stage division chart of the Akita Basin showing division of lithostratigraphic units, major depositional systems, stacking patterns, uplift and subsidence patterns, tectonic regimes, and volcanic regimes. Compiled from [22, 27-33].

In the center of the Akita Basin as well as of the Niigata Basin, thick piles of basalts accumulated in grabens [34-35]. The Middle Miocene (16-13 Ma) northeast Japan was characterized by a distinct bimodal volcanism in which basalts dominated in back-arc side whereas fore-arc side was dominated by felsic rocks [23]. Figure 3 shows a tectonostratigraphic stage division chart of the Akita Basin. The Akita Basin was further subdivided into the coastal basin (basin center) and Yokote and Shinjo marginal basins by an intervening ridge, which have uplifted to form the Dewa Hills at present (Figure. 2). During rapid rifting, syn-rift successions filled horsts and grabens formed in the Akita Basin. Two grabens; the Aosawa graben [34] along the Akita coast and the Kuroko graben in the backbone range [36] were formed in the Akita Basin (Figure. 3). The Dewa Hills and Oga Peninsula constituted the intervening structural highs (horsts). The Aosawa graben is a large-scale graben formed by pull-apart tectonism, and was filled with thick piles (~2,000 m) of graben-fill basalt lavas (Aosawa Formation) [34]. The Kuroko graben is composed of component half grabens, which were filled with thick piles (~1,000 m) of felsic volcanic and pyroclastic rocks interbedded with mudstones (Oishi Formation) [31, 36]. In contrast, the Nishikurosawa and Sugota formations, which are only 30–400 m thick, are distributed on the Oga and Dewa ridges, respectively, with unconformable contacts with the underlying volcanic successions, and consist of shallow-marine sandstones, calcareous sandstones, and conglomerates, which were deposited on a shelf environment (Figure. 3)[22, 27]. These lithofacies and thickness contrasts are likely attributed to differential subsidence between horsts and grabens during the rifting under extensional tectonics [37]. Within the Aosawa and Kuroko grabens, syn-rift volcanism commenced at around 16.5 Ma and lasted until 13.5 Ma, prolonged than that in outer arc and in other inner arc areas in northeast Hounshu [33-34, 38]. Syn-rift volcanism (16.5 – 13.5 Ma) is further divided into early syn-rift volcanism (16.5 ~ 15 Ma) characterized by basaltic volcanism in the Aosawa graben and late syn-rift volcanism (15 ~ 13.5 Ma) characterized by felsic volcanism in the Kuroko graben [33, 38]. The Kuroko ore deposits were formed at the end of syn-rift stage at around 13.5 Ma, which may have been caused by the tectonic conversion from a back-arc to an island-arc setting [33, 36]

3.3. Evolution of the Sea of Japan rift system

The Sea of Japan rift system referred herein includes all component rift systems such as the Japan and Yamato basins, and the Eastern Japan Sea Rift System associated with the opening of the Sea of Japan. Evolution of the Sea of Japan rift system is summarized as follows. Incipient rift system commenced at around 35 Ma within a narrow zone along the present eastern margin of the Sea of Japan. Geochemical features of volcanic rocks similar to continental rifts suggest a rifting in the Eurasian continental arc. During the late incipient rifting in northeast Japan, the Japan Basin started opening at around 24 Ma followed by the opening of the Yamato Basin at around 21 Ma. The spreading of the Sea of Japan may have occurred in two stages. During the first stage, the Japan and Yamato basins spread from 24 – 21 Ma to 18 ~ 17 Ma. During the second stage, the upper basalts of the Yamato Basin may have accumulated at around 16 Ma [14]. In the meantime, rapid intra arc rifting started in the Eastern Japan Sea Rift System and surrounding areas at around 18 Ma and subsidence rate attained maximum (> 1km/m.y.) at around 16 – 15 Ma with the formation of graben fill basalts in the Akita and Niigata basins. Synrift volcanism and areas of intense subsidence migrated toward east and lasted until 13.5 Ma mainly in the

Kuroko graben along the eastern margin of the Eastern Japan Sea Rift System with minor amount of basalt activity in the Aosawa graben. The overall evolution of the Sea of Japan rift system may be interpreted as temporal progression from core-complex mode to wide-rift mode to narrow-rift mode by a simplified model of crustal extension [39]. Incipient rift system (35 – 24 Ma) may be assigned to core-complex mode while spreading of the Sea of Japan (24 – 15 Ma) is assigned to wide-rift mode. During the late stage of wide-rift mode (18-15 Ma), rifting extended to the Eastern Japan Sea Rift System. Late syn-rift system (15 – 13.5 Ma) developed mainly in the Kuroko graben at the eastern margin of the Eastern Japan Sea Rift System and may be assigned to narrow-rift mode because the lithosphere became cold and strong [37, 39].

4. Intra-arc basin development in response to tectonic events: A case study from the Ou Backbone Range

This section focuses on the post-rift tectonic events based on a case study of intra-arc basin development from the Yuda Basin in the Ou Backbone Range (Figure. 2).

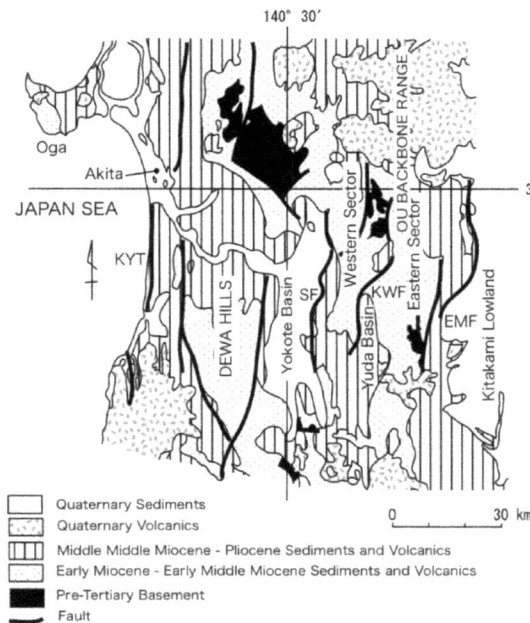

Figure 4. Index map of the Akita Basin and the Ou Backbone Range showing regional geology. SF: Senya Fault, KWF: Kawafune-Warikurayama Fault, EMF: Eastern Marginal Fault, KYT: Kitayuri Thrust.

4.1. Geology of the Yuda Basin

The Yuda Basin is located in the axis of the northern part of the Ou Backbone Range (Figure. 1). This portion of the Backbone Range is divided into western and eastern sectors, the altitudes of which are up to 1,000 m and 9,00 m, respectively. The two sectors are bounded by three fault systems; the Senya (bounding the western margin of the western sector), the Kawafune-Warikurayama (dividing the two sectors) and the Eastern Marginal Fault systems, from west to east (Figure. 4). The Kawafune-Warikurayama and Eastern Marginal Fault systems were originally formed as normal faults bounding eastern margins of half grabens during Early - Middle Miocene rifting/backarc opening. These normal faults have been reactivated (inverted) as thrusts during post-rift stages, resulting in a "pop-up" uplift of the present Ou Backbone Range (Figure. 4) [40]. The Yuda Basin is a depression between the eastern and western sectors (Figure. 4), and is 15 km long in a N-S direction and 5 km wide in the E-W direction. Pre-Tertiary basement rocks and Early – Middle Miocene syn-rift volcanic rocks are distributed in the axis of the western and eastern sectors, while Middle Miocene to Pleistocene marine and non-marine deposits are distributed in the surrounding lowlands including the Yuda Basin (Figure. 4).

The stratigraphy of the study area is divided into the Oishi, Kotsunagizawa, Kurosawa, Sannai, Hanayama and Yoshizawa Formations in ascending order (Figure. 5). The basin structure shows a simple synclinal structure bounded by the Kawafune-Warikurayama Fault at its western margin (Figure. 6). The deposits become younger toward the basin center, reflecting a synclinal structure.

The Oishi Formation consists of syn-rift volcanics, volcaniclastic rocks and mudstone. This formation had been formed by syn-rift felsic volcanism within half-grabens bounded by the Kawafune-Warikurayama and Eastern Marginal Fault systems in the Kuroko graben during the syn-rift stage (16 – 13.5 Ma) (Figure. 3).

The Kotsunagizawa Formation consists of mudstone, fine-grained sandstone and felsic tuffs. The formation conformably overlies the Oishi Formation and intercalates the Okinazawa basalt member, consisting of basaltic tuff breccia and lapilli stone in the middle part of the formation (Figures 5, 7). The uppermost several meters of the formation grades from mudstone to sandstone upward and is overlain by the Ochiai volcanic breccia bed (OB: 7 m thick)[43]. The age of the Kotsunagizawa Formation spans from 13.5 Ma to ca. 12 Ma based on biostra-tigraphy and fission-track datings (Figure, 8)[30-31, 45].

The Kurosawa Formation consists of shallow marine sandsone and tuffaceous mudstone with intercalations of felsic tuffs and conglomerate. Three key tuff beds, the Tsukano Tuff beds (TN: 20–50 m thick), Ohwatari Tuff beds (OW: 25 m thick) and the Torasawa Tuff bed (TS: 3–15 m thick) are intercalated in the middle and upper parts of the formation, respectively [30, 43] (Figures. 5-7). The Kurosawa Formation unconformably overlies the Kotsunagizawa Forma-tion in the eastern margin of the Yuda Basin. The age of the Kurosawa Formation in the Yuda Basin was dated to be 9-6.5 Ma based on fission-track dating with a notable age gap of 2 – 3 Ma at the unconformity between the Kotsunagizawa and the Kurosawa formations (Figure. 8) [31, 43]. The Kurosawa Formation is, however, continuous from the Kotsunagizawa Formation in the western margin of the Yuda Basin (Core 41PAW-1 in Figure. 7) and in the western sector (Figure. 7).

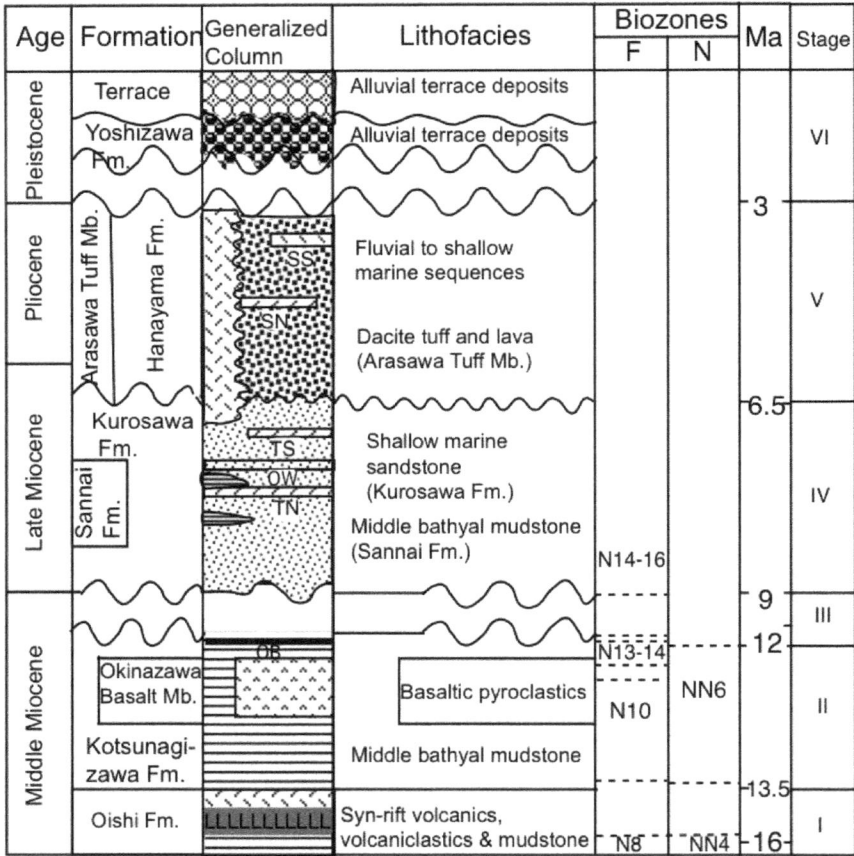

Age	Formation	Generalized Column	Lithofacies	Biozones F	Biozones N	Ma	Stage
Pleistocene	Terrace		Alluvial terrace deposits				VI
	Yoshizawa Fm.		Alluvial terrace deposits				
						3	
Pliocene	Arasawa Tuff Mb. / Hanayama Fm.	SS SN	Fluvial to shallow marine sequences / Dacite tuff and lava (Arasawa Tuff Mb.)				V
						6.5	
Late Miocene	Kurosawa Fm. / Sannai Fm.	TS OW TN	Shallow marine sandstone (Kurosawa Fm.) / Middle bathyal mudstone (Sannai Fm.)	N14-16			IV
						9	III
Middle Miocene	Okinazawa Basalt Mb.	OB	Basaltic pyroclastics	N13-14 / N10	NN6	12	II
	Kotsunagi-zawa Fm.		Middle bathyal mudstone			13.5	
	Oishi Fm.		Syn-rift volcanics, volcaniclastics & mudstone	N8	NN4	16	I

Figure 5. Generalized stratigraphy of the Yuda Basin with approximate ages and tectonic stages (I – VI). Key tuff bed names: SS: Sasoh tuff, SN: Sawanakagawa tuff, TS: Torasawa tuff, OW: Ohwatari tuff, TN; Tsukano tuff. F: planktonic foraminifer zonation after [41]. N: nannofossil zonation after [42]. Modified from [43].

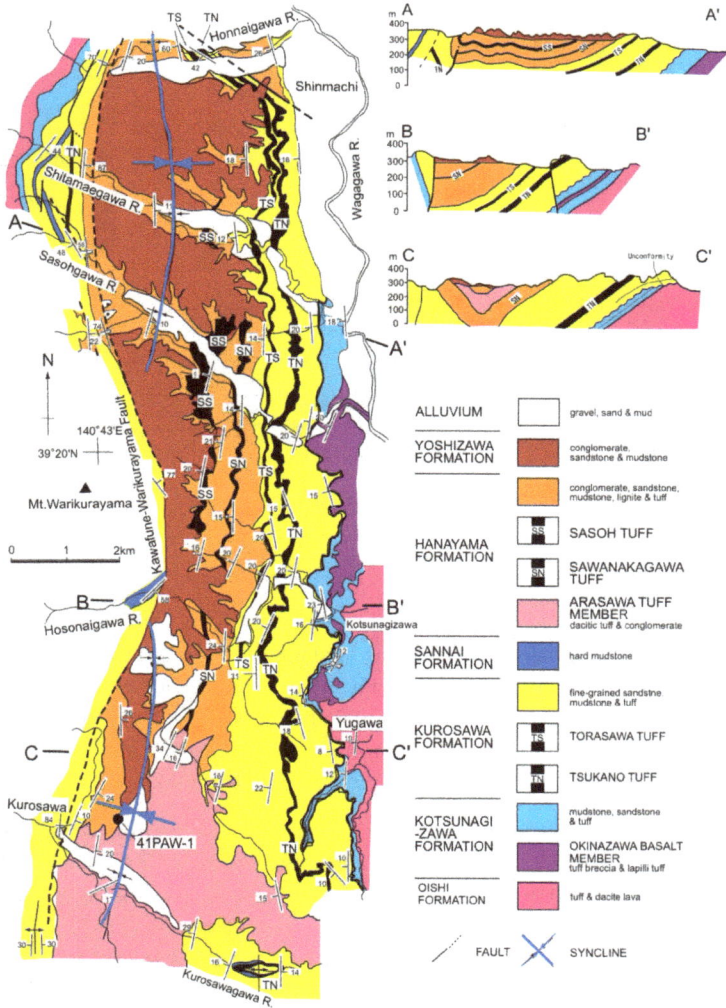

Figure 6. Detailed geological map and geological cross-sections of the Yuda Basin. A-A', B-B' and C-C' are lines of cross-sections. 41PAW-1 denotes the location of the drilling hole [44] in the southwestern part of the Yuda Basin. Modified from [31]

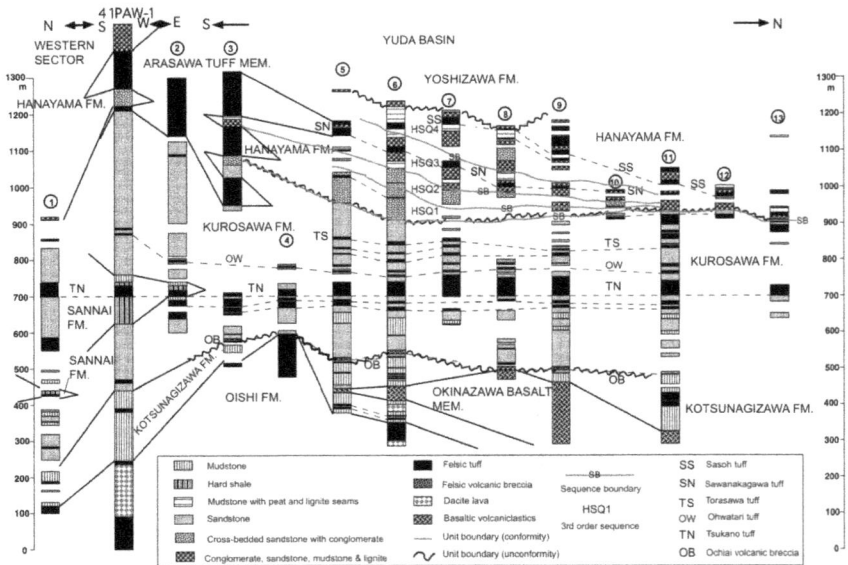

Figure 7. Correlation of lithologic columns in the Yuda Basin (modified from [43]). Numbers in open circles denotes locations of columns shown in Figure 10.

The Sannai Formation consists of alternating hard mudstone and black shale associated with felsic tuffs. The formation interfingers with the Kurosawa Formation (Figure. 7) and is distributed in the western sector and in the southern margin of the Yuda Basin, while it is absent in the eastern margin of the Yuda Basin (Figure. 7). This formation is equivalent to the Onnagawa Formation, a siliceous shale unit in the center of the Akita Basin (Figures. 2&3).

The Hanayama Formation distributed along the syncline in the axis of the Yuda Basin is divided into a main part and the Arasawa Tuff Member distributed in the southern Yuda Basin (Figure. 7). The main part of the formation unconformably overlies the Kurosawa Formation, and consists of fluvial, marsh, deltaic and shallow marine deposits [30-31]. The Hanayama Formation includes four 3rd order depositional sequences overlapped by numerous high frequency sequences (Figure. 7). Two key tuff beds, the Sawanakagawa Tuff beds (SN: 4–30 m thick) and the Sasoh Tuff bed (SS: 10–50 m thick) are intercalated in the middle and upper Hanayama Formation (Figures. 5-7)[30]. The Arasawa Tuff Member chiefly consists of dacite pumice tuff associated with dacite lava and conglomerate. This member interfingers with upper part of Kurosawa Formation and also with the main part of the Hanayama Formation in the southern Yuda Basin (Figures. 5-7). The age of the main part of the Hanayama Formation

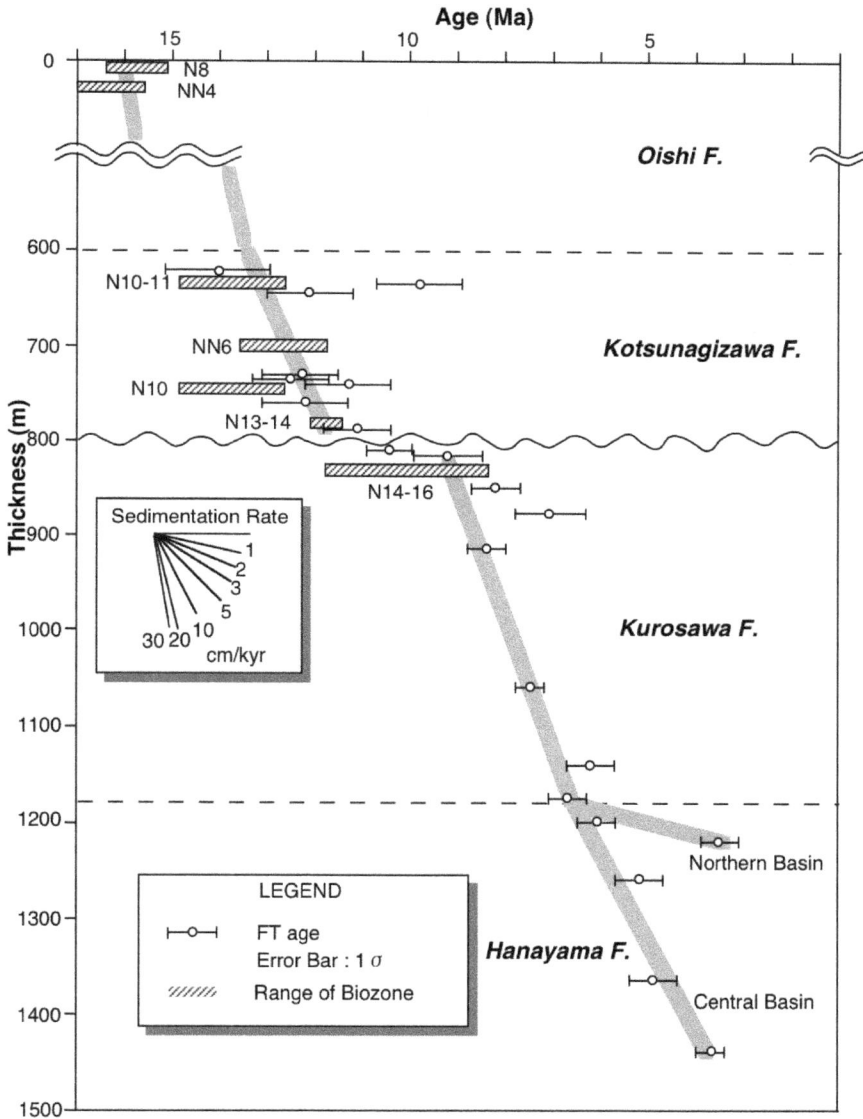

Figure 8

Figure 8. Age-thickness diagram in the Yuda Basin after [31].

was estimated from 6.5 to less than 3 Ma based on fission-track dating (Figure 8) [30, 45] although the upper boundary has not yet been precisely dated.

The Yoshizawa Formation overlies the Hanayama and Kurosawa formations with a notable angular unconformity and is distributed in the western part of the Yuda Basin along the Kawafune-Warikurayama Fault (Figure. 4). This formation consists of alternation of conglomerate, sandstone and mudstone associated with lignite seams and peat. This formation may have been deposited on alluvial fans that were formed along the eastern margin of the western sector [30, 45]. The Yoshizawa Formation may be early Pleistocene in age based on the stratigraphic relationships between this formation and the Hanayama Formation and fluvial terraces [30].

4.2. Styles and origins of unconformities in the Yuda Basin

Three major unconformities were formed around 10 Ma, 6.5 Ma and after 3 Ma in the eastern margin of the Yuda Basin. These unconformities have significant implications for post-rift tectonics not only in the Ou Backbone Range but also in Northeast Japan and thus the styles and origins of these unconformities are described and discussed herein

4.2.1. 10 Ma unconformity

The unconformity between the Kotsunagizawa Formation and the Kurosawa Formation at around 10 Ma is a partial unconformity within the basin and shows significant horizontal change in terms of the style [31]. The unconformity apparently eroded the Kotsunagizawa Formation at two hills in the eastern margin of the Yuda Basin (Figures. 5-7). At the road cut section north of the Kotsunagizawa (Figure. 6), the unconformity eroded the upper part of the Kotsunagizawa Formation and the Okinazawa basalt member and the shallow marine sandstone of the Kurosawa Formation overlies the Okinazawa basalt member with the basal conglomerate bed composed of boulders of basalts eroded from the underlying Okinawaza basalt member (Figure. 9)[43]. In this section, the upper 40 meters of the Kotsunagizawa Formation and the Ochiai volcanic breccia bed were eroded as well (Figure 9). The lower Kurosawa Formation in this section yields fission-track ages of 9.2 and 7.1 Ma (Figure. 10), significantly younger age than the top of the Kotsunagizawa Formation dated at around 12-11 Ma (Figure 8)[31]. This suggests an age gap of 3 – 2 Ma interval at the unconformity (Figure 8) [43].

The amount of erosion varies over a short distance, but tends to increase where the boundary shifts to the east, towards the uplift axis of the eastern sector (Figures. 6 & 10). The maximum amount of the erosion attains more than 100 m at another hill west of Yugawa (Figure. 6), where the unconformity eroded the entire Kotsunagizawa Formation and the Kurosawa Formation directly overlies dacite tuff breccia of the Oishi Formation (Figure. 10). At the boundary, the Kurosawa Formation onlaps the Oishi Formation with lower angle dip of 8° W than the underling Oishi Formation with a dip of 16° W (Figure. 6)[43]. This indicates angular unconformity between the Oishi and Kurosawa formations. The basal 5 m of the Kurosawa Forma-

Figure 9. A photo showing the unconformity between the Kotsunagizawa and Kurosawa formations in column K of Figure 10. White circles denote basal conglomerates derived from the Okinazawa Basalt Member of the Kotsunagizawa Formation. Modified from [43].

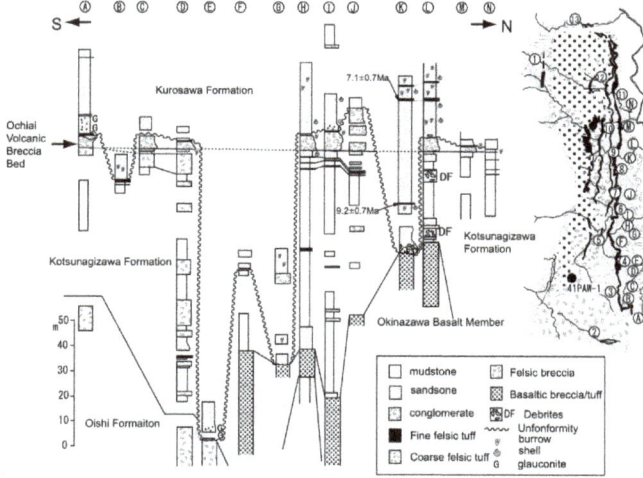

Figure 10. Detailed correlated columns showing the unconformity between the Kotsunagizawa and Kurosawa formations in the Yuda Basin. Inset shows locations of lithologic columns in Figure 7. Modified from [31].

tion in this section contains glauconite sandstone (Figure. 10), indicating that significant hiatus or low sedimentation rate event took place at the time of the unconformity[46].

In other parts of the eastern margin of the Yuda Basin, however, this boundary is a paracon-formity. Bathyal mudstones in the upper part of the Kotsunagizawa Formation grade upward into shallow marine sandstones in the top ~10 m of the formation. The Ochiai volcanic breccia bed [43] of up to 15 m in thickness rests on sandstones at the top of the formation and is a distinct key correlation bed in the uppermost Kotsunagizawa Formation (Figure. 10). The breccia bed was covered by a fine tuff bed (<1 m in thickness), which is intensely burrowed and bioturbated. This fine tuff bed is sharply overlain by shallow-marine, massive sandstone of the basal Kurosawa Formation (Figure. 10). The contact between intensely burrowed fine tuff at the top of the Kotsunagizawa Formation and the shallow marine sandstone of the basal Kurosawa Formation can be traced at the same stratigraphic position <1 m above the Ochiai breccia bed over 10 km from the south to the north except for the two hills where the uncon-formity incised deeply the underlying strata (Figure. 10). Although no significant erosion at the boundary was suggested, the significant age gaps between two formations indicate hiatus or slow deposition at the boundary.

In the western margin of the Yuda Basin (Core 41PAW-1 in Figures. 6, 7) and in the western sector (Figure. 7), the boundary between the Kotsunagizawa and Kurosawa formations is continuous. Correlation between the Yuda Basin and the western sector (Figure. 7) indicates that the equivalent time interval of unconformity in the eastern Yuda Basin corresponds to the deposition of the lower Kurosawa Formation in the western sector. Based on the above consideration, there was a westward submarine paleo-slope in the basin during ~12–9 Ma. In summary, the unconformity dated around 10 Ma was attributed to the uplift and westward tilting of the eastern sector, starting around 12 Ma and followed by the cessation of subsidence until 9 Ma and by subsequent onlapping of the Kurosawa Formation from west because of the resumption of the subsidence. The overall observation suggests that the eastern sector started to uplift and emerge earlier than the western sector. However, the western sector may have also slightly uplifted in this stage because the middle bathyal Kotsunagizawa Formation changed upward into the upper bathyal to outer shelf Kurosawa Formation (Figure 7).

4.2.2. 6.5 Ma unconformity

The base of the main part of the Hanayama Formation was defined by the first occurrence of gravelly sandstone deposited in fluvial channels. The relationship between the Kurosawa and Hanayama formations is thus unconformity formed by fluvial erosion. Nakajima et al. [30] interpreted the unconformity as a sequence boundary formed by a relative sea-level lowering. Although the erosion by the unconformity is not significant (Figure. 7) and no age gap was suggested between two formations [30-31], this unconformity may have reflected major tectonic changes at around 6.5 Ma for the following reasons [43]. After the 6.5 Ma unconformity was formed, 4th or 5th order high frequency depositional sequences comprising of fluvial, deltaic and shallow water deposits with significant amount of conglomerate were deposited [30]. This contrasts with the Kurosawa Formation, which consist totally of shallow marine sandstones. This suggests that supply of coarse sediments began to exceed the accommodation space possibly because of the uplift of surrounding backbone range. In addition, this uncon-formity marks the major changes from uniform sedimentation rate in north-south direction in the Kurosawa Formation to differential sedimentation rate in the Hanayama Formation (Figure. 7). Those changes in depositional style may have resulted from tectonic conversion from extension to compression as discussed later.

4.2.3. < 3 Ma unconformity

The angular unconformity between the Hanayama and Yoshizawa formations is the most significant structure among three unconformities described herein. The Kurosawa and Hanayama formations form a syncline with dips of 20 – 30 ° and are unconformably overlain by almost flat Yoshizawa Formation (Figure. 6). The formation of syncline and subsequent unconformity may have reflected uplift of the eastern and western sectors of the Ou Backbone Range during and after deposition of the Hanayama Formation. The unconformity was formed after 3 Ma although the precise age has not been dated.

4.3. Evolution of the Yuda Basin since Middle Miocene

Evolution of the Yuda Basin since Middle Miocene is summarized here, based on correlation of tuff beds and unconformities (Figures 7 & 10) and basin subsidence analysis (Figure. 11). Tectonic history of the Yuda Basin and the surrounding Ou Backbone Range was divided into six tectonic stages according to the basin subsidence pattern (Figure. 12)[31].

4.3.1. Stage I (Syn-rift stage; 16-13.5 Ma)

This stage was characterized by a rapid subsidence (600 m/m.y.) with accumulation of thick volcanic and volcaniclastic successions in the Yuda Basin (Figure. 11). The amount of tectonic subsidence attains no less than 1,000 m even if altitude of sea-level rise at around 15 Ma [47] is subtracted (Figure 11). The Eastern Marginal Fault and Kawafune-Warikurayama Fault systems may have been activated as normal faults and formed half grabens in the eastern and western sectors, respectively as a result of crustal stretching under extensional tectonics during Early-early Middle Miocene rifting (Figure 12) [40].

4.3.2. Stage II (Post-rif transition stage; 13.5–12 Ma)

This stage was characterized by the cessation of syn-rift volcanism and accumulation of hemipelagic sediments at a slower rate (10 cm/k.y.) in the Yuda Basin (Figures. 8 & 12). The subsidence reconstruction of the Yuda Basin (Figure. 11) indicates subsequent reduction of tectonic subsidence rate although precise estimation is difficult due to unreliable estimation of paleo-depth under bathyal environment.

4.3.3. Stage III (partial inversion stage: 12–9Ma)

This stage was represented by temporal uplift of the Ou Backbone Range and associated unconformity in the Yuda Basin. The amount of tectonic uplift in the Yuda Basin at around 12-11 Ma was estimated at more than 500 m. This uplift was followed by a cessation of

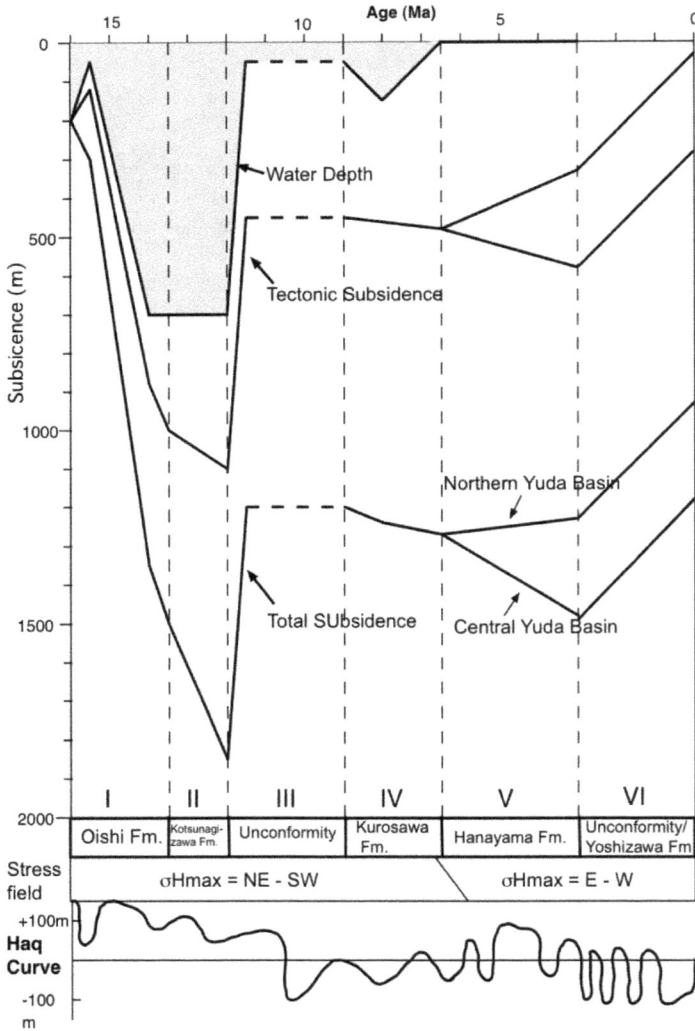

Figure 11. The total and tectonic subsidence curves of the Yuda Basin with tectonic stages (I-VI), inferred stress field [30] and a eustatic curve proposed by [47]. Modified from [31, 43].

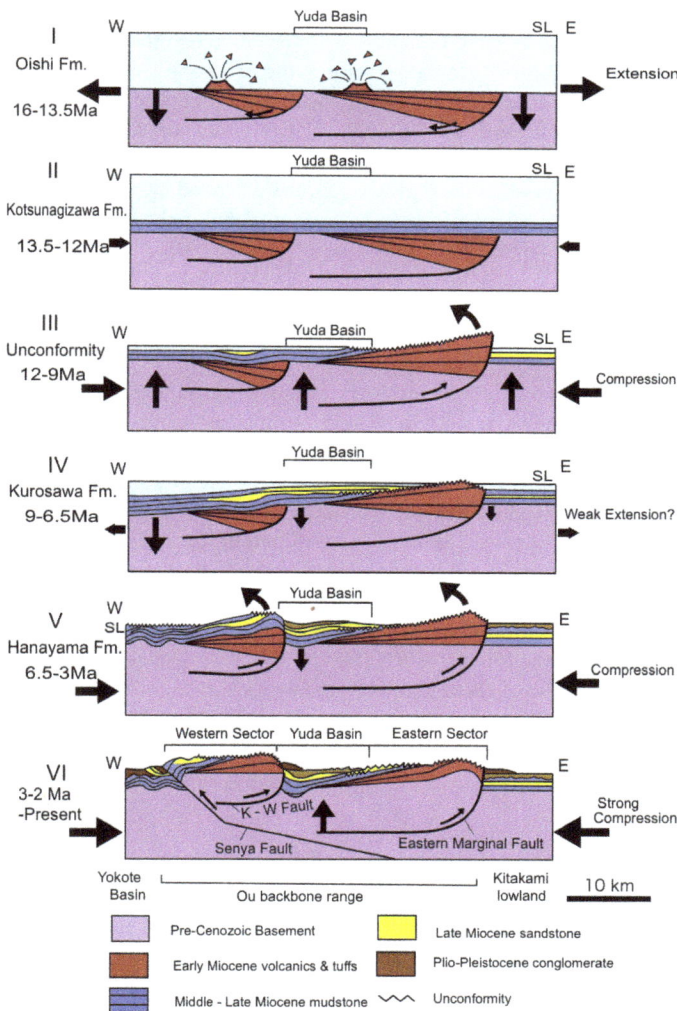

Figure 12. A cartoon illustrating the tectonic evolution of the Ou Backbone Range (modifed from [31]). K-W Fault: Kawafune-Warikurayama Fault. See text for explanation.

subsidence and uplift within the basin until about 9 Ma (Figure. 11), leading to a hiatus in the eastern margin of the basin. Correlation of key beds and unconformities (Fig. 7) clearly demonstrates that intensely uplifted areas at the eastern margin of the basin were more subject to intense truncation. The uplift was more intense in the eastern sector than in the western sector, and was accompanied by the westward tilting of the eastern sector (Figure 12). The eastern sector began to emerge and became a sediment source to the surrounding basins by subaerial erosion. The uplift of the eastern sector of the backbone range may be attributed to a subsequent inversion of a half graben formed during Stage I because of an increase in horizontal compressional stress (Figure. 12)[31]. The western sector of the backbone range also uplifted from middle bathyal to shelf environments, although the bounding normal fault does not seem to have been inverted. This stage was also characterized by reduced volcanic activity around the basin as suggested by scarcity of tuffs intercalated in the lower Kurosawa Formation in the western sector (Figure. 7).

4.3.4. Stage IV (subsidence stage: 9–6.5 Ma)

This stage was characterized by slow subsidence and deposition of sand in shallow marine environments in the Yuda Basin. The subsidence resumed at around 9 Ma and shallow-marine sandstones were deposited over the unconformity in the eastern margin of the Yuda Basin, while the eastern sector of the backbone range remained as a sediment source. The lower Kurosawa Formation thins in intensely uplifted and truncated areas such as west of Yugawa (Figures. 6 & 7), which suggests that the Kurosawa Formation onlaps these uplifted topographic highs. While subsidence rates in the eastern part of the Yuda Basin were uniform in north-south direction (Figure. 7) within the magnitude of eustatic sea-level fluctuation (Figure. 11), the western sector subsided more rapidly. Total tectonic subsidence in the western Yuda Basin was estimated at least 600 m during this stage from the thickness of the Sannai and upper Kurosawa Formations in the core 41PAW-1 (Figure. 7)[31]. This stage was also characterized by increased felsic volcanism, as represented by increased felsic tuffs such as the Tsukano and Torasawa Tuff beds in the Kurosawa Formation in the Yuda Basin (Figure. 7). Occurrence of Northeast-Southwest-trending minor normal faults in the Kurosawa Formation indicates Northwest-Southeast-trending extensional stress field [30], which resulted in regional slow subsidence.

4.3.5. Stage V (basin inversion and compression stage: 6.5 – 3~2 Ma)

Stage V represents differentiation of uplifted and subsided areas within the Yuda Basin. Subsidence pattern changed from the preceding stage IV, and the northern Yuda Basin turned into an uplifted area (Figure. 11). Moreover, conglomeratic deposits within the basin indicate uplift of the surrounding mountains. The high frequency depositional sequences were developed in this stage, suggesting that supply of coarse sediments began to exceed the accommodation space. The western sector may have also uplifted because marine environments gradually retreated from the Yuda Basin by ~4 Ma [30], which indicates the emergence of the western sector by that time. The Northeast-Southwest trending minor normal faults found in the Kurosawa Formation disappeared at the base of the Hanayama Formation (~6 Ma) in the Yuda Basin [30]. After 6 Ma, only North-South-trending reverse faults were formed in the Yuda Basin. The observation indicates that stress changed from extension to E-W compression at the beginning of this stage. The compressive stress field resulted in basin inversion and uplift of both the eastern and western sectors of the Ou Backbone Range (Figure. 12).

4.3.6. Stage VI (Intense compression stage; 3~2 Ma - Present)

Stage VI represents the uplift of the whole Backbone Range and the formation of an angular unconformity between the Hanayama and Yoshizawa Formations within the Yuda Basin (Figures 11, 12). The uplift of the Backbone Range during this stage resulted from "pop-up" of the sectors of the Backbone Range by activation of the Senya Fault system as well as reactivation of the Eastern Marginal Fault and Kawafune-Warikurayama Fault systems under intense compression (Figure. 12)[40].

5. Late Cenozoic tectonic events in northeast Japan

In this section, Late Cenozoic tectonic events deduced from developments of other fore-arc, intra-arc and back-arc basins across the northeast Japan transects were correlated with those in the Yuda Basin (Figure 13). Based on the results, Late Cenozoic tectonics in northeast Japan was clarified and was divided into seven stages from 0 to VI as described below. Then, the author will discuss the origin of regional tectonic events. The reviews presented herein could provide a revised Late Cenozoic tectonics model in northeast Japan.

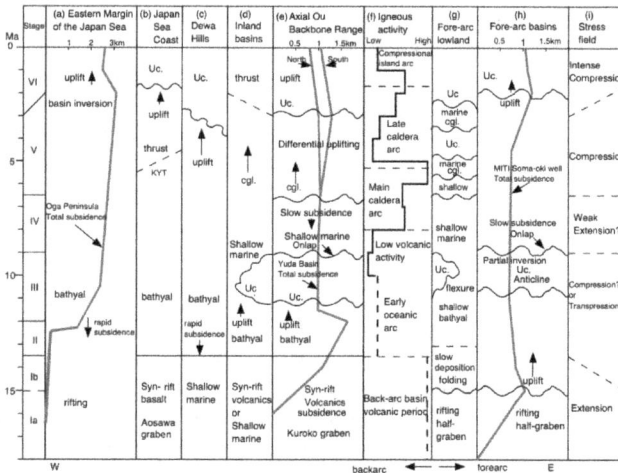

Figure 13. A compiled chart showing correlation of tectonic movements across northeast Japan transect since 18 Ma (revised after [31]). Stages I-VI are tectonic stages defined in this study. Uc.: unconformity, cgl.: conglomerate. Compiled after (a) [48] revised after new age model of [22], [49]; (b) [27-29]; (c) [27-29, 32]; (d) [32, 51-54]; (e) [31, 38, 55-56]; (f) [16, 33, 57-58]; (g) [56, 59-65]; (h) [31, 66-68]; (i) [31]

5.1. Stage 0 (Incipient rift stage; 35 – 20 Ma)

Stage 0 represents formation of the incipient rift system along the eastern margin of the Sea of Japan as already noted in section 3. Rift basins in this stage were relatively small and marine incursion was limited within a narrow zone along the present Sea of Japan coast [21]. The timing of marine incursion varied among rift basins [21], suggesting that individual marine basin was relatively short lived. The incipient rift basins had not directly developed into later syn-rift large basins, but were interrupted by regional unconformity during the late Stage 0

5.2. Stage I (Syn-rift stage: 20 - 13.5 Ma)

Syn-rift stage has been divided into two substages: IA and IB on the basis of tectonic conversion from wide rift mode to narrow rift mode [39], associated with a stress change from regional extension to coexistence of extension and compression at around 15 Ma.

5.2.1. Substage IA (20 – 15 Ma)

After the incipient rifting stage, dacitic volcanics with some amount of basaltic volcanics together with conglomerates and sandstones deposited in terrestrial environments at around 20 Ma in the Akita Basin [22, 34]. These stratigraphic units may represent slow subsidence [22] prior to rapid rifting after 18 Ma as described in section 3.2. They were unconformably overlain by non-marine to marine successions deposited during rapid rifting. The stratigraphic units deposited during the rapid rifting (ca. 18 – 15 Ma) represent regional marine transgression in northeast Japan as a result of rapid subsidence under extensional tectonics associated with rifting [26, 69]. The equivalent stratigraphic units during Substage IA also deposited within rotated half grabens in the eastern margin of the Sea of Japan [49]. Rapid subsidence associated with rifting and half grabens also took place in the fore-arc side of northeast Japan [56, 66, 70] and in the Kanto Plain (Figure 1)[5, 71]. In terms of igneous activity, this stage was assigned to a backarc basin volcanic period (Figures, 3, 13)[33, 58] and was characterized by intense basaltic volcanism within rift grabens in the Akita and Niigata basins [34, 36, 58].

5.2.2. Substage IB (15 – 13.5 Ma)

Substage IB is characterized by shrinkage of rift zones and by uplift with a notable unconformity in fore-arc side and fore-arc basins in northeast Japan. Formation of half grabens in the Kanto Plain (Figure 1) was suddenly terminated by rapid uplift with formation of a notable unconformity (the Niwaya unconformity) at 15.3 – 15.2 Ma [5, 71]. Marine fine-grained sediments with low sedimentation rates unconformably overlie the successions in rotated half grabens [71]. This change in the style of subsidence and deposition was attributed to tectonic conversion from extensional to strong compression stress, followed by relatively quiet tectonics [71]. Sedimentation rates in the post rift successions over the Niwaya unconformity in the northern Kanto Plain had been suppressed until 14 – 12.5 Ma, suggesting compression lasted until ca. 13 Ma [64]. The Joban forearc basin (Figure 1) was also inverted and uplifted at around 15 Ma (Figure 13) with a notable unconformity being formed [68]. This tectonic change was accompanied by NW – SE trending folding along the coast of Sendai [56] and in the Joban Basin [68], which suggests that the stress field in the forearc switched to a compressional or transpressional regime at around 15 Ma (Figure 13). The shallow marine successions in Substage IB in the Sendai Plain (Figure 1) comformably overlie the underlying conglomerates of ca. 16 Ma [60, 63]. However, sedimentation rates had been suppressed until about 13 Ma in the glauconite bed at the base of the shallow marine successions [60], similar to the observation in the northern Kanto Plain. Rifting associated with activities of half grabens also ceased at about 15 Ma in the Niigata Basin [69] and in the Uetsu district [26] within the Eastern Japan Sea Rift System (Figure 2). This suggests that extensional tectonics ended at ca. 15 Ma in the back-arc side as well as fore-arc side of northeast Japan. However, rifting associated with activities of half grabens lasted until 13.5 Ma in the Kuroko graben, as suggested by the rapid subsidence of the Yuda Basin (Figures 11 & 12) and voluminous felsic volcanism in the Kuroko graben [31, 33, 36]. Sato [72] also concluded that synsedimentary faults bounding the Aosawa and Niigata grabens had been active until 13.5 Ma based on the isopach map of Substage IB. These observations clearly indicate that extensional tectonics continued within some parts of the Eastern Japan Sea Rift System, particularly in the Kuroko graben as a narrow rift mode [39] during Substage IB. For this reason, both strong compression and extension coexisted in northeast Japan during Substage IB. The origin of strong compression in the fore-arc side of northeast Japan may be attributed to rapid counter-clockwise rotation of northeast Japan at around 15 Ma as a result of the opening of the Sea of Japan (See section 2). The rapid counter-clockwise rotation of northeast Japan might accelerate the relative convergence rate of the Pacific Plate at the Japan Trench, which must have resulted in increased compressional stress. Collision and transpressional movements of northeast and southwest Japan arcs along the TTL (Figure 1) as a result of differential rotation of both arcs [5] might also contribute to strong compression along the border of the two arcs such as in the Kanto Plain (Figure 1).

5.3. Stage II (Post-rift transition stage: 13.5 – 12 Ma)

Stage II was characterized by the cessation of all rifting including the Aosawa and Kuroko grabens, and by marine transgression associated with rapid subsidence of the Dewa and Oga ridges in the Akita Basin (Figure 3). Although this stage had been previously regarded as quiet thermal subsidence stage under neutral stress regime [56, 72, 73], the Dewa ridge subsided rapidly from inner sublittoral to middle bathyal environments at 13.5 Ma, followed by rapid subsidence of the Oga ridge from upper bathyal to lower bathyal environments at 12.3 Ma [22, 27]. This subsidence mode cannot solely be attributed to a post-rift thermal subsidence, because the subsidence rate far exceeded the estimated rate (~70 m/m.y.) of the thermal subsidence [31], and the timing of rapid subsidence was out of phase between the Dewa and Oga ridges. In terms of volcanism, Stage II was represented by a major change from back-arc basin stage to island-arc stage [33]. These changes in tectonic and volcanic styles suggest a stress change from extension to compression [31]. Subsidence resumed and sedimentation rates increased at about 13 Ma in the northern Kanto Plain and in the Sendai Plain, where deposition had been suppressed during the previous Substage IB because of strong compression [60, 64].

5.4. Stage III (Partial inversion stage: 12 – 9 Ma)

Stage III was represented by the temporal uplift and associated unconformity caused by partial inversion in the Backbone Range. The temporal uplift and associated unconformity at around 10 Ma occurred not only in the Yuda Basin, but also in other sections of the Backbone Range from south to north along the axis of the Backbone Range [38, 55, 56]. Moreover, contemporary uplift and associated unconformity also occurred in inland basins and forearc lowlands along the western and eastern margins of the Backbone Range, respectively (Figure 13). For example, northwestern margin of the western sector at the eastern margin of the Yokote Basin (Figure 4), rapidly uplifted (>700 m) from middle bathyal to terrestrial environments with a notable unconformity at ~9 Ma [54]. Although the southern part of the western sector had not been inverted as described in Section 4.3.3, sedimentary successions show westward progradation from the backbone range to the Dewa Hills as a local response to excess sediment supply over the rate of creation of accommodation space (Figure 3). The contemporaneous uplift and resultant unconformity also took place at ~10 Ma in the margins of the Yonezawa and Aizu Basins (Figure 2), ~200 km south of the Yokote Basin [51, 53]. Notably, the amount of erosion at the unconformities increased toward the east (toward the Backbone Range), while deposition continued in the west of the basins [51, 53]. An angular unconformity was also formed at 11.5 – 9 Ma in response to the NW-SE trend flexure activity in the Sendai Plain (Figure. 1), east of the Backbone Range [62, 63]. Sedimentary successions in the northern Kanto Plain, east of the Backbone Range showed upward shallowing successions since 10 Ma, which was followed by an unconformity at ~9 Ma [64]. A regional angular unconformity (~11–9 Ma) occurred in the sedimentary sequence in the Joban forearc basin [67]. The basement subsidence reconstruction at the MITI Soma-oki well (Figure. 1) in the Joban forearc basin suggests uplift until this time interval (Figure 13). Seismic reflection profiles across the well show that erosional surfaces within this interval truncated the top of the Soma-oki anticline and the subsequent sedimentary sequence of ~9 Ma onlaps the anticline (Figure 13)[66]. This indicates activity along a N-S trending anticline in this stage. These observations on partial inversion in this stage have been attributed to an increase in horizontal compression stress [31]. In contrast to intense tectonic movements in both the fore-arc side and Backbone Range, the Akita coastal area and the Dewa Hills are considered to have remained bathyal environments (Figure 3). Siliceous shale of the Onnagawa Formation began to deposit in the Akita Basin at 12.3 Ma and yields high TOC-content and constitute major hydrocarbon source rocks (Figure 3)[22]. However, paleogeography of the Niigata Basin (Figure 2) reconstructed from wells [74] suggests that uplift of the eastern part of the Niigata Basin at the base of siliceous shale (~12 Ma) resulted in westward shift of the basin. The eastern margin of the Sea of Japan showed only minor deformation during this stage [49]. These observations suggest attenuation of compression stress and its related deformation toward the Sea of Japan [31]. This time interval (~10-8 Ma) was also characterized by the minimum of volcanic activity in northeast Japan (Figure 13)[16, 33, 57, 75-76]. The reduced volcanic activity during Stage III seems to have been attributed to an increase in horizontal compression stress [31, 33], because compressional stress may prevent the ascent of magma through the upper brittle crust to the surface [16].

Contemporaneous tectonic events at ~10 Ma have been reported not only from northeast Japan but also from Hokkaido (Figure 1) and from further south. Regional unconformity was formed in the western part of Hokkaido at about 12 Ma [77]. Duplex structures developed in the fore-arc basin off Hidaka, north of the Sanriku fore-arc basin (Figure 1) and have been attributed to westward shift of the outer Kurile arc and resultant uplift of the Hidaka Range (Figure 1) [78]. Concurrent deformation (uplift, faulting and folding) took place in the southern Japan Sea and the southern Korea at ~11–10 Ma [48, 79]. Synchronous tectonism with regional deformation and unconformity generation took place during the latest Middle Miocene to Late Miocene in the East China Sea and Ryukyu arc area (Figure 1)[80-82]. In the latter region the Lower-Middle Miocene Yaeyama Group dated to be as young as ~13–12 Ma [83] was folded and truncated by an erosional surface, and then covered by the Upper Miocene-Pliocene Shimajiri Group, which dates from N16 zone (11–8 Ma) [84]. Arc-continent collision had also started by 9 Ma around Taiwan [85]. The Middle Miocene unconformity at around 10 Ma was traced further south to the Pattani Trough in the Gulf of Thailand [86]. The simultaneous occurrence of tectonic events at around 10 Ma in the broad zone in the eastern margin of Asia suggests that compressional tectonics of this age may have had a more regional influence in eastern Asia than previously supposed [31].

The origin of regional compressional tectonics at around 10 Ma is still speculative at the moment. Compressional stress field with NE-SW maximum horizontal stress in northeast Japan in Stage III was attributed to westward shift of the outer Kurile arc and resultant collision of the western and eastern Hokkaido along the Hidaka Range [33]. However, the observations that deformation in northeast Japan tended to reduce towards the west while it was persistent towards the south to the northern Kanto Plain and the Joban fore-arc basin suggest that the Pacific Plate might be a key control and that the westward shift of the outer Kurile arc and resultant collision of the western and eastern Hokkaido was a consequence of regional compression rather than a cause. The late Neogene change in relative motion of the Pacific-Antarctic Plates started at about 12 Ma [87]. The increase in the spreading rate at the Pacific-Antarctic Ridge at 12 Ma might accelerate subduction of the Pacific Plate toward northeast Japan [87]. This might be a possible origin for a regional compression intensified at about 12 Ma. A change in Pacific-Antarctic Plates motion would also have affected the motion of the adjacent plates such as Philippine Sea and Indo-Australian Plates. For example, contemporary increase in spreading rates of Indo-Australian Plate at ~10 Ma at the eastern part of the Southeast Indian Ridge and a consequent peak in basin inversion in the West Indonesian Tertiary basins was suggested by [88].

5.5. Stage IV (Subsidence stage: 9 – 6.5 Ma)

At around 9 Ma, subsidence resumed on the fore-arc side, as well as in the axial Backbone Range (Figure 13). A rapid retrogradation occurred at around 9 Ma in the backbone range because of the resumption of subsidence (Figure 3). In the Joban fore-arc basin, a sedimentary sequence dated at ~9 Ma onlaps an eroded anticline structure with a notable hiatus between 11 and 9 Ma [66-67]. This suggests resumption of subsidence at ~9 Ma after the preceding partial inversion stage. Simultaneous resumption of subsidence and deposition occurred in the forearc lowland (Sendai Plain) and inland basins where an unconformity had been formed during the preceeding stage III (Figure. 13)[51, 53, 63]. This stage was also characterized by a period of intense felsic volcanism, with increased caldera formation at ~8 Ma on the Ou Backbone Range (Figure. 13)[33, 57-58]. Concurrent upward lithostratigraphic change from siliceous shale of the lower part to alternation of shale and tuffs of the upper part of the Onnagawa Formation (Figure 3) was suggested by analysis of wireline logs in oil-producing wells along the Akita coast [89].

Resumption of regional subsidence with increased volcanic activity in this stage may be attributed to weak extensional stress (Figure. 12). The occurrence of Northeast-Southwest trending minor normal faults in the Kurosawa Formation within the Yuda Basin [30] is consistent with this interpretation. Felsic magma trapped in the magma chamber or laccolith within the upper crust during the previous compressional stage may have been released to the surface under an extensional stress field. Caldera formation may have caused local domelike uplift in the Backbone Range during this stage [56, 72]. However, the origin of reduction in horizontal compression stress during Stage IV is unclear because no relative plate motion changes have been reported at around 9 Ma [87].

5.6. Stage V (basin inversion and compression stage: 6.5 – 3~2 Ma)

Stage V represents the differentiation of uplifted and subsided areas associated with crustal deformation in the back-arc region because of basin inversion and compressive stress field (Figure 13)[31]. The sedimentary successions in the Akita Basin (Figure 3) show basin-scale westward progradational stacking patterns including upward shallowing cycles, which consist of slope to basin-floor, trough-fill-turbidite, shelf, nearshore, delta, and fluvial systems, and likely reflect an increased accumulation rate caused by large amount of sediment supply from the uplifted backbone range and the Dewa Hills in response to the increase in compressional stress. The activity of the Kitayuri Thrust (Figures 3 & 4) extending north–south along the Akita coast started at around 5 Ma and resulted in the deposition of trough-fill turbidite system (Katsurane Facies) on the footwall trough of the thrust as a response to syn-depositional faulting and folding under compressional stress [25, 29, 90]. West of the Shinjo Basin (Figure 2), the Dewa Hills uplifted and emerged first in the southern part at about 5 Ma, followed by emergence of the northern part at about 4 Ma and by emergence of central part at around 3 Ma [32]. Four third-order depositional sequences consisting of shallow-marine to fluvial successions (i.e. Shinjo Group) developed in the Shinjo Basin, which represent gradual retreat of marine environments from the Shinjo Basin in response to the successive uplift of the Dewa Hills [32]. Four third-order depositional sequences accompanied by high-frequency depositional sequences consisting of shallow-marine, deltaic and fluvial successions developed in the Yuda Basin from 6.5 Ma to 3 Ma. The third and fourth 3rd-order depositional sequences in the Yuda Basin are correlated with the first and second 3rd-order depositional sequences in the Shinjo Basin, respectively (Figure 3). The correlation indicates marine incursion in the center of the Backbone Range until around 4 Ma, followed by separation from the Sea of Japan by emergence of the western sector of the Backbone Range [30](Figure 3). In other inland basins (e.g. Yonezawa Basin; Figure 2), the sedimentation of conglomerate increased after ~6 Ma

(Figure 13)[53], which suggests uplift of the surrounding mountains at that time. The origin of uplift of the Backbone Range and the Dewa Hills during this stage may be attributed to basin inversion due to increased compressional stress (Figure 13). A change in regional stress field from tension into E-W compression at 7–6 Ma was suggested by the earlier stress field studies in northeast Japan [91-93]. Similar basin inversion and change in depositional style at 6.5 Ma have been reported from the Neogene Niigata-Shin'etsu Basin in central Japan [94]. A notable angular unconformity was formed at the eastern margin of the Niigata Basin at around 7 ~ 6.5 Ma [95-96]. However, half grabens in the eastern margin of the Sea of Japan had not been inverted until early Pliocene [49]. In the fore-arc lowlands, major unconformities were formed at around 6.5 Ma, 5.5 Ma and 3.5 Ma in Stage V (Figure 13)[63, 65]. The unconformity at 6.5 Ma in the fore-arc lowlands can be correlated with that in the Backbone Range and in the Niigata Basin, which suggests a regional tectonic event.

This Late Miocene tectonic change associated with compressional deformation had a greater regional influence than seen the northeast Japan Arc alone. Ingle [48] pointed out that acceleration of uplift and deformation commenced at ~5 Ma in both northeast Japan and the Kurile Arcs (Sakhalin). Itoh et al. [97] demonstrated that Late Miocene uplift and deformation widely took place in the backarc side of the southwest Japan. The compressional deformation and uplift also occurred at 6.5 Ma in Taiwan [98]. The origin of these regional tectonic events has been attributed to resumption of subduction of the Philippine Sea Plate at ~7 Ma [97, 99-100]. However, contemporaneous motion change of the Pacific Plate commenced at 6 Ma [87, 101-102], suggesting more a regional tectonic event within circum Pacific region. For examples, transpressional tectonics along the San Andreas fanult, California, and the Alpine fault, New Zealand commenced at 6 Ma in relation to this change in the Pacific Plate motion [101]. This change in the Pacific Plate motion might also change the motion and subduction of the Philippine Sea Plate.

5.7. Stage VI (Intense compression stage; 3-2 Ma–Present)

Stage VI represents intense crustal deformation associated with the uplift and emergence of all present land areas because of the increased compressive stress [31, 56]. Major angular unconformities were formed at the base of Stage VI in the Yuda Basin and in the eastern margin of the Backbone Range [103], indicating intense uplift of the Backbone Range (Figure 12). The Akita coastal plain emerged at 1.7 Ma, resulting in westward shift of a sedimentary basin and submarine-fan deposition in Oga, followed by gradual fill of the basin with coarse sediments and by emergence of the basin-fill successions [29](Figure 3). However, the timing of basin inversion and of anticline growth varied from Earlly Pliocene to < 1 Ma according to structures both in the center of the Akita Basin [104] and in the eastern margin of the Sea of Japan [49]. Coeval deformation also occurred in the central and southwest Japan [94, 105]. The cause for the increased compressive stress during Stage VI has been attributed either to a change in the Pacific Plate motion [72] or to a change in the Philippine Sea Plate motion at 3 Ma [106].

6. Conclusion

In this chapter, Late Cenozoic tectonic events in northeast Japan were reviewed. Both rifting process and post-rifting tectonics in northeast Japan were much more complex than those proposed in previous tectonic models [72]. Processes of the intra-arc rifting and opening of the Sea of Japan were interpreted as progression from core-complex mode (incipient rift system) to wide-rift mode (opening of the Sea of Japan and rapid intra-arc rifting) to narrow-rift mode (Late syn-rift system)[39]. A transition from extensional tectonics to compressional tectonics in fore-arc side of northeast Japan at the end of the wide rift mode may be related to the effect of lateral motions of the island arc; rotation of northeast Japan accelerated relative convergence rate of the Pacific Plate, thereby promoting compressional stress. A case study of intra-arc development from the Ou Backbone Range revealed three steps of uplift in 12 – 9 Ma, 6.5 – 3-2Ma, and 3-2 Ma - Present. These uplift events were correlated with regional tectonic movements not only in northeast Japan but also in other regions and were clarified as regional tectonic events. The origins of post-rift tectonic events in northeast Japan were inferred to have most likely attributed to changes in the Pacific Plate and Philippine Sea Plate motions.

The present review suggests that the tectonic mode in northeast Japan arc transformed from extension / crustal stretching to compression / crustal shortening much earlier (15 ~ 13.5 Ma) than previous models (3.5 Ma). Moreover, this change in tectonic mode was not straightforward but progressed forward and backward. Reactivation of normal faults bounding half grabens as reverse faults may have started earlier in Middle/Late Miocene. For this reason, the history of active faults may have been longer than previous esti-mates. Activities of active faults and uplift rates estimated assuming constant rate of crustal shortening after 3-2 Ma need to be reassessed. This also indicates that horizontal shorten-ing rate estimated at around 3~5 mm/yr by assuming a constant rate after 2.4 Ma [4] might be overestimated. If so, only several % of plate convergence is accommodated within the northeast Japan arc as long-term deformation. This means that the 2011 great earthquake was inevitable consequence of accumulated elastic strain in northeast Japan arc. This review thus provides important implications for assessing activities of inland active faults, and for recurrence of great subduction zone earthquakes.

Acknowledgements

The author would like to thank Prof. Emeritas Kiyotaka Chinzei, Dr. Tohru Danhara, Dr. Hideki Iwano and Dr. Shunji Moriya for past collaboration and thoughtful discussion. The author also acknowledges fruitful discussions in rifting processes and tectonics of the Eastern Japan Sea Rift System by Dr. Osamu Takano, Dr. Masahiko Yagi and Prof. Atsushi Yamaji. The review comments by the editor Dr. Yasuto Itoh and tolerant editing by Ana Pantar greatly benefited the chapter.

Author details

Takeshi Nakajima*

Address all correspondence to: takeshi.nakajima@aist.go.jp

Institute for Geo-resources and Environment, Geological Survey of Japan, AIST, Japan

References

[1] Iio H., Matsuzawa, T. The generation process of the Tohoku earthquake: Why did the magnitude 9 event occur? Journal of the Geological Society of Japan 2012; 118(5) 248-277. (in Japanese with English abstract)

[2] Ikeda, Y. A discrepancy between geologic and geodetic strain rates. Gekkan Chikyu 2003; 25(2) 125-129. (in Japanese)

[3] Ikeda, Y., Okada, S., Tajikara, M. Long-term strain buildup in the Northeast Japan arc-trench system and its implications for gigantic strain-release events. Journal of the Geological Society of Japan 2012; 118(5) 294-312. (in Japanese with English abstract)

[4] Sato, H. Degree of deformation of late Cenozoic strata in the Northeast Hounshu Arc. The Momoirs of the Geological Society of Japan 1989; 32 257-268. (in Japanese with English abstract)

[5] Takahashi, M. Tectonic boundary between Northeast and Southwest Japan Arcs during Japan Sea opening. Journal of the Geological Society of Japan 2006; 112(1) 14-32. (in Japanese with English abstract)

[6] Tamaki, K., Suehiro, K., Allan, J., Ingle Jr., J.C., Pisciotto, K.A. Tectonic synthesis and implication of Japan Sea ODP Drilling. In: Tamaki, K., Suehiro, K., Allan, J., McWilliams, M. st al. (eds) Proceedings of the Ocean Drilling Program, Scientific Results 127/128(2) College Station, TX: Ocean Drilling Program; 1992. p1333-1348.

[7] Tamaki, K. Geological structures of the Japan sea and its tectonic implications. Bulletin of the Geological Survey of Japan 1988; 39 269-365.

[8] Kaneoka, I., Takigami, Y., Takaoka, N., Yamasshita, S., Tamaki, K. ^{40}Ar-^{39}Ar analysis of volcanic rocks recovered from the Japan Sea floor: constraint of the formation of the Japan Sea. In: Tamaki, K., Suehiro, K., Allan, J., McWilliams, M. st al. (eds) Proceedings of the Ocean Drilling Program, Scientific Results127/128(2) College Station, TX: Ocean Drilling Program; 1992. p819-836.

[9] Otofuji, Y., Matsuda, T. Paleomagnetic evidence for the clockwise rotation of Southwest Japan. Earth and Planetary Science Letters 1983; 70 373-382.

[10] Otofuji, Y., Matsuda, T., Nohda, S. Paleomagnetic evidence for Miocene counter-clockwise rotation of Northeast Japan – Rifting process of the Japan arc. Earth and Planetary Science Letters 1985; 75 265-277.

[11] Tosha, T., Hamano, Y. Paleomagnetism of Tertiary rocks from the Oga Peninsula and the rotation of Northeast Japan. Tectonics 1988; 7 653-662.

[12] Otofuji, Y., Itaya, T., Matsuda, T. Rapid rotation of Southwest Japan – paleomagnetism and K-Ar ages of Miocene volcanic rocks of southwest Japan. Geophysical Journal International 1991; 105 397-405.

[13] Baba, A.K., Matsuda, T., Itaya, T., Wada, Y., Hori, N., Yokoyama, M., Eto, N., Kamei, R., Zaman, H., Kidane, T., Otofuji, Y. New age constraints on counter-clockwise rotation of NE Japan. Geophysical Journal International 2007; 171 1325-1341.

[14] Nohda, S. Formation of the Japan Sea basin: Reassessment from Ar-Ar ages and Nd-Sr isotopic data of basement basalts of the Japan Sea and adjacent regions. Journal of Asian Earth Sciences 2009; 34 599-609.

[15] Lallemand, S, Jolivet, L. Japan Sea: a pull apart basin. Earth and Planetary Science Letters 1985; 76 375-389.

[16] Jolivet, L., Tamaki, K. Neogene kinematics in the Japan Sea region and the volcanic activity of the Northeast Japan arc. In: Tamaki, K., Suehiro, K., Allan, J., McWilliams, M. st al. (eds) Proceedings of the Ocean Drilling Program, Scientific Results127/128(2) College Station, TX: Ocean Drilling Program; 1992. P1311-1331.

[17] Jolivet, L., Tamaki, K., Fournier, M. Japan Sea, opening history and mechanism: a synthesis. Journal of Geophysical Research 1994; 99 22237-22259.

[18] Jolivet, L., Shibuya, H., Fournier, M. Paleomagnetic rotations and the Japan Sea opening. In: Taylor, B., Natland, J. (eds) Active Margins and Marginal Basins of the Western Pacific. Geophysical Monograph 88. Washington: AGU; 1995. P355-369.

[19] Kano, K., Uto, K., Uchiumi, S., Ogasawara, K. Early Miocene unfonformity in the Makarov and Chekhov areas, southern Sakhallin Island, Russia, and its implication. Journal of Geography 2000; 109 262-280. (in Japanese with English abstract)

[20] Kano, K., Yoshikawa, T., Yanagisawa, Y., Ogasawara, K., Danhara, T. An unfonformity in the early Miocene syn-rifting succession, northern Noto Peninsula, Japan: Evidence for short-term uplifting precedent to the rapid opening of the Japan Sea. The Island Arc 2002; 11 170-184.

[21] Kano, K., Uto, K., Ohguchi, T. Stratigraphic review of Eocene to Oligocene successions along the eastern Japan Sea: Implication for early opening of the Japan Sea. Journal of Asian Earth Sciences 2007; 30 20-32.

[22] Kano, K., Ohguchi, T., Yanagisawa, Y., Awata, Y., Kobayashi, N., Sato, Y., Hayashi, S., Kitazato, H., Ogasawara, K., Komazawa, M. Geology of the Toga and Funakawa

District. Quadrangle Series, 1:50,000, Geological Survey of Japan, AIST 2011; p127 (in Japanese with English abstract).

[23] Ohki, J., Watanabe, N., Shuto, K., Itaya, T. Shifting of the volcanic fronts during Early to Late Miocene in the Northeast Japan arc. The Island Arc 1993; 2 87-93.

[24] Suzuki, U. Geology of Neogene Basins in the eastern part of the Seaof Japan. The Memoirs of the Geological Society of Japan 1989; 32 143-183. (in Japanese with English abstract)

[25] Sato, H., Yoshida, T., Iwasaki, T., Sato, T., Ikeda, Y., Umino, N. Late Cenozoic development of the back-arc region of central northern Houshu, Japan, revealed by recent deep seismic profiling. Journal of the Japanese Association of Petroleum Technologists 2004; 69 145-154. (in Japanese with English abstract)

[26] Yamaji, A. Rapid intra-arc rifting in Miocene Northeast Japan. Tectonics 1990; 9 365-378.

[27] Sato, T., Baba, K., Ohguchi, T., Takayama, T. Discovery of Early Miocene calcareous nannofossils from Japan Sea side, Northern Honshu, Japan, with reference to paleoenvironment in the Daijima and Nishikurosawa ages. Journal of Japanese Association of Petroleum Technology 1991; 56 263-279. (in Japanese with English abstract)

[28] Sato, T., Yamasaki, M., Chiyonobu, S. Geology of Akita Prefecture. Daichi 2009; 50 70-83. (in Japanese)

[29] Sato, T., Sato, N., Yamasaki, M., Ogawa, Y., Kaneko, M. Late Neogene to Quaternary paleoenvironmental changes in the Akita area, Northeast Japan. Journal of the Geological Society of Japan 2012; 118 62-73. (in Japanese with English abstract)

[30] Nakajima, T., Danhara, T., Chinzei, K. Late Cenozoic basin development of the Yuda Basin in the axial part of the Ou Backbone Range, northeast Japan. Journal of the Geological Society of Japan 2000; 106 93-111. (in Japanese with English abstract).

[31] Nakajima, T., Danhara, T., Iwano, H., Chinzei, K. Uplift of the Ou Backbone Range in Northeast Japan at around 10 Ma and its implication for the tectonic evolution of the eastern margin of Asia. Palaeogeography Palaeoclimatology Palaeoecology 2006; 241 28-48.

[32] Moriya, S., Chinzei, K., Nakajima, T., Danhara, T. Uplift of the Dewa Hills recorded in the Pliocene paleogeographic change of the western Shinjo Basin, Yamagata Prefecture. Journal of the Geological Society of Japan 2008; 114 389-404. (in Japanese with English abstract)

[33] Yoshida, T. Late Cenozoic Magmatism in the northeast Honshu Arc, Japan. Earth Science (Chikyu Kagaku) 2009; 63 269-288. (in Japanese with English abstract)

[34] Yagi, M., Hasenaka, T., Ohguchi, T., Baba, K., Sato, H., Ishiyama, D., Mizuta, T., Yoshida, T. Transition of magmatic composition reflecting an evolution of rifting activity – a case study of the Akita-Yamagata basin in Early to Middle Miocene, North-

east Honshu, Japan -. Japanese Magazine of Mineralogical and Petrological Sciences 2001; 30 265–287. (in Japanese with English abstract)

[35] Sato, S., Sato, T. Basalts of Nishikurosawa Stage of early to middle Miocene in the Akita and Niigata oil fields, northeast Japan. Journal of Japanese Association of Petroleum Technology 1992; 57 91-102.

[36] Yamada, R., Yoshida, T. Volcanic sequences related to Kuroko mineralization in the Hokuroku district, Northeast Japan. Resource Geology 2004; 54 399-412.

[37] Ingersoll, R.V., Busby, C.J. Tectonics of sedimentary basins. In: Busby, C.J., Ingersoll, R.V. (eds) Tectonics of Sedimentary Basins, Oxford: Blackwell; 1995. p1-51.

[38] Yamada, R., Yoshida, T., Relationship between the evolution of the Neogene volcanism and the Kuroko mineralization in the vicinity of the Hokuroku district, NE Japan–A chronological study–. Shigen Chishitsu 2002; 53 69-80 (in Japanese, with English abstract)

[39] Buck, W.R. Modes of continental lithospheric extension. Journal of Geophysical Research 1991; 96(B12) 20161-20178.

[40] Sato, H., Hirata, N., Iwasaki, T., Matsubara, M., Ikawa, T., Deep seismic reflection profiling across the Ou Backbone Range, Northern Honshu Island, Japan. Tectonophysics 2002; 355 41-52.

[41] Blow, W.H. Late middle Eocene to Recent planktonic foraminiferal biostratigraphy. Proceedings of 1st International Conference of Planktonic Microfossils. 1969; 1 p199-421.

[42] Martini, E., Standard Tertiary and Quaternary calcareous nannoplankton zonation. In: Farinacci, A. (ed.) 2nd Internatioanl Conference of Planktonic Microfossils: proceedings of the 2nd international conference of Planktonic Microfossils, 1970, Roma (Tcnoscienza); 1971; p738-785.

[43] Nakajima, T. Description of unconformities and pyroclastic key beds in the Yuda Basin, Iwate Prefecture, and their implications for the uplift of the Ou Backbone Range. Earth Science (Chikyu Kagaku) 2012; 66(3) 69-83. (in Japanese, with English abstract)

[44] Takeuchi, T., Inoue, T., Matsukuma, H., Yamaoka, K, Ueda, R., Oikawa, A., Ohtagaki, T., Abe, M., Honma, T., Koumura, A., Moriai, Y., Koguma, H., Kaneno, Y., Hirayama H., Baba, A., Akiyama S., Nishide, S, Report of the regional survey in the fiscal year 41,the Waga-Omono region. Tokyo; Ministry of International Trade and Industry; 1967. 21pp (in Japanese).

[45] Nakajima, T., Danhara, T. Fission-track ages of Middle Miocene to Pliocene tuffs and a lava from the Yuda Basin, Iwate Prefecture, Northeast Japan. Journal of the Geolocical Society of Japan 1999; 105 668-671 (in Japanese, with English abstract).

[46] Bornhold, B.D., Giresse, P. Glauconitic sediment on the continenatal shelf of Vancouver Island, British Columbia. Journal of Sedimentary Petrology 1985; 55 653-664.

[47] Haq, B.U., Hardenbol, J., Vail, P.R. Mesozoic and Cenozoic chronostratigraphy and eustatic cycles. In: Wilgus, C.K., Hastings, B.S., Kendall, C.G.ST.C., Posamentier, H., Ross, C.A., Van Wagoner, J.C. (Eds.) SEPM Special Publication 1988; 42 p 71-108.

[48] Ingle, J.C.Jr. Subsidence of the Japan Sea: stratigraphic evidence from ODP sites and onshore sections. In: Tamaki, K., Suyehiro, K., Allan, J., McWilliams, M., et al. (eds). Proceeding of ODP, Scientific Results, 127/128(2), College Station, TX: Ocean Drilling Program: 1992; p1197-1218.

[49] Okamura, Y., Watanabe, M, Morijiri,, R., Satoh, M. Rifting and basin inversion in the eastern margin of the Japan Sea. The Island Arc 1995; 4 166-181.

[50] Sato, T., Takayama, T., Kato, M., Kudo, T. Calcareous microfossil biostratigraphy of the uppermost Cenozoic Formations distributed in the coast of the Japan Sea–Part 3: Akita area and Oga Peninsula. Journal of Japanese Association of PetroleumTechnology 1988; 53 199-212 (in Japanese, with English abstr.).

[51] Aita, Y., Taketani, Y., Okada, H., Hasegawa, S., Murayama, T., Nemoto, N. Miocene microbiostratigraphy and paleoceanographic study of the Aizu area, Northeast Japan. Bulletin oh the Fukushima Museum 1998; 13 1-119 (in Japanese, with English abstract)

[52] Sato, H., Hirata, N. Deep structures of active faults and development of the Japanese Islands. Kagaku 1998; 68 63-71 (in Japanese).

[53] Yanagisawa, Y., Yamamoto, T. Geology of the Tamaniwa district. Quadrangle Series, 1:50,000, Geological Survey of Japan 1998; p94. (in Japanese with English abstract)

[54] Nakajima, T., Danhara, T., Iwano, H. Earliest Late Miocene compressional tectonics deduced from the basin development of the Ou Backbone Range and Yokote Basin, northeast Japan. Abstract of the 109th Annual Meeting of the Geological Society of Japan 2002; p12. (in Japanese).

[55] Usuta, M. Geotectonic history of the southern part of Akita Prefecture, Northeast Japan. The Memoirs of the Geological Society of Japan 1989; 32 57-80. (in Japanese with English abstract)

[56] Sato, H. Late Cenozoic tectonic evolution of the central part of Northern Honshu, Japan. Bulletin of the Geological Survey of Japan 1992; 43 119-139. (in Japanese with English abstract).

[57] Yoshida, T., Aizawa, K., Nagahashi, Y., Sato, H., Ohguchi, T., Kimura, J., Ohira, H.. The formation of late Cenozoic caldera swarm in the island arc volcanic period of Northeast Honshu arc. Earth Monthly. Special Volume 1999; 27 123-129. (in Japanese)

[58] Yoshida, T., Nakajima, J., Hasegawa, A., Sato, H., Nagahashi, Y., Kimura, J., Tanaka, A., Prima, O.D.A., Ohguchi, T. Evolution of Late Cenozoic magmatism in the NE

Honshu Arc and its relation to the crust-mantle structures. Daiyonki Kenkyu 2005; 44 195-216. (in Japanese with English abstr.act)

[59] Yanagisawa, Y. Diatom biostratigraphy of the Neogene Sendai Group, northeast Hounshu, Japan. Bulletin of the Geological Survey of Japan 1990; 41(1) 1-25. (in Japanese with English abstr.act)

[60] Shimamoto, M., Ota, S., Hayashi, H., Sasaki, O., Saito, T. Planktonic foraminiferal biostratigraphy of the Miocene Hatatate Formation in the southwestern part of Sendai City, Northeast Japan. Journal of the Geological Society of Japan 2001; 107(4) 258-268. (in Japanese with English abstr.act)

[61] Suto, I., Yanagisawa, Y., Ogasawara, K. Tertiary geology and chronostratigraphy of the Joban area and its environs, northeastern Japan. Bulletin of the Geological Survey of Japan 2005; 56(11/12) 375-409. (in Japanese with English abstr.act)

[62] Fujiwara, O, Yanagisawa, Y., Shimamoto, M., Fuse, K., Danhara, T. Stratigraphic position and significance of the unconformity recognized from the upper Miocene Natori Group in the SW region of Sendai city, NE Honshu. Abstract of the 112nd Annual Meeting of the Geolical Society of Japan 2005; p.70 (in Japanese).

[63] Fujiwara, O., Yanagisawa, Y., Irizuki, T., Shimamoto, M., Hayashi, H., Danhara, T., Fuse, K., Iwano H. Chronological data for the Middle Miocene to Pliocene sequence around the southwestern Sendai Plain, with special reference to the uplift history of the Ou Backbone Range. Bulletin of the Geological Survey of Japan 2008; 59(7/8) 423-438. (in Japanese with English abstract)

[64] Geological Society of Japan., editor. The Geology of Japan 3, Kanto District. Tokyo: Asakura shoten; 2008. (in Japanese)

[65] Yanagisawa, Y., Oishi, M. Diatom age of the Upper Miocene Hishinai Formation in the Western Area of Kitakami City, Iwate Prefecture, Northeast Japan. Bulletin of the Iwate Prefectural Museum 2009; 26 1-10. (in Japanese with English abstract)

[66] Kato, S., Akiba, F., Moriya, S. The Upper Cretaceous-Cenozoic stratigraphy and geologic structure in the offshore Soma area, northeast Japan. Journal of the Geological Society of Japan 1996; 102(12) 1039-1051. (in Japanese with English abstrract).

[67] Kameo, K., Sato, T. Recent development of calcareous nannofossil biostratigraphy and its application – Neogene and Quaternary stratigraphy of offshore Joban based on calcareous nannofossils-. Journal of the Japanese Association of Petroloreum Technology 1999; 64 16-27. (in Japanese with English abstract)

[68] Iwata, T., Hirai, A., Inaba, T., Hirano, M. Petroleum system in the Offshore Joban Basin, northeast Japan. Journal of the Japanese Association of Petroloreum Technology 67(1), 62-71 (in Japanese with English abstract).

[69] Kudo, T., Yanagisawa, Y., Iwano, H., Danhara, T. Chronostratigraphy of the Lower to Middle Miocene Successions and Basin Development in the Kamo Area, Niigata

Sedimentary Basin, Northeast Japan Arc. Journal of Geography 2011; 120(4) 654-675. (in Japanese with English abstract).

[70] Nakamura, K. Inversion tectonics and its structural expression. Journal of the Tectonic Research Group of Japan 1992; 2 1-16. (in Japanese with English abstract)

[71] Takahashi, M., Hayashi, H., Kasahara, K., Kimura, H. Geologic interpretation of the seismic reflection profile along the western margin of the Kanto Plain – With special reference to the Yoshimi metamorphic rocks and western extension of the Tonegawa Tectonic Line-. Journal of the Geological Society of Japan 2006; 112(1) 33-52. (in Japanese with English abstrract).

[72] Sato, H. The relationship between late Cenozoic tectonic events and stress field and basin development in northeast Japan. Jour.nal of Geophysical Research 1994; 99 22,261-22,274.

[73] Yamaji, A., Sato, H. Miocene subsidence of the Northeast Honshu Arc and its mechanism. The Memoirs of the Geological Society of Japan 1989; 32 339-349 (in Japanese with English abstract).

[74] Sato, T., Kudo, T., Kameo, K. On the distribution of source rocks in the Niigata oil field based on the microfossil biostratigraphy. Journal of the Japanese Association of Petroloreum Technology 1995; 60(1) 76-86. (in Japanese with English abstract)

[75] Fujioka, K., History of the Explosive Volcanism of the Tohoku Arc from the Core Sediment Samples of the Japan Trench. Kazan 1983; 28(1) 41-58. (in Japanese with English abstract)

[76] Cambray, H., Cadet, J-P., Pouclet, A. Ash layers in deep-sea sediments as tracers of arc volcanic activity: Japan and Central America as case studies. The Island Arc 1993; 2 72-86.

[77] Geological Society of Japan, editor. The Geology of Japan 1, Kanto District. Tokyo: Asakura shoten; 2010. (in Japanese)

[78] Osawa, M., Nakanishi, S., Tanahashi, M., Oda, H. Structure, tectonic evolution and gas exploration potential of offshore Sanriku and Hidaka provinces, Pacific Ocean, off northern Honshu and Hokkaido, Japan. Journal of the Japanese Association of Petroloreum Technology 2002; 67(1) 37-51 (in Japanese, with English abstr.).

[79] Chough, S.K., Berg, E. Tectonic history of Ulleung basin margin, East Sea (Sea of Japan). Geology 1987; 15 45-48.

[80] Ishiwada, Y., Aida, H., Atake, M., Araki, H., Iijima, A., Ikeda, A., Okuda, Y., Kikuchi, Y., Kojima, K., Saito, T., Sato, Y., Tanaka, S., Tono, S., Hirayama, J., Honza, E., Miyazaki, H., Morishima, H., harada, Y., editors. Oil and Natural Gas Resources in Japan. Tokyo; Natural Gas Association; 1992. (in Japanese)

[81] Nakagawa, H. Geologic events in late Cenozoic around the Miyako-jima, Ryukyu Islands. Abstrract of the 2003 Technical Meeting of the Japanese Association of Petroleum Technology 2003; p23. (in Japanese)

[82] Lee, G.H., Kim, B., Sun, K., Sunwoo, D. Geologic evolution and aspects of the petroleum geology of the northern East China Sea shelf basin. American Association of Petroleum Geologists, Bulletin 2006; 90(2) 237-260.

[83] Fujimoto, M., Ishida, H., Iwata, T., Nakagawa, H. A review of stratigraphy of MITI Miyakojimaoki well based on fission-track ages. Abstract of the 2001 Technical Meeting of the Japanese Association of Petroleum Technology 2001; p58. (in Japanese)

[84] Aiba, J., Sekiya, E. Distribution and characteristics of the Neogene Sedimentary Basins around the Nansei-Shoto (Ryukyu Islands). Journal of the Japanese Association of Petroloreum Technology 1979; 44 97-108. (in Japanese with English abstract)

[85] Sibuet, J.-C., Hsu, S.-K., Debayle, E. Geodynamic context of the Taiwan Orogen. In: Clift, P., Kuhnt, W., Wang, P., Hayes, D. (Eds.), Continent-ocean interactions within East Asian marginal seas. AGU Geophysical Monograph Ser. 149, 127-158.

[86] Fujiwara, M. The Middle Miocene Unconformity (MMU) in the southern part of the Pattani Trough in the Gulf of Thailand. Journal of the Japanese Association of Petroloreum Technology 2011; 76(6) 545-555. (in Japanese with English abstract)

[87] Cande, S. C., Raymond, C.A., Stock, J., Haxby, W.F. Geophysics of the Pitman fracture zone and Pacific-Antarctic Plate motion during the Cenozoic. Science 1995; 270 947-953.

[88] Honda, H., Nagura, H. A note on the Tertiary history of Indo-Austaralian plate-movements and the West Indonesian Tertiary stratigraphy.. Journal of the Japanese Association of Petroloreum Technology 2000; 65 270-277 (in Japanese, with English abstract).

[89] Honda, H., Yoshitake, N., Yamamoto, S., Ashida, T., Kawamoto, T., Tsuji, T., Todoroki, T. Physical diagenesis and rock properties of the On'nagawa siliceous mudstone in the central and southern parts of the Neogene Akita sedimentary basin. Journal of the Japanese Association of Petroloreum Technology 2013; 78(1) 61-67. (in Japanese with English abstract)

[90] Nakajima, T., Danhara, T., Iwano, H., Yamashita, T. A fission-track age from the Katsurane Facies in the Tentokuji Formation at Hanekawa, Akita, northeast Japan. Journal of the Geological Society of Japan 2003; 109(4) 252-255. (in Japanese with English abstract)

[91] Takeuchi, A., Stress field and tectonic process during the Neogene and later period in the northern part of Nagano Prefecture, central Japan.. Journal of the Geological Society of Japan 1977; 83 679-691 (in Japanese with English abstr.act).

[92] Tsunakawa, H. Neogene stress field of Japanese arcs and its relation to igneous activity. Tectonophysics 1986; 124 1-22.

[93] Otsuki, K. Neogene tectonic stress fields of northeast Honshu Arc and implications for plate boundary conditions. Tectonophysics 1990; 181 151-164.

[94] Takano, O. Changes in depositional systems and sequences in response to basin evolution in a rifted and inverted basin: an example from the Neogene Niigata-Shin'etsu basin, Northern Fossa Magna, central Japan. Sedimentary Geology 2002; 152 79-97.

[95] Kobayashi, I. Watanabe K. Geologic event in the eastern margin of Niigata Tertiary sedimentary basin, especially on Mio-Pliocene unconformity. Report of the Geology and Mineralogy Department, Faculty of Science, Niigata University 1985; 5 91-103 (in Japanese with English abstr.act).

[96] Kudo, T., Uchino, T., Komatsubara, T., Takahashi, Y., Yanagisawa, Y. Geology of the Kamo district. Quadrangle Series, 1:50,000, Geological Survey of Japan, AIST 2011; p162. (in Japanese with English abstr.act).

[97] Itoh, Y., Nakajima, T., Takemura, A. Neogene deformation of the back-arc shelf of Southwest Japan and its impact on the palaeoenvironments of the Japan Sea. Tectonophysics 1997; 281 71-82.

[98] Lin, A.T., Watts, A.B., Hesselbo, S.P. Cenozoic stratigraphy and subsidence history of the South China Sea margin in the Taiwan region. Basin Research 2003; 15 453-478.

[99] Kamata, H., Kodama, K. Tectonics of an arc-arc junction: an example from Kyushu Island at the junction of the Southwest Japan Arc and the Ryukyu Arc. Tectonophysics 1994; 233 69-81.

[100] Niitsuma, N. Japan Trench and tectonics of the Japanese Island Arcs. The Island Arc 2004; 13 306-317.

[101] Wessel, P., Kroenke, L.W. Ontong Java Plateau and late Neogene changes in Pacific plate motion. Journal of Geolhysical Research 2000; 105(B12) 28,255-28277.

[102] Wessel, P., Kroenke, L.W. Reconciling late Neogene Pacific absolute and relative plate motion changes. Geochemistry, Geophysics, Geosystems 2007; 8(8) doi: 10.1029/2007GC001636.

[103] Oishi, M., Yoshida, H., Kim, K.N. The Plio-Pleistocene formations of the Waga and Geto River area in the Kitakami Lowland, Northeast Japan. Research Report of the Iwate Prefectural Museum 1998; 14 5-20. (in Japanese with English abstr.act).

[104] Hiramatsu, C., Miwa, M. Occurrence patterns of the Pliocene to Pleistocene planktonik foraminiferal marker beds with relation to the timing of hydrocarbon trap formation, offshore and coastal areas of Akita Prefecture, Japan. Journal of the Japanese Association of Petroloreum Technology 2005; 70(1) 104-113. (in Japanese with English abstr.act).

[105] Taira, A. Tectonic evolution of the Japanese Island Arc system. Annual Revue of Earth and Planetary Sciences 2001; 29 109-134.

[106] Takahashi, M. Tectonic Development of the Japanese Islands controlled by Philippine Sea Plate Motion. Journal of Geography 2006; 115(1) 116-123.

Chapter 7

Foreland Basins at the Miocene Arc-Arc Junction, Central Hokkaido, Northern Japan

Gentaro Kawakami

Additional information is available at the end of the chapter

1. Introduction

A foreland basin can be defined as a linear depression formed from the flexure of foreland lithosphere under the load of an orogenic wedge. Various factors control the basin geometry and the stratigraphic architecture of the basin fill [1].

During the Miocene age, a series of foreland basins were formed in Hokkaido, the northern island of Japan, due to an arc-arc collision event [2–4]. The basin area has a total length of 400 km, and it is characterized by several rows of deep depressions stretched in a north-south direction and separated by topographic highs (Figure 1). The degree of total crustal shortening and exhumation by the collision event is larger in the southern region of the collision zone [5], and thus, the basin geometry and the stratigraphic/sedimentologic architecture of the basin fills vary among the depressions. Namely, the depression in the southern region has suffered progressive deformation due to thrust propagation, and the depocenter has migrated foreland-ward, a feature noted in many foreland basins [1]. In contrast, the depression in the northern region shows restrictive deformation and foreland-ward migration of depocenter. Thus, the initial geometry and stratigraphic architecture of the basin fill are well preserved even in the proximal margin of the basin. On the other hand, unlike the northern and southern regions, the depression located near the center of the foreland basin area has been affected by the existence of a projection on the foreland lithosphere. As a result, the depression shows very narrow basin geometry and restrictive foreland-ward migration of the depocenter.

Such spatial variations are caused mainly by regional differences in the degree of tectonic disturbance, as there is no radical difference in the geodynamic states of the underlying lithosphere throughout the foreland area. Considering the above factors, this study focuses on how the difference in the degree of tectonic deformation affects the foreland basin geometry, basin-filling process, and the resultant stratigraphy in a geological record. This paper firstly reviews the stratigraphic architecture and depositional system in each depression, and then discusses the relationship between basin evolution and tectonics.

2. Geological setting

The present Hokkaido Island is located on a junction between the Northeast Japan Arc and the Chishima (Kuril) Arc (Figure 1), which originated from an interaction between the Eurasian and Okhotsk plates that commenced in the Late Eocene [4]. A right-lateral oblique collision between the two arcs during the Miocene age formed a west-vergent fold-thrust belt and a subsiding foreland area nearly 400 km long in central Hokkaido [4].

Because of the right-lateral oblique collision and the subsequent westward migration of the Chishima forearc sliver [4], the collision event has endured longer in the south. In the northern region, the thrust activity had declined by the late Miocene, but in the southern region, it continues up untill the present-day. Therefore, the degree of total crustal shortening and exhumation by the collision event is larger in the southern region [5]. For instance, the amount of crustal shortening in the Sorachi-Yezo Belt varies from 12 km in the north, to 50–64 km in the south [5, 6].

An exhumed basement in the eastern orogen (the eastern zone of the Sorachi-Yezo Belt and the Hidaka Belt) consists of Cretaceous to Early Paleogene accretionary complexes (Figure 1) formed by the arc-trench system along the eastern margin of the Eurasian continent [7]. A metamorphic core of granulite facies crops out only in the southern area of the orogen (the Hidaka metamorphic rocks) [8]. The cooling ages of the metamorphic core indicate the middle to late Miocene ages (20–10 Ma), and are synchronized with the evolution of the foreland basin ([9] and references therein).

Before the foreland basin subsidence, relatively shallow sea environments were widespread in the area during the Early to early-Middle Miocene ages corresponding roughly to the sea-level rise during the Mid-Neogene Climatic Optimum [10]. However, the detailed tectonic control of this shallow marine basin remains unclear. It is evident that initial deep depressions were formed around 15–16 Ma, due to rapid subsidence (Figure 2). On close inspection, the central part of the foreland basin area is constricted, and this is probably owing to the existence of a projection on the foreland lithosphere (the Kabato basement high (KBH) of Rebun-Kabato Belt in Figure 1). The Rebun-Kabato Belt is defined by the Cretaceous arc volcanics/plutonics and forearc basin-fills [11] which are overlain by the thick sedimentary succession of Paleogene–Quaternary in the Ishikari Lowland.

Slightly right-stepping deep depressions developed in both the northern and southern parts of the foreland basin area. These deep depressions, known as the Tenpoku, Haboro, Ishikari, and Hidaka basins (from north to south) [3, 12] (Figure 1), are filled with Miocene–Pliocene deposits up to 6000 m thick (Figure 2), sourced from the eastern orogen [9, 12–16]. Most of the basin fills consist of parallel-sided, non-channelized turbiditic sand and basin-plain mud interbeds ("basinal turbidites": [17]) and coarser-grained turbiditic deposits of immature

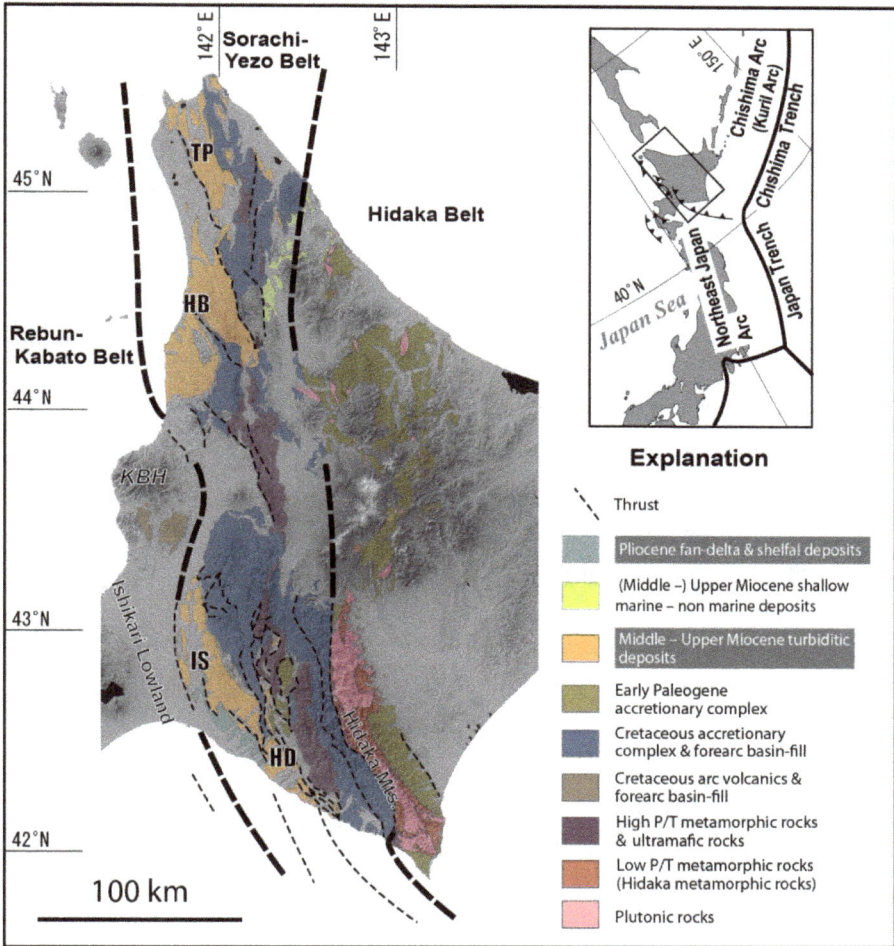

Figure 1. Simplified geologic map of central Hokkaido, northern island of Japan. The Miocene foreland basins were formed in the western zone of the Sorachi-Yezo Belt. The central part of the foreland basin area is constricted by the projection (Kabato basement high: KBH). Tenpoku Basin (TP), Haboro Basin (HB), Ishikari Basin (IS), and Hidaka Basin (HD) are major depressions separated by topographic highs (after [3]).

facies. These are settled in various areas from the outer to the inner foredeep settings. In this paper, the terms "outer", "axial" and "inner foredeep" follow the previous work [17, 18], (i.e., the outer foredeep is the most distal part of the foredeep above the foreland ramp; the axial foredeep is the central and deepest part of foredeep with relative flat basin floor; and the inner foredeep is the proximal part of foredeep along the thrust front characterized by steep slope and rough topography due to thrust propagation). The turbiditic deposits are covered with siliceous/diatomaceous and/or shelfal muddy deposits of the late Miocene to Pliocene ages (Figure 2).

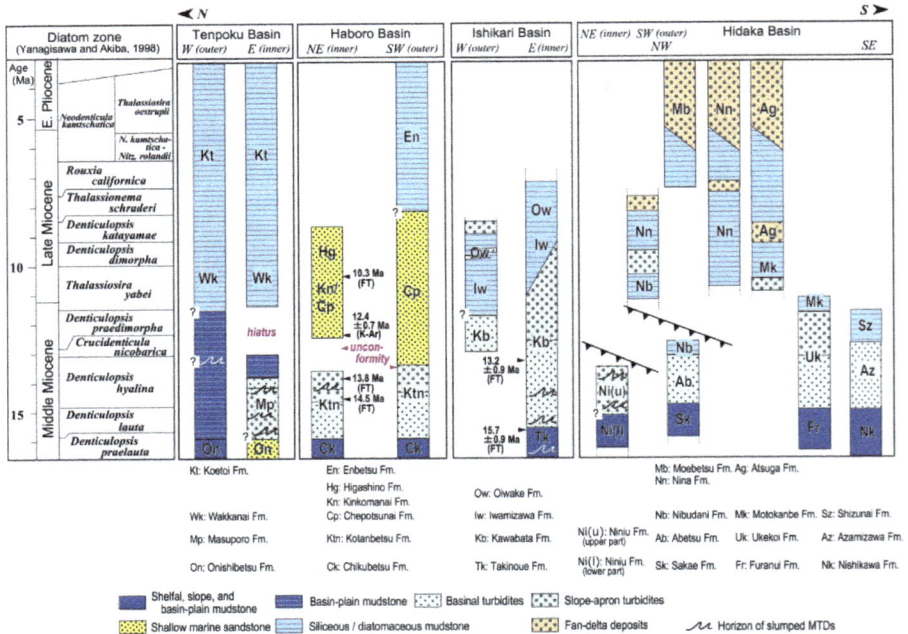

Figure 2. Stratigraphy of the middle Miocene to early Pliocene deposits in central Hokkaido. Depositional ages have been determined by diatom biostratigraphy and chronostratigraphic data. Diatom biostratigraphic zonation after [19]. Columns show the successions of inner (orogen-ward, proximal) and outer (foreland-ward, distal) areas of each depression.

3. Basin geometry and stratigraphy

3.1. Tenpoku Basin

The northernmost depression is 80 km wide and at least 60 km long (wider than other depressions later described, see Table 1), and is known as the Tenpoku Basin. The western half and the northern part of the basin extend to the Japan Sea (Figure 3). Because the fold-thrust propagation and resultant basin deformation is restrictive, a Middle to Late Miocene basin fill, 1000–2000 m thick, crops out only in the easternmost area. The Miocene axial foredeep is located around the present-day coastline along the Japan Sea, where, according to well data, the basin fill reaches its maximum thickness (ca. 4000 m) [20].

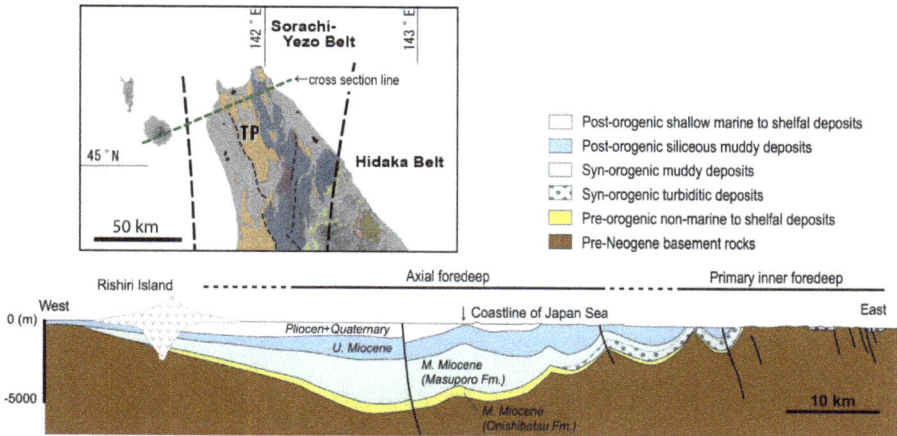

Figure 3. Cross section of the Tenpoku Basin (after [21]). Primary basin geometry is well preserved because of restricted later tectonic disturbances. See Figure 1 for legend in the index map.

In the easternmost outcrop, a middle Miocene turbiditic succession (the Masuporo Formation) is characterized by abundant mass-transport deposits (MTDs), such as slumped sand/mud interbeds and chaotic sand to gravel beds, bearing many intrabasinal blocks [20, 22–25] (Figure 4). It is noteworthy that the MTDs at the base of the Masuporo Formation rest directly on the shallow marine sandy deposits of the early Middle Miocene (the Onishibetsu Formation), which settled prior to the foreland basin subsidence. In the uppermost horizon of the turbiditic succession, mud-prone basinal turbidites and basin-plain mudstones are predominant, and the succession thus shows an overall fining-upward trend.

The basin fill fines also foreland-ward drastically, and the axial foredeep is filled mainly with basin-plain muddy deposits (2000–3000 m thick) [26]. As an exception, a 200 m thick slumped interval occurs in the upper part of this muddy succession [26]. This interval is characterized by muddy chaotic deposits containing granule-grade grains, although no detailed sedimentary features are described.

The entire part of the basin was covered by basin-plain muddy deposits in the late-Middle Miocene, after which siliceous/diatomaceous muddy deposits (>1000 m thick) were accumulated basin-wide during the late Miocene.

Although the sediment dispersal pattern in the Tenpoku Basin is not clearly understood, a clastic composition of basin fill, which is rich in granite and hornfels clasts, indicates a sediment supply from the Hidaka Belt in the east.

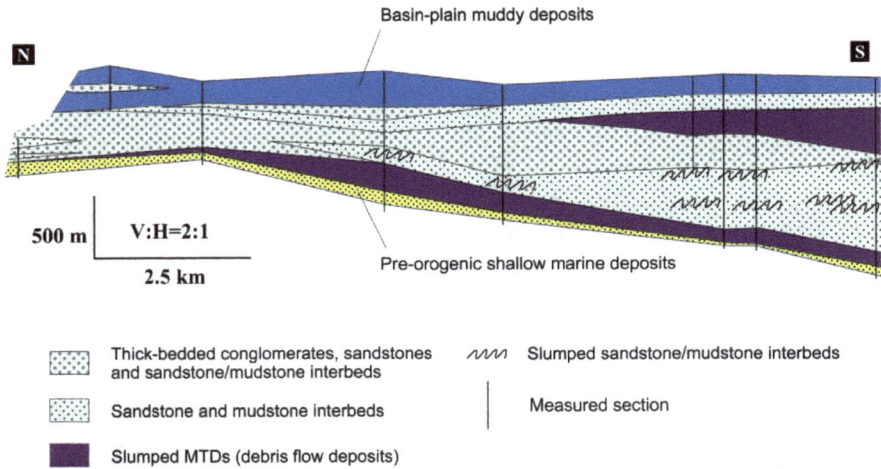

Figure 4. Basin-axial profiles of the middle Miocene turbidites at the eastern margin of the Tenpoku Basin (after [22]) (above), and representative photo of the slumped MTDs (below). The cliff is about 12 m high.

3.2. Haboro Basin

The stratigraphic and sedimentary architecture of the Haboro basin fill, (at least 50 km wide and 90 km long), has been well reconstructed by many studies (e.g. [2, 3, 14, 27]). A large part of the accommodation space is filled with 2000–3000 m thick middle Miocene turbidites (the Kotanbetsu Formation) composed of lower basinal turbidites and upper coarse-grained immature turbiditic deposits (slope-apron turbidites, to be discussed later) (Figure 5). The basinal turbidites are invariably underlain by relatively thin (<100 m) shelfal muddy deposits of the upper part of the Chikubetsu Formation. The blackish muddy deposits at the top of the Chikubetsu Formation indicate a condensed horizon formed during rapid basin subsidence [27]. At the southwestern margin of the basin, the turbiditic deposits show lateral onlap onto the early Miocene shallow-marine/non-marine deposits fringing the KBH. On the other hand, the basin fill was exposed subaerially around 12 Ma at the eastern margin of the basin.

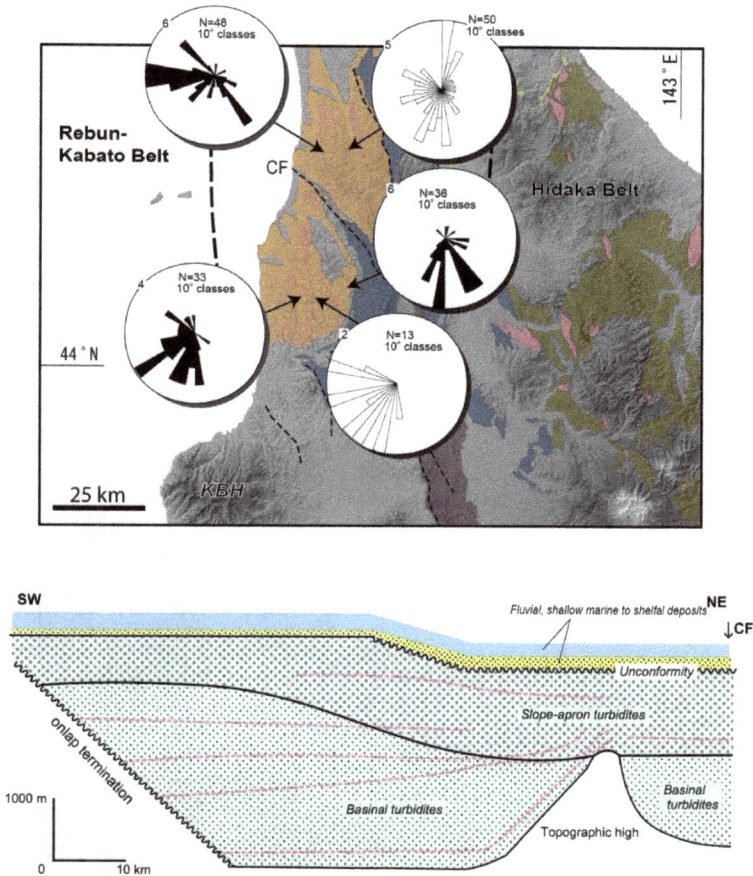

Figure 5. Rose diagrams showing paleocurrent directions of the middle Miocene turbidites in the Haboro Basin (above: compiled from [2, 14]), and transverse profile of the basin fill (below: modified from [2]). The solid and open petals of rose diagram mean the direction measured from sole-marks and clast fabric (solid) and cross lamination (open). Pink-colored dashed lines are stratigraphic markers (ash turbidite beds). CF: Chikubetsu Fault. See Figure 1 for legend of the index map.

A north-south stretched topographic high lying parallel to the basin-axis separates an initial depression into two segments [2, 3, 14, 28]. Basinal turbidites buried the segments progressively from the inner (eastern orogen-ward segment) to the outer (western foreland-ward segment), and flattened the irregular bottom of the basin (Figure 5). Slumped MTDs are developed, especially in the inner segment [27]. Subsequently deposited coarse-grained turbidites characterized by amalgamated and channelized sandy/gravelly beds are prograded on the basinal turbidites [2, 3]. These turbiditc deposits contain abundant large granite clasts indicating a sediment supply from the Hidaka Belt in the east. Sole marks within the basinal turbidites in the central to southern outcrops reveal southwest-to-south directed flows. The coarse-grained turbidites found at the stratigraphically uppermost part of the northern outcrop reveal west-directed flow (Figure 5).

3.3. Ishikari basin

The Ishikari Basin is characterized by its very narrow basin geometry (Figure 6). At present, the Middle Miocene basin fill reveals a north-south stretched elongated distribution (<15 km wide and 60 km long). Although the original dimension of the basin is uncertain because of the post-dated basin deformation, the strongly concentrated paleoflow data in the basin-axis direction indicates a confined basin floor (Figure 7). The primary western margin of the basin, where a thrust-front advanced during the Pliocene age [5], was bordered by horst structures formed from the basement rocks and the overlying Upper Oligocene to Lower Miocene strata.

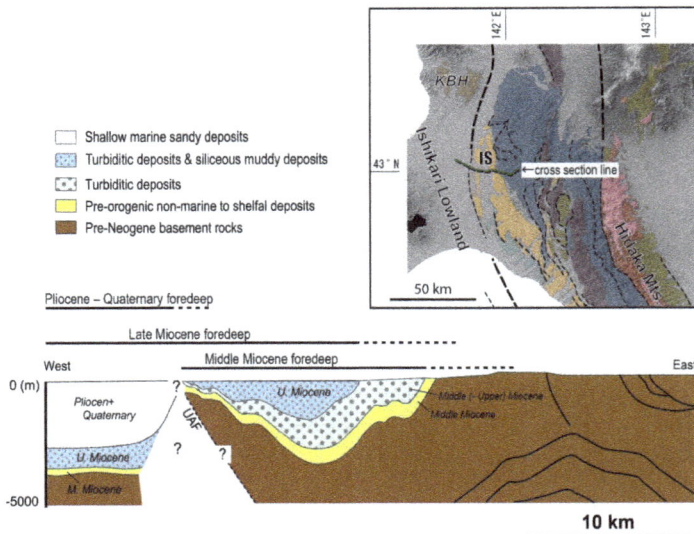

Figure 6. Cross section of the Ishikari Basin (after [6]) suggesting very narrow basin geometry concordant with strongly concentrated paleoflow data shown in Figure 7. UAF: Umaoi active fault. See Figure 1 for legend of the index map.

The basin is filled mainly with lower basinal turbidites and upper coarse-grained slope-apron turbidites of the middle to late Miocene Kawabata Formation, 3500 m thick. The basinal turbidites buried the irregular basin floor and they onlap to the slope in the western and southern margins of the basin. The slope-apron turbidites are longitudinally prograded southward onto the basinal turbidites (Figure 8).

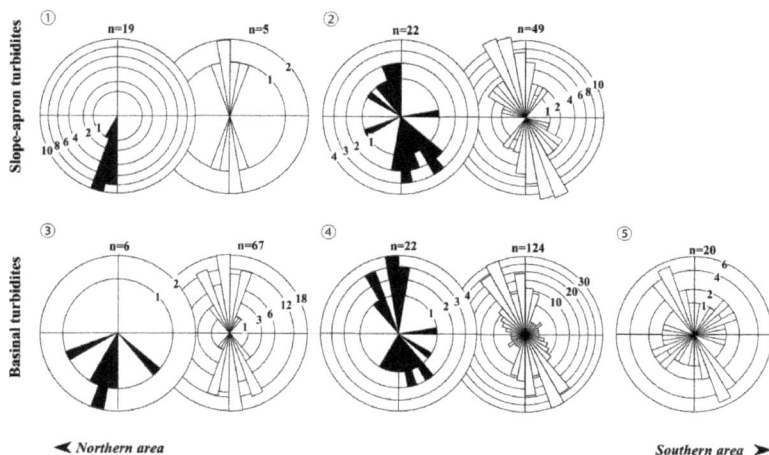

Figure 7. Rose diagrams showing paleocurrent directions of the middle to late Miocene tubidites in the Ishikari Basin. The solid and open petals of diagram mean the direction measured from flute marks and clast fabric (solid) and from groove marks and parting lineation (open). Note the diagrams are drawn with square root scaling. See Figure 8 for approximate localities of the measured data.

The basinal turbidites are underlain by shelfal, slope, and basin-plain muddy deposits (the early Middle Miocene Takinoue Formation). The muddy deposits are generally several hundreds of meters thick, but reach a thickness of 1000 m in local areas, where thick chaotic intervals (70–420 m thick), consisting of slump and debris flow deposits, occur [29]. The debris flow deposits notably contain abundant boulder-size serpentinite and sandstone blocks in addition to slate and chert cobbles.

Three chaotic intervals (slumped MTDs) are also encased within the basinal turbidites, and these consist of slump and debris flow deposits similar to the intervals in the Takinoue Formation. They lack serpentinite blocks, but contain granite cobbles-boulders, indicating a sediment supply from the Hidaka Belt [9]. Subsequently, clastic compositions changed synchronous with the change in sedimentary style from basinal turbidites to coarse-grained slope-apron turbidites. Abundant granite and hornfels grains in the basinal turbidites decrease upward. This is counterbalanced by an increase of green rocks, chert, and coeval andesitic-rhyolitic volcanic grains in the slope-apron turbidites (Figure 9). Such change in clastic composition indicates that a sediment provenance advanced from the Hidaka Belt to the Sorachi-Yezo Belt.

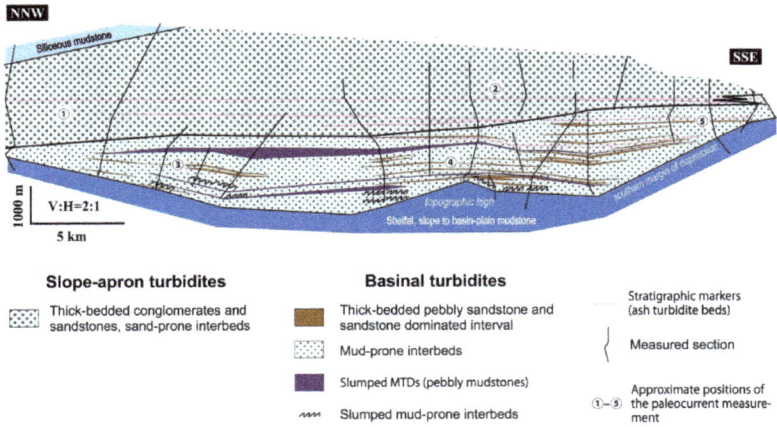

Figure 8. Basin-axial sedimentary profile of the Ishikari basin fill.

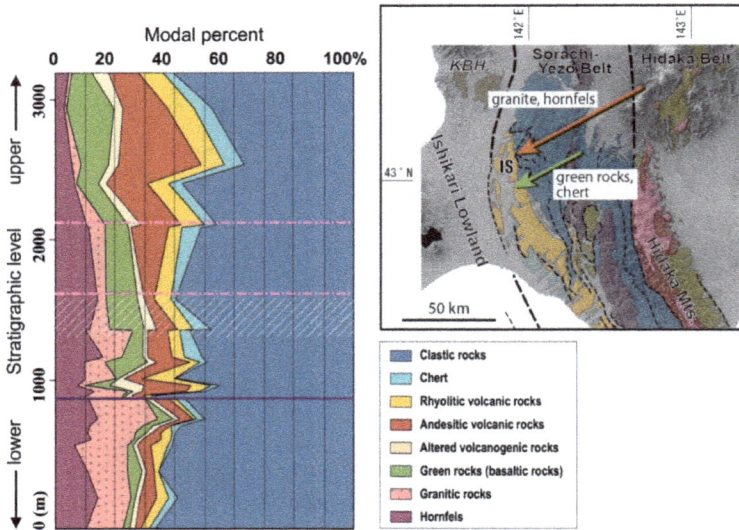

Figure 9. Modal evolutions of the lithic fragments in the coarse-grained sand to granule-grade turbiditic beds filling the Ishikari Basin, measured by point-counting method for thin section. White diagonal hatch indicates the horizon of the change in sedimentary style from basinal to slope-apron turbidite system. Stratigraphic level is based on the thickness.

3.4. Hidaka Basin

The Hidaka Basin, (>30 km wide and 70 km long), is fragmented by thrust propagation. The deformation is especially intensive in the eastern area. Depositional ages of the basin fill are progressively younger to the west (Figure 10).

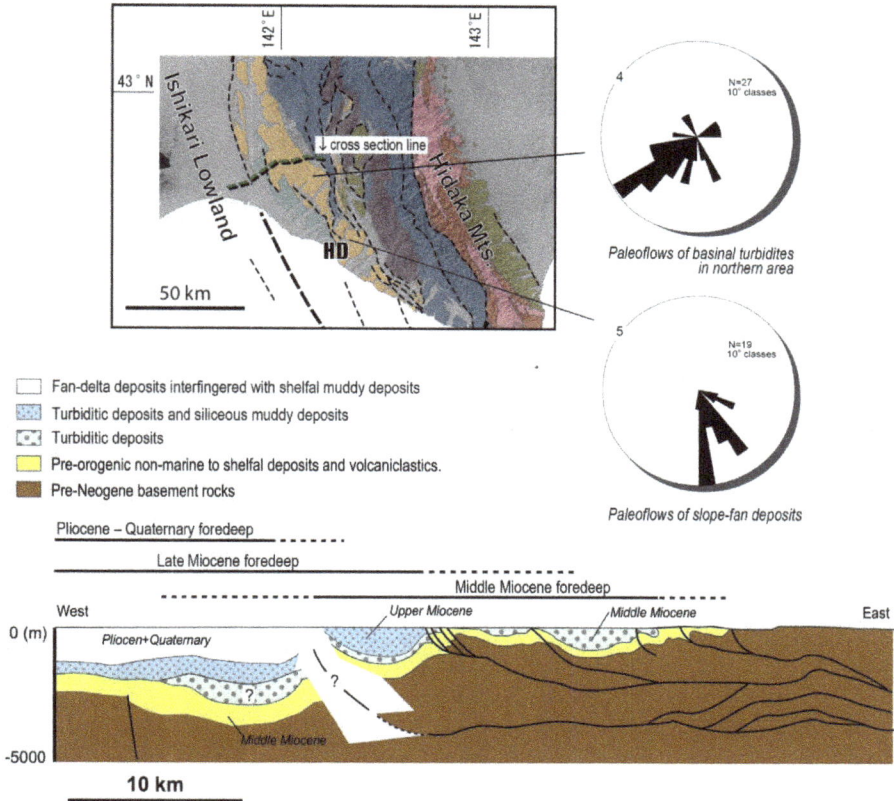

Figure 10. Cross section of the Hidaka Basin (after [6]). Paleoflow data are measured from flute marks and AMS (anisotropy of magnetic susceptibility) fabrics of the Middle Miocene turbidites (compiled after [30, 31, 33]). See Figure 1 for legend of the index map.

In the eastern margin of the basin, pre-orogenic shallow marine to shelfal muddy deposits of the early-Middle Miocene age are widely distributed (the lower part of the Niniu Formation and its correlatives, several hundred meters thick). These muddy deposits are locally overlain by turbiditic deposits accompanied with slumped MTDs (the upper part of the Niniu Formation, >500 m thick). A lenticular gravelly body (slope-fan deposits, which are discussed later),

1000 m thick and 10 km width, is locally developed (the eastern part of the Abetsu and the Ukekoi Formations). The gravelly deposits pass rapidly down-dip to the south into the basinal turbidites [30].

Locality	Dimension	Stratigraphy and sedimentary features			Tectonic deformation	Remarks
	Width and length at middle Miocene age	Pre-orogenic (underlying deposits)	Syn-orogenic a: primary inner foredeep, b: axial foredeep, c: transition from axial to inner foredeep	Post-orogenic (overlying deposits)		
Tenpoku	80 km, >60 km	Shallow marine sand-stone	a: MTD dominant, coarse-grained turbiditic deposits (1000–2000 m thick) b: Muddy turbidites and basin-plain muddy de-posits (2000–3000 m thick)	Siliceous / diatoma-ceous mudstone (1000–2000 m thick)	Restrictive Declining thrust activity in early Late Miocene	Slumped interval (200 m thick) in the basin-plain muddy deposits of axial foredeep
Haboro	>50 km, 90 km	Shelfal mudstone	a: MTD dominant, coarse-grained turbiditic deposits, details unknown b: Basinal turbidites (1500–3000 m thick) c: Slope-apron turbidites (1000–1500 m thick)	Shallow marine sand-stone (<500 m thick) and subsequently aggraded diatoma-ceous mudstone (500–1000 m thick)	Moderate The western margin of the basin is bordered by KBH Declining thrust activity during late Miocene ?	Initially segmented axial foredeep
Ishikari	>>15 km, 60 km	Shelfal mudstone	a: Not preserved b: Slope to basin-plain mudstones (<500 m) and overlying basinal turbidites (300–2400 m thick) c: Slope-apron turbidites (2000–2800 m thick) locally accompanied with siliceous / diatomaceous mudstones	–	Moderate–high The western margin of the basin is bordered by KBH Continuing thrust activity to the present?	Very narrow geometry Longitudinal sediment transport Pebbly mudstones in the basinal turbidite succession
Hidaka	>30 km, >70 km	Shelfal mudstone	a: MTD dominant, coarse-grained turbiditic deposits (>500 m thick), locally developed slope-fan deposits (1400 m thick) b: Basinal turbidites (middle Miocene, 1500 m thick), shelf-type fan-delta (Pliocene, 1000–1800 m thick) c: Slope-apron turbidites (1100–2600 m thick)	–	High Continuing thrust activity to the present	Depocenter migration

Table 1. Summary of the features of the depressions formed in the Miocene foreland basin area, Hokkaido, northern Japan.

In the central zone of the outcrops, middle to late Miocene turbiditic deposits (3000 m thick) are distributed (the western part of the Abetsu and the Ukekoi Formations, and the Azamizawa Formation). Basinal turbidites developed in the middle Miocene age show a slightly southward-fining megatrend. Paleocurrent directions show dominant southwestward flows in the northern area (Figure 10), although the basin-wide tendency is undefined. The basinal turbidites are overlain by late Miocene siliceous sandy mudstones and coarse-grained slope-apron turbidites [30, 31], but the detailed sedimentary style of the late Miocene deposits remains unknown. The sandy mudstones yield fossils of shallow marine shells [30].

In the westernmost zone of the outcrops, Pliocene sandy to gravelly deposits interfinger with and/or are prograded onto shelfal sandy mudstones [32]. The sand and gravelly beds often show large-scale cross bedding. The distribution area of the Pliocene deposits is roughly separated from the Miocene turbidites by a thrust fault parallel to the basin axis. The total thickness of the Pliocene deposits reaches 1500–2000 m.

The detrital composition of the Middle Miocene to Pliocene succession in the Hidaka Basin records an exhumation history of deep-seated lower crustal rocks of the Chishima Arc (the Hidaka metamorphic rocks) [9, 15, 30].

4. Sedimentary processes

4.1. Primary wedge-top to inner foredeep

A sedimentary succession within the primary inner foredeep can be observed in the eastern Tenpoku Basin (Figure 4). The sedimentary fill typically includes syn-orogenic packages of chaotic MTDs. The MTDs rest directly on the pre-orogenic shallow marine sandstones and often bear abundant intrabasinal blocks. Thus, the primary inner foredeep setting was quite unstable because of rapid syn-depositional subsidence and thrusting. The laterally discontinuous distribution of MTDs and sandy/gravelly turbiditic deposits with locally presented sand/mud interbeds suggests the ponding of frequently generated dense flows in small accommodation spaces of the irregular basin floor. As discussed in the next subsection, a large collapse event also resulted in a thick accumulation of cohesive debris flow deposits in the axial foredeep setting.

In the Hidaka Basin, a thick gravelly body is locally present in the boundary zone between the inner and the axial foredeep. Its basin-axial paleoflow direction suggests the development of a sediment conduit controlled by the thrust fault itself (cf. [34]). The gravelly body rapidly passes down-dip into the basinal turbidites of the axial foredeep and is interpreted as slope-fan deposits fed by a northern point source [30].

The sedimentary features of primary wedge-top basin (i.e., basins on thrust sheets and located at proximal side of inner foredeep) during the early syn-orogenic phase of foreland basin evolution are not evident. Non-marine to shallow marine deposits of middle to late Miocene age occur adjacent to the Tenpoku and Haboro Basins, but most of the deposits were formed during the post-orogenic phase in the northern collision zone.

4.2. Axial foredeep

In the Tenpoku Basin, much of the sandy to gravelly deposits were trapped in the accommodation space at the inner foredeep. As a result, the axial foredeep was filled mainly by basin-plain muddy deposits more than 2000 m thick. A 200-m-thick chaotic interval, containing exceptionally of sandy to granular grains in muddy matrices, occurs in the basin-plain muddy succession, suggesting large mass failure events in the proximal setting (i.e., the inner foredeep).

Except for the Tenpoku Basin, the axial foredeep is filled with a 1000–3000 m thick turbiditic succession mainly consisting of basinal turbidites (Figure 11a). The basinal turbidites are underlain by shelfal, slope, and/or basin-plain, muddy sediments with a condensed interval of blackish mud at their top, showing gradual but larger amount subsidence compared to the primary inner foredeep setting.

The basinal turbidites consists of monotonous interbeds of parallel-sided sand/mud beds. Spatiotemporally, they show no apparent coarsening or fining facies trends, other than a sandier upward trend in their basal section. The monotonous interbeds incorporate with isolated beds and packages of thick-bedded pebbly sandstone several tens of meters thick. These package-forming beds are more or less graded and unchannellized, and rarely show bed amalgamation. The concentration of paleoflows along the basin axis suggests that flows entering from lateral entry point(s) were deflected and transformed to basin-axial flows, which may explain the lack of an ordered facies trend. The basinal turbidites are transported by efficient turbidity currents in the confined basin plain (*sense* [35]).

Slumped MTDs intervene in the basinal turbidites of the axial foredeep and are particularly well developed in the Ishikari Basin. The MTDs mainly consist of debris flow deposits (pebbly mudstone) that contain many intrabasinal blocks and extrabasinal cobbles-boulders outsized for the succession (Figure 11b). These debris flow deposits are traceable basin-wide, suggesting large collapse events at the basin margin settings (i.e., the inner foredeep and wedge-top). MTDs of a similar scale do not exist in the slope-apron turbidites of the upper stratigraphic level.

4.3. Transition from axial to inner foredeep

Temporally, an axial foredeep setting evolves into an inner foredeep setting owing to thrust propagation. Other than in the Tenpoku Basin, the basinal turbidites of the axial foredeep are overlain by a turbiditic succession consisting of poorly sorted and coarser-grained sand to gravel beds interfingered with monotonously interbedded sandstone and sandy mudstone. As a whole, these coarse-grained deposits form a gravelly wedge prograded on the basin floor of the inner to the axial foredeep settings. In the Haboro Basin, the gravelly wedge prograded foreland-ward, while the wedge in the narrow Ishikari Basin shows axial progradation southward.

The thickly bedded sandstones and disorganized cobble to boulder conglomerates show an erosive base and frequent bed amalgamation (Figure 11c, 11d). These coarser-grained beds

Figure 11. Field occurrence of the Miocene turbidites in the Ishikari Basin. (A) Monotonous, mud-prone interbeds of basinal turbidites, (B) extrabasinal-cobble bearing pebbly mudstone in the basinal turbidite succession, (C) sand-prone interbeds and an amalgamated sand bed more than 5 m thick in the slope-apron turbidites, arrow: interval of aligned mudclasts (D) amalgamated stack of disorganized conglomerates and coarse-grained sandstones interpreted as fills of multiple chutes.

show variable facies such as disorganized, massive, cross-stratified, crudely laminated to well laminated, or crudely graded to graded bedding. Some beds consist of gravel, sand, and minor thin mud layers partitioned by surfaces with abrupt grain-size reduction (tripartite beds). The sand beds frequently contain plant debris and the associated mudstones are sandy and bioturbated. Thus, these deposits were transported by poorly-efficient dense flows, and filled multiple chutes developed at relatively proximal and shallower settings than the former basin-plain setting (e.g., the slope-apron setting: [36]). The paucity of muddy deposits can probably be attributed to the finer-grained dilute portions of the flows being stripped off and bypassing the area because of consistent infilling of the confined basin (e.g., [37]). In addition, as discussed in the Chapter 5, changes in proximity of sediment source and/or in lithologic compositions in the orogen also caused a decrease in the fine-grained fraction.

4.4. Basin migration and Pliocene foredeep

Pliocene gravelly deposits crop out on the northwestern side of the Hidaka Basin and inter-finger with shallow marine to shelfal deposits. Their distribution, distinct from that of the Miocene basin fill, suggests a major foreland-ward shift of the depocenter. Although the sedimentary processes of the Pliocene deposits are poorly known, the spatiotemporal distribution of the gravelly body suggests cyclic progradation of shelf-type fan-delta systems toward the west [32] or south to southwest (Takano, personal communication).

4.5. Post-orogenic sedimentation

In the Tenpoku Basin, the MTD dominated coarse clastic wedge is overlain by mud-prone interbeds and subsequently developed basin-plain muddy deposits, suggesting rapidly declining tectonic activity in the northern area of the collision zone by the early Late Miocene. Subsequently accumulated siliceous/diatomaceous muddy deposits buried the "abandoned foredeep." A recent study suggested that the accommodation space of the late Miocene age was maintained as a pull-apart depression along a right-stepping dextral fault system transformed from the former transpressional thrust system [38]. The siliceous/diatomaceous deposits are interpreted as siliceous muddy turbidites [39]. The declining thrust activity resulted in starvation of terrigenous sediment input to the Tenpoku Basin. Siliceous tests, originating from siliceous phytoplankton blooms in the photic zone, were alternately transported by laterally induced muddy turbidity currents [38, 39].

Notably, diatom productivity dramatically increased in the North Pacific and paleo-Japan Sea during the late Miocene [40]. As a result, syn-orogenic turbiditic deposits in the Ishikari and Hidaka Basins are also accompanied by similar muddy deposits during the late Miocene to Pliocene ages.

5. Tectonically controlled basin geometry and stratigraphic architecture

It is supposed that along the collision zone in central Hokkaido, the spatial variation in geodynamic properties of the foreland lithosphere is small, because of the assumption of a uniform geological composition, geotectonic history, and geothermal structure throughout the region. Thus, the variations in basin geometry and stratigraphic/sedimentary architecture for each depression are attributed to the regional difference in the degree of horizontal compression and resultant basin deformation.

The wider geometry of the Tenpoku Basin is the result of restrictive basin deformation. Many of the turbiditic deposits are confined to the inner foredeep and the axial foredeep is filled with muddy deposits. This condition is common to other depressions at the initial stage of basin evolution. Conversely, the thick accumulation of turbiditic deposits in the axial foredeep requires persistently high-relief basin physiography from the hinterland to the basin plain and high-volume sediment input.

Consistent basin infilling and/or thrust propagation results in shallowing of the foredeep and a transition in sedimentary style [37]. In addition, synchronous changes in the detrital composition in the Ishikari and Hidaka Basins suggest a close relationship between the sedimentary style in the foredeep and structural deformation in the hinterland (see discussion of [41]). In the Ishikari Basin, the compositional change in the detritus suggests lateral growth of the orogen (extension from the Hidaka Belt to the Sorachi-Yezo Belt). The proximity between the depocenter and the newly emergent source area resulted in an increase in the coarser-grained fractions and generation of relatively poorly-efficient flows. In contrast, the evolution of the detritus composition in the Hidaka Basin implies exhumation of deep-seated crustal rocks. Thus, the increase in coarse-grained deposits through poorly-efficient dense flows is attributed to an increase in the distributional area of crystalline rocks in the Hidaka Belt (Hidaka metamorphic rocks). Although the detrital compositional signal is unclear in the Haboro Basin, subaerial erosion of the basin fills in the eastern area suggests syn-depositional thrusting near the basin margin.

A similar succession, consisting of lower basinal turbidites and upper coarser-grained turbiditic deposits, is well documented from the Miocene foredeep turbidites in the Northern Apennines [e.g., 42]. In that area, the change in sedimentary style was controlled by the narrowing and closure (shallowing) of the foredeep due to thrust propagation [42]. In Hokkaido, in contrast, the stratigraphic architecture does not show obvious narrowing of the foredeep depressions. The coarse-grained slope-apron turbidites occur basin-wide and their thickness is approximately the same as that of the basinal turbidites. In addition, they are covered with relatively thick siliceous/diatomaceous muddy deposits. Nevertheless, the depressions appear to shallow gradually upward, as indicated by the dominance of bioturbation, plant debris, and shell fossils in the Late Miocene basin fills. Despite the migration of the depocenter in the Hidaka Basin, the depth of the depocenter gradually decreases foreland-ward until the Pliocene. Initial regional shallowing occurred around 13–14 Ma, beginning in the northern foreland area. An eustatic sea-level fall [43] is not sufficient to explain such a long-term gradual shallowing of the basin; however, a flexural rebound of underlying lithosphere [44] can explain such shallowing. The rebound was probably caused by isostatic readjustment for a thinning orogen or decreased horizontal compressional stress corresponding to a gradual or stepwise decline of thrust activities in central Hokkaido [45, 46].

6. Conclusions

This paper provides an introduction to a tectonically controlled foreland basin stratigraphy at the arc-arc collision zone of Miocene age in Hokkaido, northern Japan. Spatial differences in the degree of tectonic disturbance caused variations in stratigraphic/sedimentary architecture between the separately developed depressions in the foreland basin area. Limited tectonic activity resulted in trapping of coarse-grained deposits in the inner foredeep setting. Thick muddy deposits filled the sediment-starved and abandoned axial foredeep. Moderate to high thrust activity formed turbiditic successions several thousand meters thick in the axial foredeep. Progressed thrusting caused an increase in coarser-grained sediment-input to the foredeep and the sedimentary style changed from the basinal turbidites by efficient-flows to the slope-apron turbidites by poorly-efficient flows. The long-term shallowing of the foreland basin area can be explained by lithospheric flexural rebound caused by the isostatic readjustment for a thinning orogen and/or decreased horizontal compressional stress due to a gradual or stepwise decline of thrust activities in central Hokkaido.

Acknowledgements

The invitation of the editor, Dr. Y. Itoh, to submit this manuscript is highly appreciated. The early version of the manuscript is greatly improved by constructive comments by reviewers (Y. Itoh and O. Takano). The following people are thanked for discussion and helpful suggestion: K. Arita, K. Sawada, S. Furota, T. Nakajima, T. Watanabe, S. Ohtsu, and M. Kawamura.

Author details

Gentaro Kawakami*

Address all correspondence to: kawakami-gentaro@hro.or.jp

Geological Survey of Hokkaido, Hokkaido Research Organization, Japan

References

[1] Allen PA., Homewood PN. Foreland Basins. IAS Special Publication 8. Oxford: Blackwell Science; 1986.

[2] Hoyanagi K. Progradational lithofacies change of turbidite sequence, Middle Miocene Kotambetsu Formation, central Hokkaido, Japan. Journal of the Geological Society of Japan 1989; 95: 509-525.

[3] Hoyanagi K. Coarse-grained Turbidite Sedimentation resulting from the Miocene Collision Event in Central Hokkaido, Japan. In: Taira A., Masuda F. (eds) Sedimenta-

ry Facies in the Active Plate Margin. Tokyo: Terra Scientific Publishing Company (TERRAPUB); 1989. P689-709.

[4] Kimura G. Collision orogeny at arc-arc junctions in the Japanese Islands. The Island Arc 1996; 5: 262-275.

[5] Kato N., Sato H. Active tectonics in the axial zone of Hokkaido –an example from the Umaoi Hills area–. Chikyu Monthly 2002; 24: 481-484.

[6] Kazuka T., Kikuchi S., Ito T. Structure of the foreland fold-and-thrust belt, Hidaka Collision Zone, Hokkaido, Japan: re-processing and re-interpretation of the JNOC seismic reflection profiles 'Hidaka' (H91-2 and H91-3). Bulletin of the Earthquake Research Institute, University of Tokyo 2002; 77: 97–109.

[7] Ueda H., Kawamura M., Niida K. Accretion and tectonic erosion processes revealed by the mode of occurrence and geochemistry of greenstones in the Cretaceous accretionary complexes of the Idonnappu Zone, southern central Hokkaido, Japan. The Island Arc 2000; 9: 237-257.

[8] Osanai Y., Komatsu M., Owada M. Metamorphism and granite genesis in the Hidaka Metamorphic Belt, Hokkaido, Japan. Journal of Metamorphic Geology 1991; 9: 111-124.

[9] Kawakami G., Ohira H., Arita K., Itaya T., Kawamura M. Uplift history of the Hidaka Mountains, Hokkaido, Japan: a thermochronologic view. Journal of the Geological Society of Japan 2006; 112: 684-698.

[10] Tsuchi R. Neogene events in Japan and the Pacific. Palaeogeography, Palaeoclimatology, Palaeoecology 1990; 77: 355-365.

[11] Nagata M., Kito N., Niida K. The Kumaneshiri Group in the Kabato Mountains: the age and nature as an Early Cretaceous volcanic arc: Monograph of the Association for the Geological Collaboration in Japan 1986; 31: 63-79.

[12] Hoyanagi K., Miyasaka S., Watanabe Y., Kimura G., Matsui M. Depositions of turbidites in the Miocene collision zone, central Hokkaido. Monograph of the Association for the Geological Collaboration in Japan 1986; 31: 265-284.

[13] Takahashi K. Sedimentological analysis on pebbles from Tertiary conglomerates in north Hokkaido, Japan. Report of the Geological Survey of Hokkaido 1974; 46: 17-43.

[14] Okada H., Tandon SK. Resedimented Conglomerates in a Miocene Collision Suture, Hokkaido, Japan. In: Koster EH., Steel RJ. (eds) Sedimentology of Gravels and Conglomerates. Memoir of Canadian Society of Petroleum Geologists 10. Calgary: Canadian Society of Petroleum Geologists; 1984. P 413-427.

[15] Miyasaka S., Hoyanagi K., Watanabe Y., Matsui M. Late Cenozoic mountain-building history in central Hokkaido deduced from the composition of conglomerate. Monograph of the Association for the Geological Collaboration in Japan 1986; 31: 285-294.

[16] Kawakami G., Arita K., Okada T., Itaya T. Early exhumation of the collisional orogen and concurrent infill of foredeep basins in the Miocene Eurasian - Okhotsk Plate boundary, central Hokkaido, Japan: inferences from K-Ar dating of granitoid clasts. The Island Arc 2004; 13: 359-369.

[17] Mutti E., Tinterri R., Benevelli G., di Biase D., Cavanna G. Deltaic, mixed and turbidite sedimentation of ancient foreland basins. Marine and Petroleum Geology 2003; 20: 733-755.

[18] Roveri M., Ricci Lucchi F., Lucente CC., Manzi V., Mutti E. Part III Stratigraphy, facies and basin fill history of the Marnoso-Arenacea Formation. In: Mutti E., Ricci Lucchi F., Roveri, M. (eds) Revisiting turbidites of the Marnoso-arenacea Formation and their basin-margin equivalents: problems with classic models. Excursion guidebook of the workshop organized by Dipartimento di Scienze della Terra, (Università di Parma) and Eni-Division Agip for the 64th EAGE Conference and Exhibition, Florence (Italy), May 27-30; 2002. III1-26.

[19] Yanagisawa Y., Akiba F. Refined Neogene diatom biostratigraphy for the northwest Pacific around Japan, with an introduction of code numbers for selected diatom biohorizons. Journal of the Geological Society of Japan 1998; 104: 395-414.

[20] Takahashi K., Fukusawa H., Wada N., Hoyanagi K., Oka T. Neogene stratigraphy and Paleogeography in the Area along the Sea of Japan of northern Hokkaido. Earth Science (Chilyu Kagaku) 1984; 38: 299-312.

[21] Hokkaido Mining Industry Promotion Committee. Petroleum and Natural Gas Resources in Hokkaido: Exploration and Development in 1977-1988. Sapporo: Hokkaido Mining Industry Promotion Committee; 1990.

[22] Mitani K., Saito N., Osanai H. Oil and natural gas: from the Masuporo anticline to northern Tenpoku oil field area. Report of the geological resources in Hokkaido 1962; 79: 1-16.

[23] Mitani K., Saito N., Matsushita K., Osanai H. Oil and natural gas: from the headwaters of the Masuporo River and its southern side area to northern Tenpoku oil field area. Report of the geological resources in Hokkaido 1963; 88: 1-16.

[24] Hirooka E. Petroleum geological study on Tenpoku district in Hokkaido, Japan. Journal of the Japanese Association for Petroleum Technology 1962; 27: 113-134.

[25] Motoyama I., Nakamura S. Radiolarian biostratigraphy of the Miocene Masuporo and Wakkanai formations of the Uruyagawa section, Wakkanai, Hokkaido, Japan, with special reference to unconformity. Journal of the Geological Society of Japan 2002; 108: 219-234.

[26] Hokkaido Mining Industry Promotion Committee. Petroleum and Natural Gas Resources in Hokkaido: Exploration and Development in 1968-1976. Sapporo: Hokkaido Mining Industry Promotion Committee; 1979.

[27] Takahashi K., Kiminami K. Sedimentation of the Miocene Kotanbetsu Formation around the Haboro dome. Earth Science (Chilyu Kagaku) 1983; 37: 250-261.

[28] Matsuno K. The deposition of the sedimentary basin of the Kotambetsu Formation. Journal of the Japanese Association for Petroleum Technology 1958; 23: 19-21.

[29] Kanno S., Ogawa H. Geology of the Tertiary System in the Momijiyama-Takinoue district, Yubari City, Japan. Journal of the Geological Society of Japan 1963; 69: 262-278.

[30] Hoyanagi K., Mito N., Yoshioka M., Miyasaka M., Watanabe Y., Matsui M. Stratigraphy and sedimentology of the Miocene sediments in the southern part of the Ishikari-Teshio Belt, Hokkaido, Japan. Earth Science (Chilyu Kagaku) 1985; 39: 393-405.

[31] Kawakami G., Shiono M., Kawamura M., Urabe A. Koizumi I. Stratigraphy and depositional age of the Miocene Kawabata Formation, Yubari Mountains, central Hokkaido, Japan. Journal of the Geological Society of Japan 2002; 108: 186-200.

[32] Sagayama T., Hoyanagi K., Miyasaka S. Diatom biostratigraphy and the stage of Neogene coarse-grained deposits in the Hidaka coastal land, central Hokkaido, Japan. Journal of the Geological Society of Japan 1992; 98: 309-321.

[33] Itoh Y., Tamaki M., Takano O. Rock Magnetic Properties of Sedimentary Rocks in Central Hokkaido: Insights into Sedimentary and Tectonic Processes on an Active Margin. In Itoh, Y., ed., Mechanism of Sedimentary Basin Formation - Multidisciplinary Approach on Active Plate Margins (this volume).

[34] Covault JA., Graham SA. Turbidite architecture in proximal foreland basin-system deep-water depocenters: insights from the Cenozoic of Western Europe. Austrian Journal of Earth Sciences 2008; 101: 36-51.

[35] Mutti E., Tinterri R., Remacha E., Mavilla N., Angella S., Fava L. An Introduction to the Analysis of Ancient Turbidite Basins from an Outcrop Perspective. AAPG Course Note 39. Tulsa, Oklahoma: The American Association of Petroleum Geologists; 1999.

[36] Soh W., Tanaka T., Taira A. Geomorphology and sedimentary processes of a modern slope-type fandelta (Fujikawa fandelta), Suruga Trough, Japan. Sedimentary Geology 1995; 98: 79-95.

[37] Sinclair HD., Tomasso, M. Depositional evolution of confined turbidite basins. Journal of Sedimentary Research 2002; 72: 451-456.

[38] Itoh Y., Kusumoto, S., Inoue, T. Magnetic properties of siliceous marine sediments in Northern Hokkaido, Japan: a quantitative tectono-sedimentological study of basins along an active margin. Basin Research in press.

[39] Fukusawa H. Sedimentary mechanism of Neogene bedded siliceous rocks – on late Miocene Wakkanai Formation of northern Hokkaido, Japan. Journal of the Geological Society of Japan 1988; 94: 669-688.

[40] Yamamoto M., Watanabe Y., Watanabe M. Paleoceanographic controls on the deposi-
      tion of Neogene petroleum source rocks, NE Japan. Bulletin of the Geological Survey
      of Japan 1999; 50: 361-376.

[41] Bábek O., Mikuláš R., Zapletal J., Lehotský T. Combined tectonic-sediment supply-
      driven cycles in a Lower Carboniferous deep-marine foreland basin, Moravice For-
      mation, Czech Republic. International journal of Earth Sciences 2004; 93: 241-261.

[42] Tinterri R., Muzzi Magalhaes P. Synsedimentary structural control on foredeep turbi-
      dites: An example from Miocene Marnoso-Arenacea Formation, Northern Apen-
      nines, Italy. Marine and Petroleum Geology 2011; 28: 629-657.

[43] Haq BU., Hardenbol J., Vail PR. Chronology of fluctuating sea levels since the Trias-
      sic. Science 1987; 235: 1156-1167.

[44] Allen PA., Allen JR. Basin Analysis: principles and applications, 2nd edition. Oxford:
      Blackwell Science; 2005.

[45] Arita K., Ganzawa Y., Itaya, T. Tectonics and uplift process of the Hidaka Mountains,
      Hokkaido, Japan inferred from thermochronology. Bulletin of the Earthquake Re-
      search Institution of University Tokyo 2001; 76: 93-104.

[46] Kawakami G., Kawamura M. 2003, Reconsideration to the collision tectonics in cen-
      tral Hokkaido from the Miocene stratigraphy. Earth Science (Chikyu Kagaku) 2003;
      57: 333-342.

Chapter 8

Tectonic Basin Formation in and Around Lake Biwa, Central Japan

Keiji Takemura, Tsuyoshi Haraguchi,
Shigekazu Kusumoto and Yasuto Itoh

Additional information is available at the end of the chapter

1. Introduction

Located on the convergent margin of the Eurasian plate, the Japanese archipelago features many tectonic, volcanic, and coastal lakes that are well suited for studies of Quaternary intraplate tectonics. A famous and often studied tectonic lake in Japan is Lake Biwa (Figure 1). Along the west coast, an extremely active fault system in Japan designated as the Biwako-seigan Fault zone runs north to south (Figure 1).

The ca. 1.5-Ma-old present Lake Biwa (82 m a s l.) on south-central Honshu Island is the largest freshwater lake in Japan, measuring 22.6 km wide by 68 km long (Figure 1). Lake Biwa is divided into two basins. The *Northern Lake* is a deep basin with a maximal depth of 104 m and average depth of 40 m. The much smaller *Southern Lake* is extremely shallow with average depth of about 3 m. Herein we summarize and discuss that the tectonic basin formation in the paleo and present Lake Biwa basin as an example of intraplate basin formation.

2. Sedimentary sequences in present Lake Biwa Basin

Lake sediments are important archives for understanding tectonic history at different scales. Several attempts to recover core sediments from Lake Biwa have been made, mainly in the 65-70 m deep depression situated in the southern part of the Northern Lake (Figure 1). Horie et al. first recovered a 6-m-long sediment core in 1965 and then an 11.5-m-long piston core in 1967 [2]. In 1971, with considerable effort, they drilled sediments in the same depression (Figure 2) and obtained core samples of about 200 m in all [3]. Finally, in 1982 and 1983, they recovered a 1400-m-long core covering the entire sediment sequence to the basement.

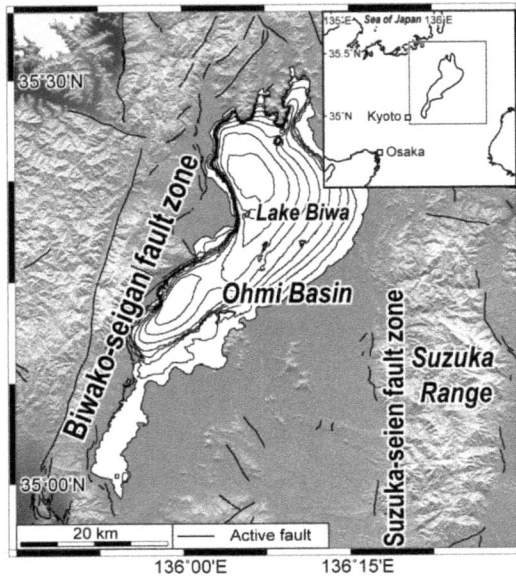

Figure 1. Geomorphology and active faults around Lake Biwa (illustrated by D. Ishimura) Surface traces of active faults are from [1]. This map is from 10m DEM of the Geospatial Information Authority of Japan. The bathymetric contour interval in Lake Biwa is 10 m.

This record confirmed that the basin is filled with lacustrine and fluvial sediments about 800 m thick with a ca 100 m thick pebbles and cobbles layer resembling debris flow deposits piled on Mesozoic- Paleozoic basement rocks [5,6]. Sediments were divided into five units based on differences in predominant grain-size distributions [9, 10]. These units have been named the P (ca 100-m-thick pebble and cobble layer) and fluvial and lacustrine sediments (Q, R, S, and T beds) from deepest to most shallow. The Q bed is a 72.3m thick unit (731.8-804.1 m below lake floor, mblf) composed of alternating layers of sand, gravel, and silt. The R bed is 149.9 m thick (581.9-731.8 mblf) and is considered to be continuous with the S Bed above it. Subunits of bluish gray nonlaminated clay and of layers of silt, sand, and sandy gravel alternate at approximately 10 m intervals throughout this unit. The S bed is 332.4 m thick (249.5-581.9 mblf) and is believed to be continuous with the overlying T Bed. It consists of thin alternations of sands and silts interspersed with sandy gravels. The T bed is 249.5 m thick (0-249.5 mblf) and is composed of bluish gray, nonlaminated clay. They contain 54 layers of volcanic ash intercalated throughout them [5]. The uppermost unit (T bed) was estimated as having been deposited continuously during the last 430 ka [5,11] (Figure 3). A horizon in Figure 3 is correlated with the bottom of T bed. In 1986, additional samples of 141-m-thick sediment were recovered about 5 km northeast of older drilling sites (Figure 1; [7]). Although the neighboring (ca. 20 km) basin of Lake Suigetsu has varved sediments of the past 150 ka[12], Lake Biwa has

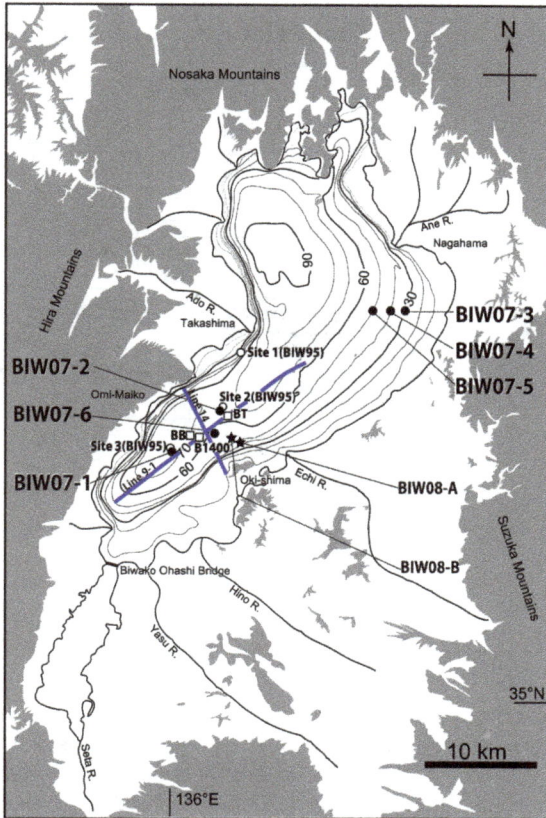

Figure 2. Map showing locations of principal coring sites in Lake Biwa [4]. BB (□ : 200 m drilling in 1971; [3]), B1400 (□: 1400 m drilling in 1982-1983; [5, 6]), BT (□: 141 m drilling in 1986; [7]), Site 1, 2, 3 (○ : BIW95; Piston cores in 1995; [8]), BIW07-1 to BIW07-6 (●:Piston cores in 2007; [4]), BIW08-A and BIW08-B (★:Drilled cores in 2008; [4]). Blue line 9-1 and 14 are the survey lines of seismic reflection shown in Figure 3 and Figure 8.

continuous sediments of a million years age. Therefore, the two lake basin records together will facilitate understanding of the Quaternary climate and tectonics at annual to orbital time scales.

Initially, the scientific value of Lake Biwa sediments was not properly acknowledged because the first attempt of fission track dating assigned an incorrect Pliocene age to the basal part. This suggested a markedly crooked sediment accumulation rate curve, casting doubt on the continuity of the Lake Biwa sediment record. In 1993, based on a stratigraphic correlation of the Biwa core with marine data, Meyers et al. [11] reported that the fission track dates were erroneous. Then in 2005, improvements on the fission track timescale identified the paleo-

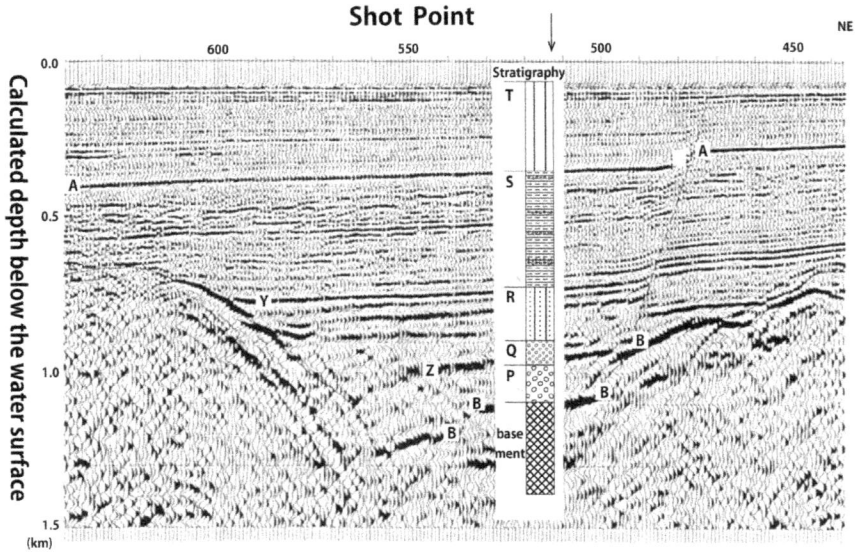

Figure 3. Multichannel seismic reflection profile along the Line 9-1. The core site is indicated by an arrow. The inset roughly correlates reflectors with the stratigraphic column [5].

magnetic reversal near the base as Jaramillo rather than Olduvai, estimating the time coverage of the Lake Biwa core as ca. 1.5 Ma ([13]; Figure 4). Figure 4 shows the nearly linear sediment accumulation from 0.57m/kyr to 0.60 m/kyr during the sedimentary record in present Lake Biwa, despite the lithologic units are different. The average sedimentation rate is calculated from the data of depth of 695.6 m with about 1211 ka.

This was evidence for the stable sedimentary environment of the basin and was evidence for the suitability of Lake Biwa as a tectonic archive. Moreover, progress in Japanese tephrochronology in recent decades has enabled the identification of several marker tephras [14] in and around the basin. Lake Biwa is therefore an ideal terrestrial site for exploration of the paleoclimate and tectonic history of eastern Asia during at least the past 1.5 Ma.

Although Lake Biwa sediments have been analyzed using various methods, high-resolution studies have not yet been conducted. In most studies, a single core was obtained at a single site. It was therefore difficult to evaluate the completeness of core recovery and the disturbance of core samples. For example, during deep drilling of 1982 and 1983, it is known that rotary coring caused a disturbance of the upper sediment samples. For a detailed study of the sedimentary record, in 1995, we recovered seven piston cores (10-15 m long) at three localities (sites 1, 2, 3) in the northern part of Lake Biwa (Figure 1). We designed the coring plan (1) to take at least two cores from each site; (2) to take cores at three locations having different

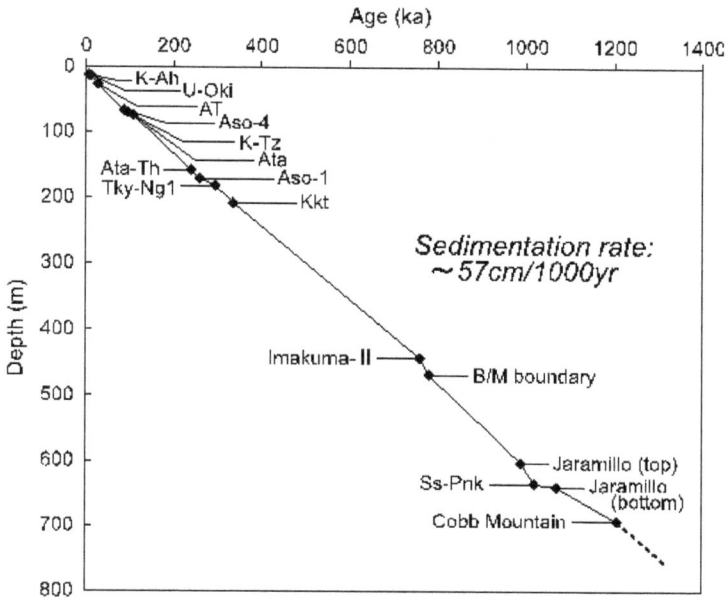

Figure 4. Summary of chronology of core B1400 based on the re-investigation of fission track ages, tephra identification and magnetostratigraphy [13]. Tephra horizons: K-Ah, U-Oki, AT, Aso-4, K-Tz, Ata, Ata-Th, Aso-1, Tky-Ng1, Kkt, Imakuma II, Ss-Pink; Paleomagnetic information: B/M (Brunhes/Matuyama) boundary, top of Jaramillo event, bottom of Jaramillo event, Cobb mountain event.

sedimentation rates, and (3) to recover the longest possible undisturbed sediment sequence. Analyses of the core samples include paleomagnetism, environmental magnetism, physical properties, organic and inorganic chemistry, pollen analysis and ^{14}C dating. We also demonstrated that magnetic susceptibility data are extremely useful to find microscopic tephra horizons, and establish correlation and age assignment of core sediments from different locations.

Drilling challenges are continuing for high-resolution studies of island arc tectonics. In 2007 and 2008, we obtained six new piston cores covering at least 50 ka, two longer cores covering 300 ka [4], and 300-km-long shallow seismic surveys. Various interdisciplinary analyses are expected to generate high-resolution records of the dynamics of the tectonic convergence. Sato et al. [15] described that two stages indicating shallow lacustrine delta formation were intercalated with sediment cores during 300 ka at the present depth of 50m. This intercalation is important evidence of tectonic deformation of the lake basin from coastal area migration.

3. Active tectonics and geophysical data related to present Lake Basin

At Lake Biwa, plate motion-related basin formation can be generalized because the Lake Biwa basin formation apparently resulted from subduction of the Philippine Sea plate into the Eurasian plate. Tectonic approaches on the Lake Biwa sediment therefore provide a case study for basin-forming mechanisms related to plate subduction activity, which is applicable to other active subduction zones throughout the world.

Lake Biwa is the largest and oldest fresh-water lake in Japan. Its surrounding geology comprises Mesozoic-Paleozoic formations, Cretaceous Granitic Rocks and Volcanic Rocks, Miocene sedimentary rocks, Plio-Pleistocene ancient lake sediments (Kobiwako Group in Paleo-lake Biwa), terrace deposits, and alluvium.

The present Lake Biwa, surrounded by several active faults, is a tectonic basin with a long history extending from about 1.5 Ma to the present day. Historical earthquake records exist around Lake Biwa region. Lake Biwa is located in the southwestern Japan, which is considered in the area under the east-west compressional stress state from geophysical, topographical and Quaternary geological information [16]. Huzita [16] reported a triangular neotectonic province designated as "Kinki Triangle" (Figure 11 inset) with the E-W compressional stress state and the basin and mountain topography from east to west in the district. In central Japan, including the mountainous region in Chubu district (east of Kinki district) and Chugoku district (west of Kinki district), we can recognize the clear conjugate fault system, and active faults with NW-SE direction of left lateral transcurrent component, and those with NE-SW direction of right lateral component, and also in Kinki district, reverse fault system with N-S direction are recognized accompanied with transcurrent fault system (Figure 5). This evidence implies that this area is influenced by the E-W compressional stress state. The force from the Pacific Plate movements and oblique subduction movements of the Philippine Sea Plate influences the Kinki District. According to the combination of both forces, E-W compressional stress occurs in southwestern Japan. Lake Biwa region is located at northern part of "Kinki Triangle" region. Most active and highest rate of activity active fault in this region is the Biwako-seigan Fault zone, which runs along the west coast of Lake Biwa with reverse faulting. Biwako-seigan Fault zone is about 59 km long with reverse fault sense of east side subsidence. The activity of southern part of the fault is calculated from about 1.4 m/kyr of average movement and recurrent interval is calculated as about 4.5 – 6.0 kyr with latest event of 1185 AD. Relative movement of single event is inferred about 6-8 m [17].

Deep seismic reflection survey and gravity measurement through the 1970s and early 1980s yielded information related to the basement topography (Figures 6-9), and revealed a tilting structure from east to west (Figure 8). However, the deepest part is located in the Northern Lake basin ("Hokko basin"), where Prof. Horie's team drilled in the early 1980s from the seismic reflection survey. A seismic reflection survey revealed the existence of active fault in the deep basin. The lithostratigraphical information indicated the subsidence history of the present lake basin during about 1.5 my. The recent subsidence rate is calculated as 0.74-m/kyr by data of the T bed (thickness 250 m, duration 430 kyr, water depth 68 m), because T bed is the sediment under pelagic environment. Shallow seismic reflection surveys conducted in the

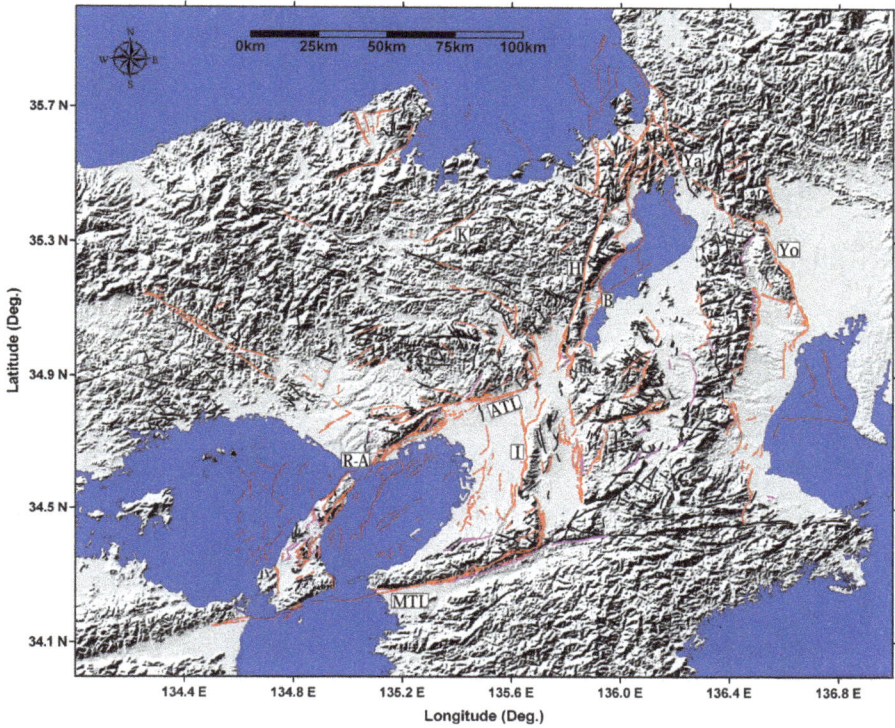

MTL: Median Tectonic Line, ATL: Arima Takatsuki Tectonic Line, R-A: Rokko-Awaji fault zone, I: Ikoma Fault zone, K: Kambayashi Fault, H: Hanaore Fault, B: Biwako-seigan Fault zone, Ya: Yanagase Fault zone, Yo: Yoro Fault zone

Figure 5. Active fault system in Kinki district [18]

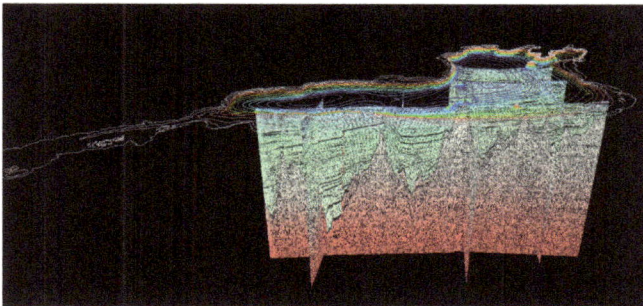

Figure 6. Seismic reflection survey (example of long section of Line 9) (drawn from [19])

1980s, 1990s and 2007 show the distribution of active fault traces and size of displacement accompanied with the activity of Biwako-seigan Fault zone along the coast of Lake Biwa.

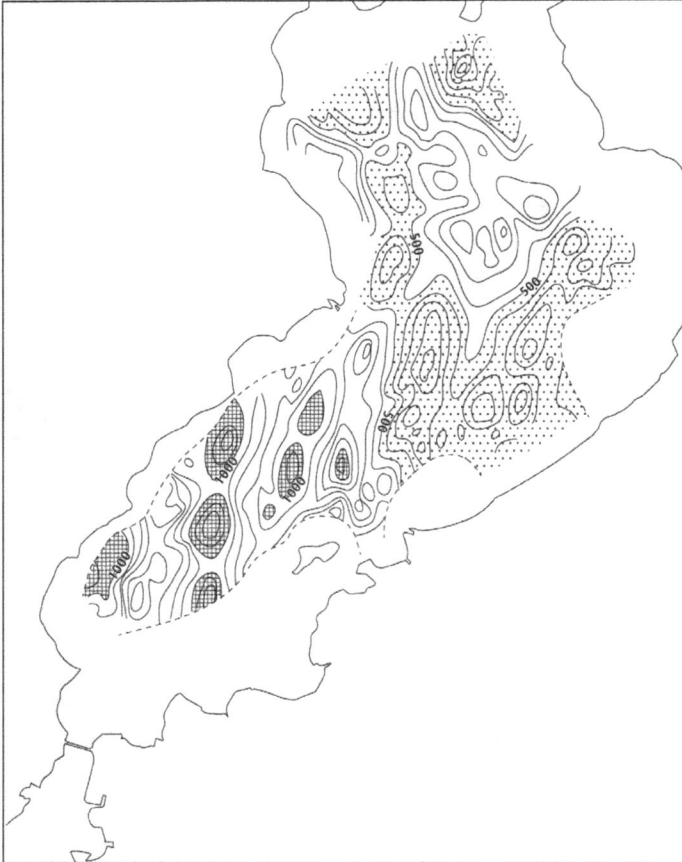

Figure 7. Basement structure revealed by seismic reflection survey [19] Meshed part: depth of basement more than 1, 000m. Dotted part: depth of basement lower than 500-m.

LINE 14 (1980)

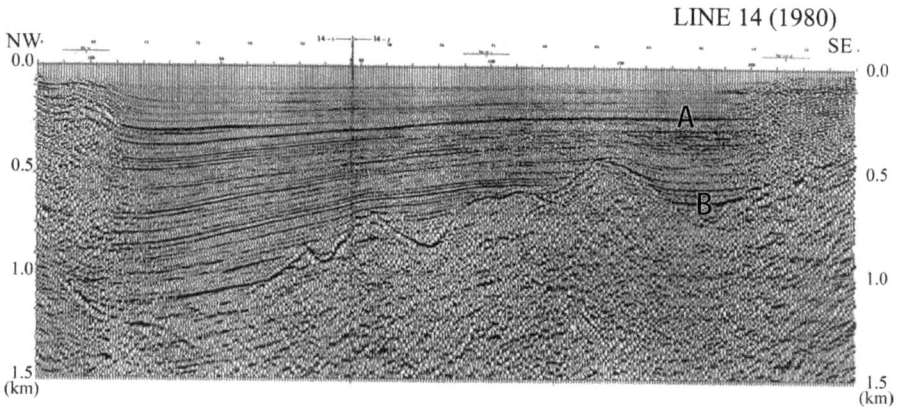

Figure 8. Tilting structure from east to west revealed by sediment structure including horizon A indicating the boundary of T and S bed [5]. Survey Line is shown in Figure 2.

A gravity survey revealed that the lowest Bouguer anomaly area is in the north lake basin. We show the Bouguer gravity anomaly map around our study area (Figure 10). This Bouguer gravity anomaly map is based on gravity mesh data reported by [20]. The Bouguer density is 2670 kg/m³.

Gravity anomaly in this area is characterized by the negative gravity anomalies caused by intra-arc sedimentary basins and isostasy attributable to the loading of mountains in central Japan. In this region, the Nohbi Plain, the Ise Plain, the Ohmi Plain including Lake Biwa, the Kyoto Basin, the Nara Basin, the Osaka Plain, the Osaka Bay, and the Sanda Basin are distributed from east to west. Each basin or plain has negative gravity anomalies correspond individually to intra-arc basins. Almost all local negative gravity anomalies in these negative anomalies are included to the active tectonic zone during the Quaternary called "Kinki-Triangle" (e.g., [21] Figure 11inset).

Two large negative gravity anomaly areas exist in the "Kinki-Triangle" area. They are the Osaka Bay area and Lake Biwa area (the Ohmi Plain). Negative gravity anomalies around the Osaka Bay and the Lake Biwa respectively reach -15mGal and -60 mGal.

Negative gravity anomalies in the Osaka Bay area are known to be explainable by sediments accumulated in and around the Osaka Bay (e.g., [22]; [23]). These negative gravities are divided by some active faults. In contrast, negative gravity anomalies in the Lake Biwa area are known to be unexplainable using the distribution of soft sediments in the lake (e.g., [24]). Nishida et al. [24] reported that depression of the Conrad surface or existence of very low density materials because of faulting is necessary to explain the gravity lows reaching -60 mGal.

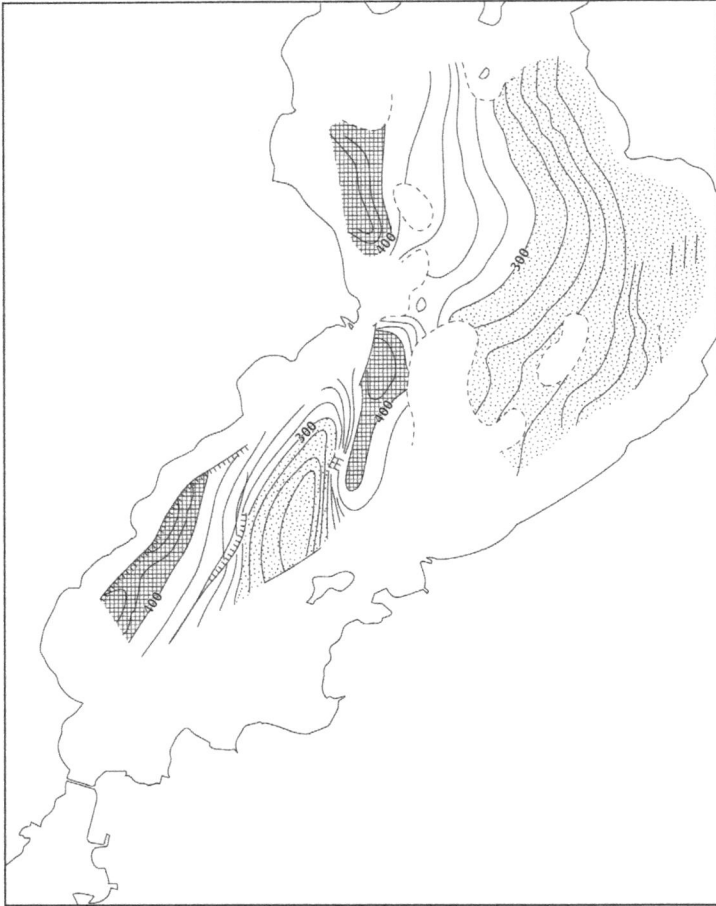

Figure 9. Active structure revealed by horizon A (Boundary of T and S bed) since about 0.43 Ma) of the seismic reflection survey [19] Meshed part: depth of A horizon more than 400 m, Dotted part: depth of A horizon lower than 300 m

4. Basin formation and migration in central Kinki since the Pliocene

Three sedimentary basins are arranged E-W latitudinally in the eastern part of "Second Setouchi Inland Depression of [25]: Lake Tokai, Paleo-Lake Biwa and Paleo-lake (bay) Osaka. All are filled with the Plio-Pleistocene deposits (Figure 11). They are respectively named Tokai,

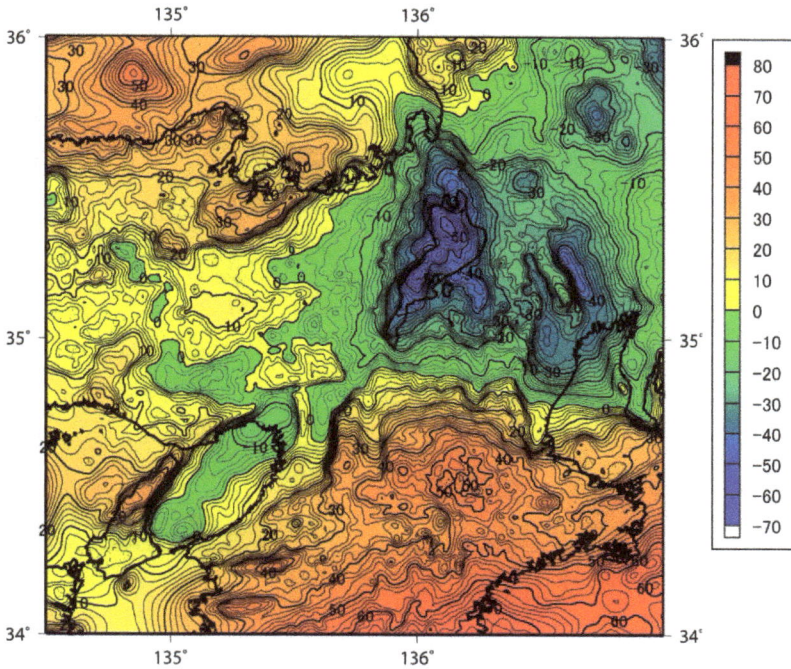

Figure 10. Bouguer gravity anomaly map. The Bouguer density is 2670 kg/m³, and contour interval is 2 mGa

Kobiwako, and Osaka Groups. Comparing of the sedimentary basin transition between Paleo-Lake Biwa and Lake Tokai, the geohistory in central Kinki Region is known subdivisible by tectonosedimentary facies of about 3.0 Ma and 1.2-1.5 Ma (Figures 12, 13) which suggests that the change of tectonic stress state of this province took place simultaneously throughout these sedimentary basins. To complete the geohistory of Paleo-lake Biwa, patterns of sedimentary basin transition must be discussed. According to Yokoyama ([26], [27]), the geohistory of Paleo-Lake Biwa has four stages: Older I, Older II, Actual I, and Actual II. First, Paleo-Lake Biwa appeared in the Iga Basin and clay-dominant sediments were deposited in the water body ("Iga-ko") (a and b in Figure 12) [28]. In the second, the sedimentary basin center shifted its place to the north from Iga Basin to Ohmi Basin, forming a stable water body ("Sayama-ko") with massive clay deposition (c and d in Figure 12). Northward shifting of the sedimentary basin was inherited (Older II Stage) (e and f in Figure 13). During the time from Older II stage to Actual I stage, the center of the sedimentary basin migrated northwestward on a large-scale. Great amounts of gravel are represented as final sediments of Older II stage (g in Figure 13). The sedimentary basin of Actual I stage shifted its place gradually to west, accompanied by upheaval of the eastern mountain area. This explanation is supported by results of lithology, paleocurrent and sedimentological studies of the deposits of the Kosei area and data from 1,000

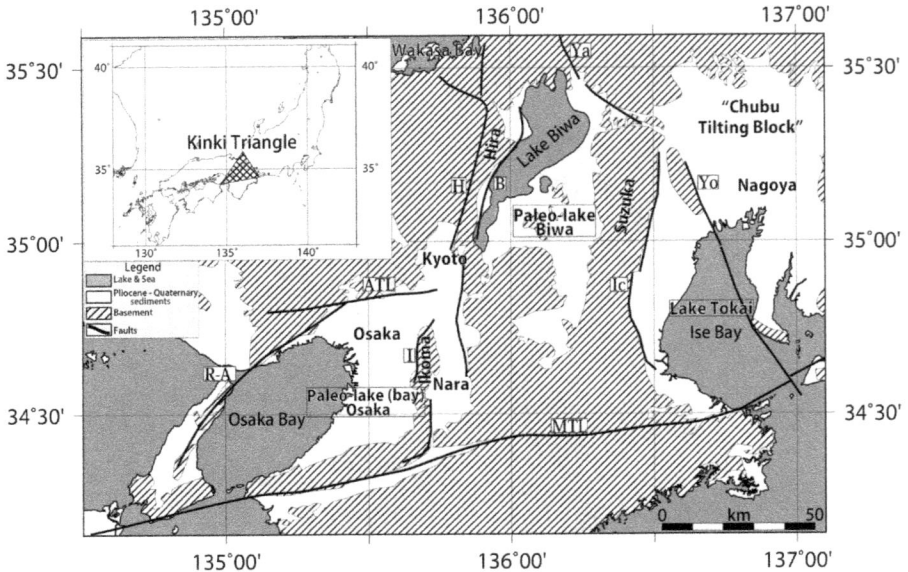

Figure 11. Basement and Pliocene-Quaternary sediment distribution in and around Lake Biwa and Ise Bay region. "Kinki Triangle" region is shown in inset figure. Paleogeographical regions such as "Chubu Tilting Block", Lake Tokai, Paleo-lake Biwa, and Paleo-lake (bay) Osaka are shown. MTL: Median Tectonic Line, ATL: Arima-Takatsuki Tectonic Line, I: Ikoma Fault, H: Hanaore Fault, B: Biwako-seigan Fault zone, Ya: Yanagase Fault, Ic: Ichishi Fault, Yo: Yoro Fault

m drilling cores at the rivermouth of the Yasu River [29]; [30]; [31]. The gravel of the uppermost part of the Kobiwako Group in Kosei Area (western side of present Lake Biwa) constitutes sediments of the last Actual I stage. These gravels were the first sediments from present western Hira Mountains. Subsequently, the crustal movements along the Katata Fault (southern part of Biwako-seigan Fault zone) became more active. Thereafter, in actual II stage, the area of actual northern lake ("Hokko") basin of present Lake Biwa began to subside rapidly.

The process of basin transition in and around Paleo-Lake Biwa is summarized as follows:

1. In the early stage, the sedimentary basin appeared in the southern area of the basin, and migrated gradually to the north in Older Stage I.

2. Subsequently, the sedimentary basin migrated northwestward on a large scale in Older Stage II.

3. The sedimentary basin migrated gradually to the west. Finally, it is divided by the structural movements in N-S trend accompanied rapid subsidence of the eastern area (rapid upheaving of western area) in Actual stage I and Actual stage II of the present Lake Biwa basin formation.

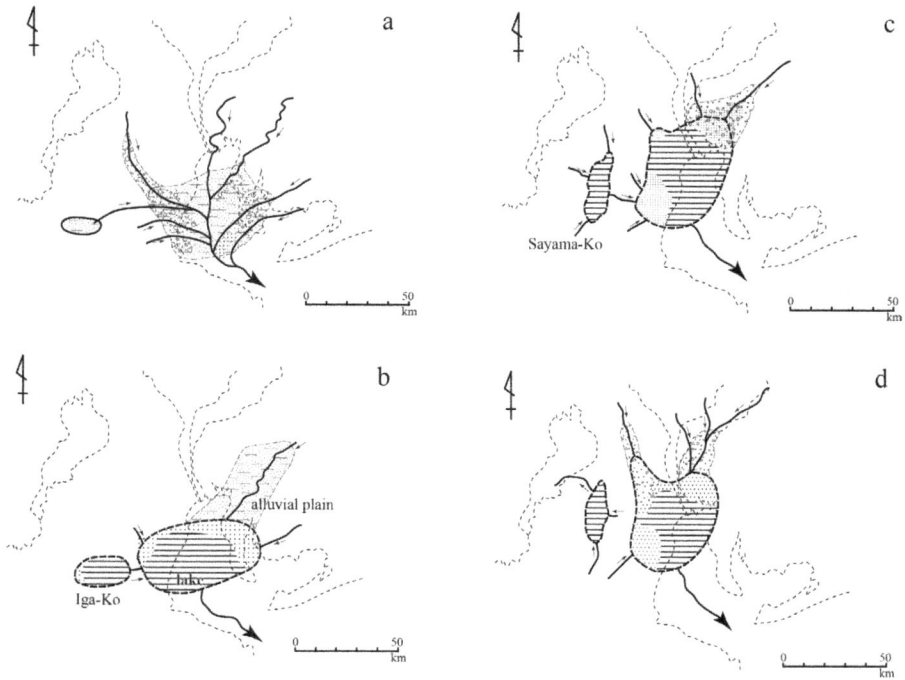

Figure 12. Paleogeographic map of the central Kinki Region including Paleo-Lake Biwa and Lake Tokai area from the early Pliocene to 1.2 – 1.5 Ma [32] a: Initial Stage I of Lake Tokai and Initial Older Stage of Paleo-Lake Biwa, b: Stage I of Lake Tokai and Older Stage of Paleo-Lake Biwa, c: d: Older stage II of Paleo-Lake Biwa and Stage II of Lake Tokai

5. Tectonic basin formation in and around Lake Biwa: Chronological development and mechanism

As described previously, the geohistory in central Kinki region had two conspicuous episodes of ca 3.0 Ma and of ca 1.2-1.5 Ma. A similar pattern of basin transition was revealed as recognized in both sedimentary basins of Lake Tokai and Paleo-lake Biwa (Figure 14), which suggests that the changes of the tectonic stress state were common throughout that province. The sedimentary basins before ca 3.0 Ma (Stage 1 of Lake Tokai and older I stage of Paleo-lake Biwa) are characterized mainly by E-W arrangement of depressional zone and their northward migration. Those characteristics might be attributable to the upheaval of the southern area under the tectonic stress state in N-S direction. However, in the middle of Stage 1, Lake Tokai received a large amount of gravel supply from the east, which was related to the movements of "Chubu Tilting Block". Kuwahara[33] discussed tectonism of eastern area of Ise Bay (the area from Nohbi Plain to mountains). He stated that this area had received tilting movement since Pliocene and this movement formed the topographic contrast between the sedimentary

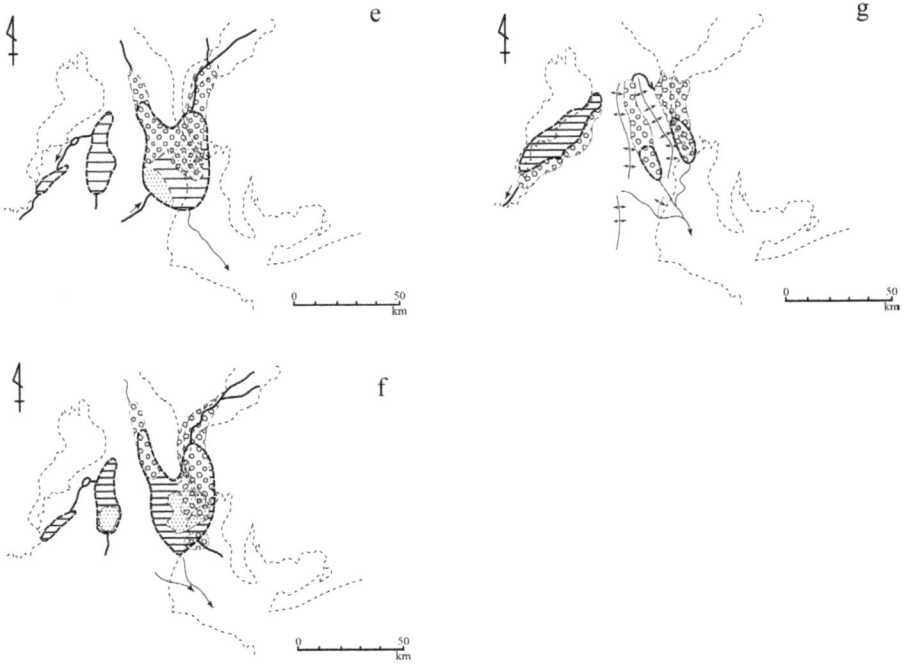

Figure 13. Paleogeographic map of the central Kinki Region including Paleo-Lake Biwa and Lake Tokai area [32] e: & f: Older stage II of Paleo-Lake Biwa and Stage II of Lake Tokai, g: final stage of Older stage II of Paleo-Lake Biwa and Stage II of Lake Tokai

basin of Lake Tokai and eastern upheaving area. He called this tilting block as "Chubu Tilting Block". A large amount of gravel supply suggests the beginning of new structural control superposing the older one. At ca 3.0 Ma, the Lake Tokai sedimentary basin migrated north-westward.

Subsequently, the sedimentary basin of Lake Tokai migrated to the west as a whole, although slightly northward, because of active movement of "Chubu Tilting Block" under the stress state of the E-W trend. However, during the same time interval, the sedimentary basin of Paleo-lake Biwa migrated gradually northward by the upheaval of southern area. At the time of ca 1.2 Ma, the sedimentary basin of Lake Tokai became to extinct, accompanying the upheaving of the Suzuka and Yoro mountains with a N-S structural trend. It was peculiar that the sedimentary basin of Paleo-lake Biwa transferred its position northwestward. Thereafter, it migrated westward gradually by tilting of the eastern area. This fact shows the origination of conspicuous movement under E-W tectonic stress in the Ohni Basin.

In this way, the sedimentary basin migration in Lake Tokai and Paleo-lake Biwa is commonly explained by the hypothesis of interaction between upheaving of the southern area and tilting

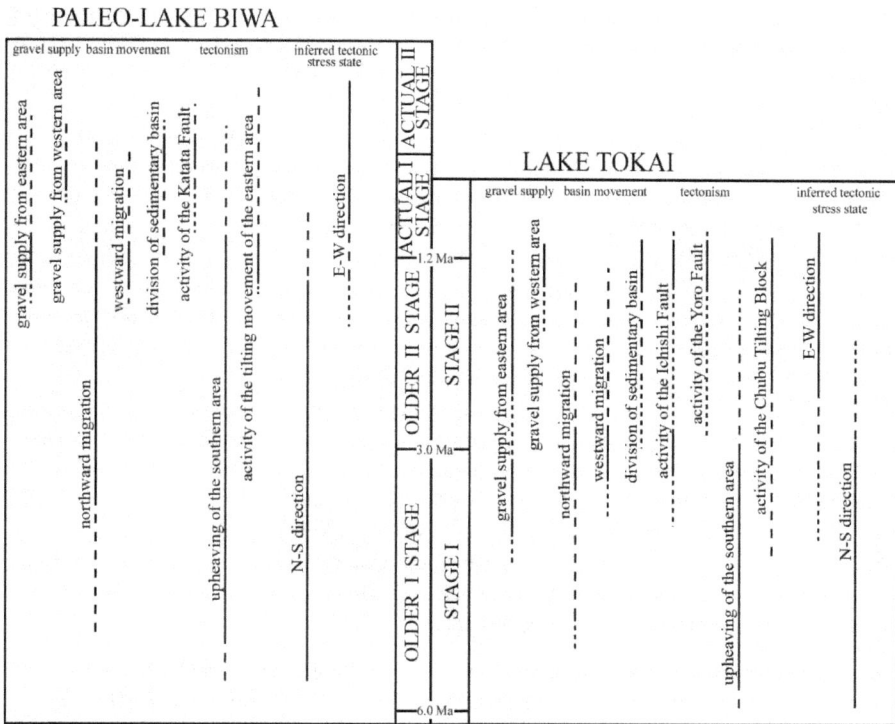

Figure 14. Tectonic implications in central Kinki Region including Paleo-Lake Biwa and Lake Tokai area [32]

of the eastern area (Figure 14). Conclusively we proposed an idea that the deposition in the central part of Kinki region since the Pliocene was under the influence of two superposed tectonic stress states represented by upheaving of the southern area and tilting of the eastern area. Hitherto, Huzita [34] have stated that the tectonic stress state change since Miocene took place in the Inner zone of southwestern Japan during the time of Pliocene and Pleistocene. This change from N-S compressional stress state to E-W one was conspicuous. Therefore, the results of the present work are not contradictory to that idea. In fact, the results confirm the chrono-logical setting and detailed process of tectonosedimentary turnover since the Pliocene.

6. Present Lake Biwa Basin and tectonosedimentary implications

Interruptions in the sedimentary record are commonly regarded as evidence of tectonic changes in a lake basin. Major unconformities exist in the core lithology and seismic profiles at the B Horizon separating the P Bed from the basement rocks and at the Z Horizon between the Q Bed and the P Bed (Figure 3). The alternating characteristics of the S, R, and Q beds indicate short-term cycles in the depositional environment. These cycles may represent periods of uplift of surrounding areas, with subsequent erosional readjustments, or periods of subsidence of the lake basin with subsequent infilling. A study of the lithology and chronology of the Kobiwako Group, the Plio-Pleistocene freshwater sediments distributed around Lake Biwa, reveals that the history of this basin is divisible into two main stages; the Older and the Actual [27]. The deposits of the Older Stage are mainly distributed southward of the present Lake Biwa. Accordingly, the sedimentary depocenter of the Older Stage was located southward of the modern Lake Biwa. Between the Older Stage and the Actual Stage, the depocenter evidently shifted northward to its present location, as indicated by the stratigraphy and chronology of the Kobiwako Group, creating the basin of modern Lake Biwa. Large amounts of gravel began to deposit at the northern part of the Older Stage Lake Biwa basin starting about 1.5 Ma. This occurrence is thought to record the beginning of the migration of the lake depocenter. The R bed corresponds to the sediments of the early stage of Actual Stage. The upper part of the S bed and the entire T bed represent the Actual Stage. The high and constant sedimentation rates within the drilling sequences in present Lake Biwa represent a time of rapid infilling. Rapid subsidence was necessary for the deposition of the continuous sequence in the present Lake Biwa with a sedimentation rate of 0.57 m/kyr [13] and inferred a subsidence rata of 0.74 m/kyr from the data of age of T bed and water depth at the 1400 m drilling site. Reconstruction of these tectonic implications cannot be done competently from data from one core site, and we must use surrounding geological data and geophysical data.

Figure 15 shows the first-order horizontal derivative of the Bouguer gravity anomalies (the horizontal gradient of gravity anomaly) more than 2 mGal/km and contour interval is 1 mGal/km. The horizontal gradient of gravity anomaly (Δg_{hg}) is defined as the following equation.

$$\Delta g_{hg} = \sqrt{\left(\frac{\partial g(x,y)}{\partial x} \right)^2 + \left(\frac{\partial g(x,y)}{\partial y} \right)^2} \qquad (1)$$

Here, g(x, y) are gravity anomaly data given on xy mesh with a constant interval.

The horizontal gradient of gravity anomaly emphasizes shorter wavelength signals of subsurface structures. Therefore, it is a good indicator of a conspicuous density change and/or a large fault. High gradient anomalies of greater than 2 mGal/km correspond well with large faults and tectonic lines such as the Rokko-Awaji fault zone, Arima-Takatsuki tectonic line, Ikoma fault, Kanbayashigawa fault, Hanaore fault, Biwako-seigan fault and others. They have NE-SW and E-W trends in their strike directions. However, distribution of high gradient anomalies around the Lake Biwa is extremely complex. It is difficult to find a specific trend in the strike directions of the horizontal gradient of gravity anomaly. As one reason, it can be inferred that they reflect subsurface structures caused by extreme crustal activities including faulting during the Quaternary.

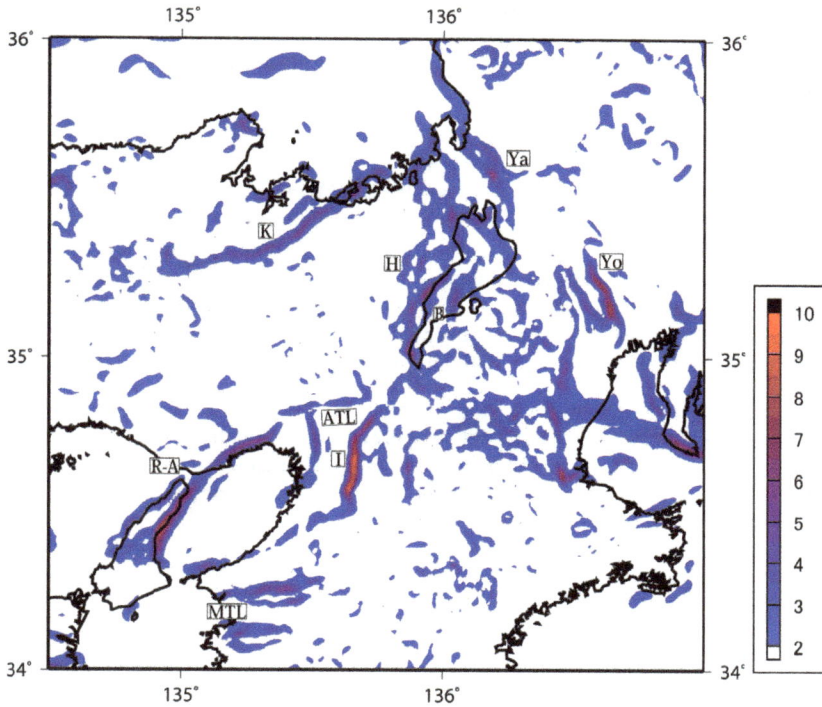

Figure 15. First order horizontal derivative of Bouguer gravity anomalies more than 2 mGal/km. Contour intervals is 1 mGal/km. MTL: Median Tectonic Line, ATL: Arima Takatsuki Tectonic Line, R-A: Rokko-Awaji Fault zone, I: Ikoma Fault zone, K: Kambayashi Fault, H: Hanaore Fault, B: Biwako-seigan Fault zone, Ya: Yanagase Fault zone, Yo: Yoro Fault zone

7. Summary

The present Lake Biwa is a tectonic basin under the E-W compressional stress state. Distribution of active faults is characterized by the western boundary of the lake (Biwako-seigan Fault zone; Hira, Katata Faults with N-S direction) and the northeast region of Lake (Yanagase Fault etc). Activity of faults (mainly Biwako-seigan Fault zone) of more than 1 m/kyr along the west side of Lake Biwa is important for the present lake basin formation.

Basement topography revealed by the seismic reflection survey shows alignment of the valley and range with a N-S direction. The "A" horizon (bottom of T bed since about 0.43 Ma) topography of the seismic reflection profile revealed its tilting topography from east to west. The sedimentary record from present Lake Biwa by deep drilling includes a shift of the lake depocenter from farther to the south to its current location. The sedimentary record shows constant sedimentation from at least 1.3 Ma. The sedimentation rate of the present Lake Biwa is about 0.57 m/kyr as revealed by age depth curve of 1400-m deep drilling core taken in 1982-1983. The subsidence rate is calculated as 0.74-m/kyr by data of the T bed (thickness 250 m, duration 430 kyr, water depth 68 m).

Paleogeographic evidence since Pliocene in central Kinki district including Paleo-Lake Biwa area shows the characteristics by which the history of this basin has been divided into two main stages; the Older and the Actual. Deposits of the Older Stage are distributed mainly southward of the present Lake Biwa. The sedimentary depocenter of the Older Stage was located south of the modern Lake Biwa. The depocenter evidently shifted northward to its present location creating the modern Lake Biwa Basin. Those changes and basin migration since the Pliocene are regarded as tectonic basin formation under the influence of superposed two tectonic stress states represented by upheaving of the southern area and tilting of eastern area. Tectonic stress change from a N-S compressional stress state to the E-W one was conspicuous.

Lake Biwa area (Ohmi Plain) is a large negative gravity anomaly area in the 'Kinki-Triangle' area with -60 mGal indicating the inhomogeneity of lower crust, or lower crust thickness or existence of very low density materials because of faulting. Distribution of high gradient anomalies around Lake Biwa is extremely complex. It is difficult to find a specific trend in the strike directions of the horizontal gradient of the gravity anomaly, perhaps because they reflect subsurface structures because of extreme crustal activities including faulting during the Quaternary.

Acknowledgements

We thank Dr. Daisuke Ishimura, Miss Yurie Yamamoto, Mr. Keitaro Yamada for help on preparing the figures.

Author details

Keiji Takemura[1], Tsuyoshi Haraguchi[2], Shigekazu Kusumoto[3] and Yasuto Itoh[4]

1 Beppu Geothermal Research Laboratory, Institute for Geothermal Sciences, Graduate School of Science, Kyoto University, Noguchibaru, Beppu, Japan

2 Graduate School of Science, Osaka City University, Osaka, Japan

3 Graduate School of Science and Engineering for Research, University of Toyama, Toyama, Japan

4 Graduate School of Science, Osaka Prefecture University, Osaka, Japan

References

[1] Nakata, T. and Imaizumi, T. eds. Digital Active Fault Map of Japan. University of Tokyo Press 2002, 2 DVD-ROM, 60p, 1 appended figure.

[2] Horie S., Mitamura, O., Kanari, S., Miyake, H., Yamamoto, A. and Fuji, N. Paleolimnological study on lacustrine sediments of Lake Biwa-ko. Contribution from the Geological Institute, Kanazawa University 1971; 18: 745-762.

[3] Horie S. Lake Biwa. Dr W Junk Publishers: 1984; 654pp.

[4] Takemura K., Iwabe, C., Hayashida, A., Danhara, T., Kitagawa, H., Hraguchi, T., Sato, T. and Ishikawa, N. Stratigraphy of marker tephras and sediments during past 50,000 years from multiple sites in Lake Biwa, Japan, The Quaternary Research Japan 2010; 49: 147-160 (in Japanese with English abstract)

[5] Takemura K. Tectonic and climatic record of the Lake Biwa, Japan, region provided by the sediments deposited since Pliocene times. Palaeogeography, Palaeoclimatology, Palaeoecology 1990; 78: 185-193.

[6] Horie S. Die Geschichte des Biwa-See in Japan: seine Entwicklung, dargestellt anhand eines 1400 m langen Tiefbohrkerns. Universitatsverlag Wagner: 1991; 349pp.

[7] Yoshikawa S and Inouchi Y. Tephrostratigraphy of the Takashima-oki boring core samples from Lake Biwa, central Japan. Chikyu Kagaku (Earth Science) 1991; 45: 81-100 (in Japanese with English abstract)

[8] Takemura K., Hayashida, A., Okamura, M., Matsuoka, H., Ali, M., Kuniko, Y. and Tori, M. Stratigraphy of multiple piston-core sediments for the last 30,000 years from Lake Biwa, Japan. Journal of Paleolimnology 2000; 23(2): 185-199

[9] Takemura, K. and Yokoyama, T. Sedimentary facies of the 1,400 m drilling sample from Lake Biwa with reference to the discontinuity in the sedimentary sequence. IPPCCE NEWSLETTER 1989; 5: 36-48.

[10] Yokoyama, T. and Takemura, K. Geologic column obtained by the deep drilling from the bottom surface of Lake Biwa, Japan. IPPCCE Newsletter 1983; 3: 21-23.

[11] Meyers, P. A., Takemura, K. and Horie, S. Reinterpretation of Late Quaternary sediment chronology of Lake Biwa, Japan, from correlation with marine glacial-interglacial cycles, Quaternary Research 1993; 39: 154-162.

[12] Nakagawa, T., Gotanda, K., Haraguchi, T., Danhara, T., Yonenobu, H., Brauer, A., Yokoyama, Y., Tada, R., Takemura, K., Staff, R. A., Payne, R., Bronk Ramsey, C., Bryant, C., Brock, F., Schlolaut, G., Marshall, W., Tarasov, P., Lamb, H., Suigetsu 2006 Project Members SG06, a fully continuous and varved sediment core from Lake Suigetsu, Japan: stratigraphy and potential for improving the radiocarbon calibration

model and understanding of late Quaternary climate changes. Quaternary Science Review 2012; 36: 164-176 doi:10.1016/j.quascirev.2010.12.013

[13] Danhara T., Yamashita, T., Iwano, H., Takemura, K. and Hayashida, A. Re-investigation of chronology for the 1400m sediment core obtained from the Lake Biwa in 1982-1983. The Quaternary Research (Jpana) 2010; 49: 101-119 (in Japanese with English abstract)

[14] Machida H and Arai F. Atlas of Tephra in and around Japan. University of Tokyo Press; 2003, 336 pp (in Japanese)

[15] Sato, T., Danhara, T., Haraguchi, T., Hayashida, A. and Takemura, K. Sedimentation rate and relative lake-level change during last 300 ky in Lake Biwa, Japan. Annual meeting of Japan Geoscience Union; 2009 (Makuhari, Chiba, Japan).

[16] Huzita, K. Role of the Median Tectonic Line in the Quaternary tectonics of the Japanese Islands. Mem. Geol. Soc. Japan 1980; (18) 129-153.

[17] The Headquarters for Earthquake Promotion, Long term evaluation of Biwako-seigan Fault zone. 2009.

[18] Okada, A. and Togo, M. Active faults in Kinki. University of Tokyo Press; 2000: 395p.

[19] Horie, S. ed. Paleolimnology of Lake Biwa and the Japanese Pleistocene; 11, 99p, Institute of Paleolimnology and Paleoenvironment on Lake Biwa of Kyoto University; 1983.

[20] Komazawa, M. Gravity grid database of Japan. Gravity CD-ROM of Japan, ver. 2, Digital Geoscience Map P-2: Geological Survey of Japan; 2004.

[21] Huzita K. The Quaternary tectonic stress states of Southwest Japan. Journal of Geosciences, Osaka City University 1976; 20: 93-103.

[22] Nakagawa, K., Ryoki, K., Muto, N., Nishimura, S., and Ito, K., Gravity anomaly map and inferred basement structure in Osaka Plain, central Kinki, south-west Japan. Journal of Geosciences, Osaka City University 1991; 34: 103-117

[23] Inoue, N., N. Kitada, Y. Itoh, K. Takemura, and Nakagawa, K. Integrated study of high resolution geophysical and geological information of Osaka Bay, Southwest Japan. Journal of Asian Earth Sciences 2003; 22: 1-11.

[24] Nishida J, Katsura I, Nishimura S. Gravity survey around Lake Biwa, Southwest Japan. Journal of Physics of the Earth 1990; 38: 1-17.

[25] Ikebe, N. Cenozoic geohistory of Japan. Proc. 8th Pacific Science Congress 1956; 2: 446-456.

[26] Yokoyama, T. Tephrochronology and paleogeography of the Plio-Pleistocene in the eastern Setouchi Geologic province, southwest Japan. Mem. Fac. Sci. Kyoto University, Geol. & Mineral. 1969; 36: 19-85.

[27] Yokoyama, T. Stratigraphy of the Quaternary System around Lake Biwa and geohistory of the ancient Lake Biwa. In: Lake Biwa. Horie, S. (ed.) The Hague, Dr. W. Junk; 1984: 43-128.

[28] Yokoyama, T., Makinouchi, T., Takemura, K., Hayashida, A., Sannomiya, T. and Yamamura, H. Stratigraphy of the Pliocene Iga-Aburahi Formation of the Kobiwako Group at the east of Iga-Ueno City, Mie Prefecture, Japan. Paleolimnology of Lake Biwa and the Japanese Pleistocene 1980; 8: 45-64.

[29] Yokoyama, T., Ishida, S., Danhara, T., Hayashida, S., Hayashi, T., Hayashida, A., Nakagawa, Y., Nakajima, T., Natsuhara, N., Nishida, J., Otofuji, Y., Sakamoto, M., Takemura, K., Tanaka, N., Torii, M., Yamada, K., Yoshikawa, S. and Horie, S. Lithofacies of the 1000 m core samples on the east coast of Lake Biwa, Japan. Paleolimnology of Lake Biwa and the Japanese Pleistocene 1976; 4: 52-66.

[30] Takemura, K., Nishimura, S., Danhara, T. and Yokoyama, T. Properties and fission track age of volcanic ashes in the 1000 m core sample of Lake Biwa with special reference to correlation by tephra among the 1000 m, 200 m boring core samples of Lake Biwa and the Kobiwako Group. Paleolimnology of Lake Biwa and the Japanese Pleistocene 1976; 4: 79-95.

[31] Takemura, K., Iida, Y. and Yokoyama, T. Mineral composition and origin of sand grains in the 1,000 m core sample and at the coast and rivers around Lake Biwa, Japan. Paleolimnology of Lake Biwa and the Japanese Pleistocene 1979; 7: 78-99.

[32] Takemura, K. The Plio-Pleistocene Tokai Group and the tectonic development around Ise Bay of central Japan since Pliocene. Memoirs of the Faculty of Science, Kyoto University, Series of Geology & Mineralogy 1985; 51: 21-96.

[33] Kuwahara, T. The Noobi Plain and its fault block movements. Quaternary Research (Japan) 1968; 7: 235-247.

[34] Huzita, K. Tectonic development of southwest Japan in the Quaternary Period. Journal of Geosciences, Osaka City University 1969; 12: 53-70.

Chapter 9

Neotectonic Intra-Arc Basins Within Southwest Japan — Conspicuous Basin-Forming Process Related to Differential Motion of Crustal Blocks

Yasuto Itoh, Keiji Takemura and
Shigekazu Kusumoto

Additional information is available at the end of the chapter

1. Introduction

Southwest Japan is a continental sliver rifted during the Miocene opening of the Japan Sea [1]. Its post-rifting tectonic architecture is closely related to the mode of convergence of oceanic plates around the eastern Eurasian margin. It seems that neotectonic deformation in southwest Japan becomes increasingly intense from west to east, which has long been understood as a contraction regime caused by the westerly subducting Pacific Plate (e.g., [2]). Collision between the Eurasian and North American Plates around the eastern margin of the Japan Sea and the Itoigawa-Shizuoka Tectonic Line (Figure 1) is also responsible for the deformation trend [3]. Another remarkable structural trend, a late Neogene strong inversion along the backarc coast of the island arc, is sometimes related to the subduction of the Philippine Sea Plate [4,5]. The Philippine Sea Plate changed its convergent direction in the Quaternary [6], and enhanced right-lateral wrenching within southwest Japan [7].

The authors concentrate upon the Quaternary morphological features in the eastern part of the island arc, and describe conspicuous basin-forming processes in the study area. In a general context, we aim at comprehension of the regional tectonic zones controlling the vigorous formation of intra-arc basins, and at identifying the neotectonic domains divided by them. Geophysical information such as gravity anomalies [8] and reflection seismic data are utilized to visualize the deep interiors of the intra-arc basins and evaluate structural trends in the damaged upper crust surrounded by the tectonic zones. Closer observation of geologic structures leads us on the path to understand the formation processes of subordinate structures in the study area.

Neotectonic Intra-Arc Basins Within Southwest Japan — Conspicuous Basin-Forming Process
Related to Differential Motion of Crustal Blocks

209

2. Description of major tectonic zones

2.1. Median Tectonic Line

The Median Tectonic Line (MTL) is the largest crustal break in southwest Japan, which bisects the island arc into the old terranes intruded by igneous rocks in the Inner Zone and partly metamorphosed accretionary complexes in the Outer Zone (Figure 1; [9]). It has a period of activity as long as 100 m.y. and highly complicated change in slip direction. In the following sections, we present a chronicle of the MTL activity based on previous research and original interpretation of geophysical data.

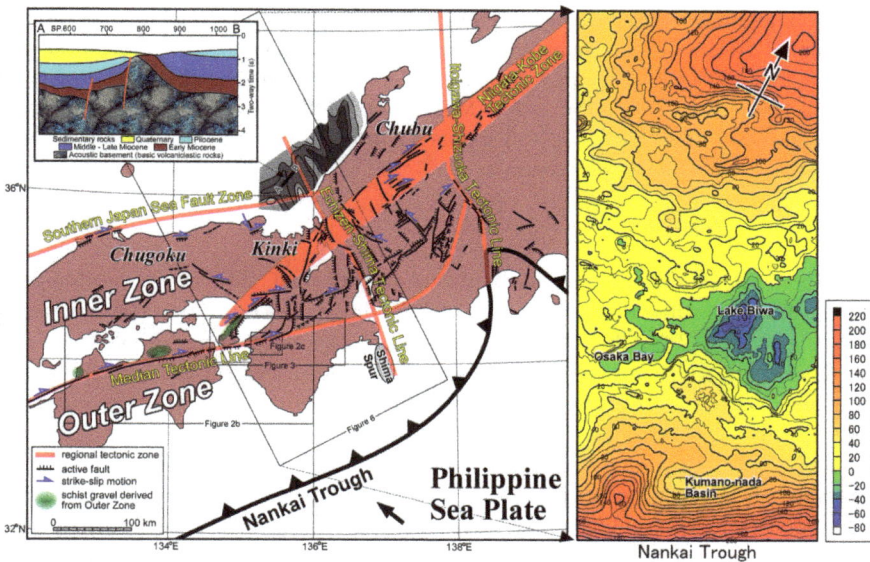

Figure 1. An index of the neotectonic regime in southwest Japan. Base map is after Huzita [2]. Gradation on backarc shelf showing onlapping sedimentation pattern and a seismic profile (shown inset) in the area are after Itoh et al. [5]. Influx of crystalline schist gravels is shown by green areas [22-24]. (Right) Bouguer gravity anomaly map. The Bouguer density is 2670 kg/m³, and contour interval is 10 mGal. Gravity map is generated based on [8]

2.1.1. Initiation of the regional fault zone

Compiling reliable paleomagnetic data, Itoh et al. [5] reconstructed the Cretaceous to early Paleogene paleogeography around the eastern Eurasian margin (Figure 2a). They pointed out that the MTL constituted a larger fault zone together with the Central Sikhote Alin Fault, and had a left-lateral slip sense as a result of the quite rapid northerly motion of the Izanagi Plate [10]. Along the fault zone, conspicuous pull-apart basins were developed and buried by the

Cretaceous silici-clastic rocks of the Izumi Group [11], which was deposited in an elongate basin (300 km long by 10~20 km wide) along the MTL with sinistral strike-slip movements (Figure 2b).

Wrenching deformation associated with the ancient left slip on the MTL is identified in the Outer Zone of the Kii Peninsula (Figure 2c). Wang and Maekawa [12] showed that the metamorphic grade in the Sanbagawa belt have an en echelon anticlinal trend along the MTL. The trend is a result of deformation after high-pressure metamorphism, the climax of which is assigned around the middle of Cretaceous [13]. It is noted that the most intensive post-metamorphic deformation zone does not coincide with the geologically determined MTL that runs upon the northern bank of the River Kinokawa. Hirota [14] found a remarkable discontinuity in the metamorphic grade around the Funaokayama bar within the River Kinokawa (Figure 2c), and regarded it as a tectonic block. Takasu et al. [15] argued, on the basis of chronological data, that the amalgamation of metamorphosed blocks had occurred during the late Cretaceous. A steep gradient in the gravity anomaly along the river also implies a concealed structure parallel to the surface MTL. Although the resolution is lower, the geomagnetic anomaly trend (Figure 3; [16]) supports a difference in upper crustal constituents along the same line as the density contrast.

2.1.2. Neotectonic activity

It is accepted that the MTL has been reactivated as a right-lateral fault since the late Neogene under the influence of the oblique subduction of the Philippine Sea Plate (e.g., [17]). Nakamura et al. [6] demonstrated that the oceanic plate shifted its convergent motion counterclockwise in the Quaternary, which resulted in vigorous slips on the MTL and westward transportation of the Outer Zone (e.g., [18]). However, when compared with the older stages, geomorphological features (e.g., [19]) suggest that the active segment of the MTL shrank during the late Quaternary. No active portion is identified in the eastern part of the Kii Peninsula (Figure 3), in which the geomagnetic anomaly contrast is also obscured. This is in contradiction to the plate subduction regime, and further study of the transient shift of MTL activity is necessary to solve this tectonic paradox. Another noteworthy point is that the MTL trace is characterized by frequent jogs and steps. A sounding survey in the Kii Channel [20] delineated a complex fault pattern that may cause great diversity in basin formation.

2.1.3. Episodic change of deformation mode

Subsurface structures delineated by reflection seismic data [21] suggest a different phase of the recent activities of the MTL. Figure 4 is a N-S (normal to the MTL) seismic profile of the northern bank of the River Kinokawa. Fault morphology is classified into high-angle flower structures, implying lateral motion, and north-dipping reverse faults, reflecting a complicated slip history. Amongst the structures, the most remarkable feature is the thrust at the bottom of the Cretaceous Izumi Group. Because it is underlain by recent sediments, a strong contraction episode in the Quaternary should be responsible for the structure.

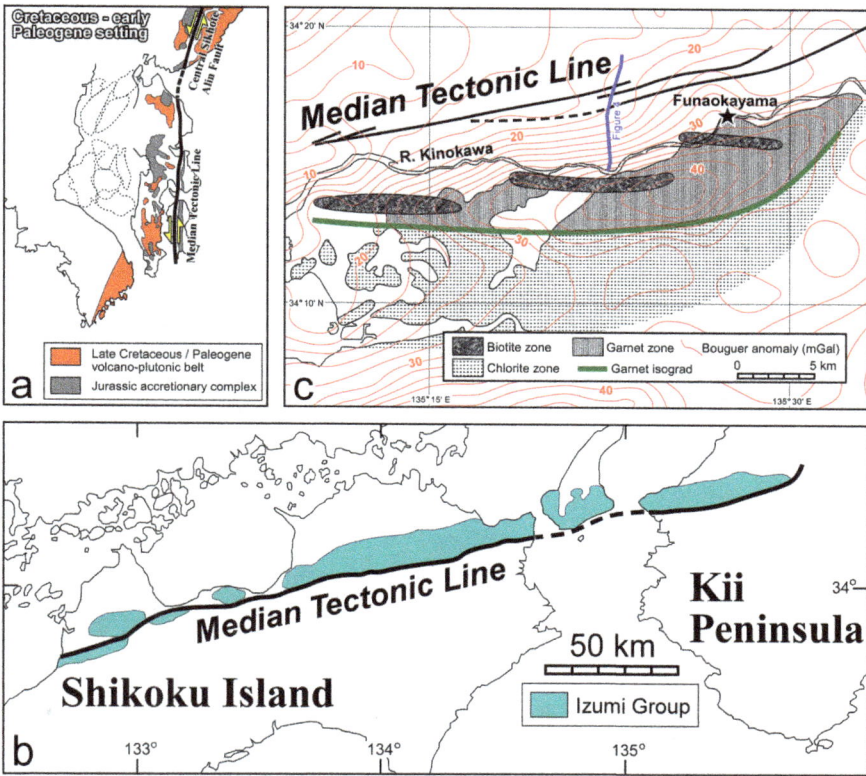

Figure 2. Incipient activity of the Median Tectonic Line (MTL). (a) Paleoreconstruction of the eastern Eurasian margin in the Cretaceous and early Paleogene stage [5]. (b) Distribution of the Cretaceous Izumi Group deposited in a series of pull-apart basins [11]. (c) Geologic features showing sinistral motion of the Median Tectonic Line in the Kii Peninsula. Metamorphic grade in the Sanbagawa belt is after Wang and Maekawa [12]. A star shows the Funaokayama bar in the River Kinokawa, where a remarkable gap in metamorphic grade was confirmed [14]. Mapped areas are shown in Figure 1

Provenance studies of the recent clastics on the northern flank of the MTL support the theory of a strong contraction phase in the Kii Peninsula. In some areas of the Pleistocene exposure, the frequent influx of schist gravels, apparently derived from the Sanbagawa belt in the Outer Zone (Figure 1), has been confirmed by many researchers (e.g., [22-24]). In contrast, Pleistocene sediments around Osaka Bay are lacking in such components in spite of the fact that the aforementioned metamorphic unit is widely distributed in the Kii Peninsula. The authors submit a hypothesis that the strong contraction phase provoked an inversion of the Cretaceous Izumi sedimentary basin along the MTL trace, and an E-W barrier (Izumi Mountains) prevented northward sediment transport through the late Quaternary.

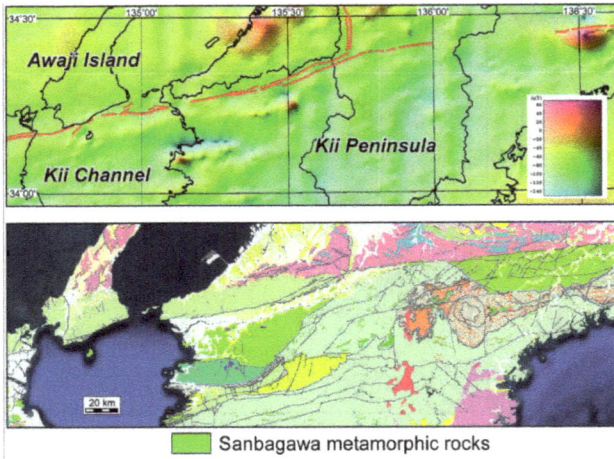

Figure 3. Recent active trace (red line) of the Median Tectonic Line around the Kii Peninsula, compiled after Yoshika-wa et al. [20,21] and an active fault database [19]. See Figure 1 for mapped area. Base maps in upper and lower frames show geomagnetic anomaly [16] and geology [9], respectively

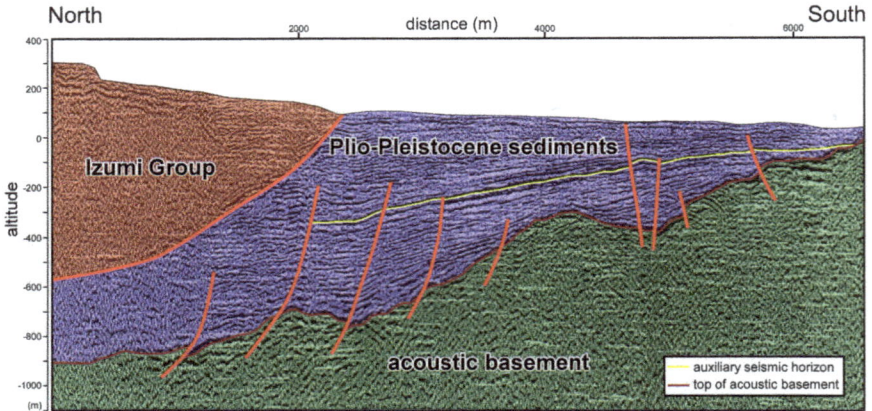

Figure 4. Geologic interpretation of a N-S depth-converted seismic profile across the Median Tectonic Line in the Kii Peninsula. Location of seismic line is shown in Figure 2. Original seismic data is after Yoshikawa et al. [21]

Visualization of subsurface structures in the forearc region implies that the contraction event has a broader impact upon the basin formation/deformation processes of southwest Japan. Takano et al. [25] stated that the Kumano-nada basin (Figure 1) suffered from an episode of contraction around the early Pleistocene, which became dormant in later periods. Thus the

MTL seems to have a much more complicated activation history than the common theory would suggest. This history may be a key to reconstructing the motion of the Philippine Sea Plate, which is not determined from global plate kinematics.

Basin morphology along the MTL shows a spatial variation under the Quaternary transient tectonic stress. Figure 5 presents a plan view of an active trace of the westernmost part of the MTL (upper) and the deep structure of recent sedimentary basins developed along the active segment, which is interpreted from reflection seismic data [26] (lower). It is obvious that the active MTL has a releasing bend around the Beppu Bay where countless secondary tensile faults develop. The volume of the pull-apart basin is estimated based on gravity data in this book (Itoh, Y., Kusumoto, S. and Takemura, K.). Deep structures interpreted from two seismic profiles with no vertical exaggeration indicate the following characteristics. (1) The youngest structural trend is a bunch of high-angle faults (with a so-called 'flower structure') implying strike-slip motion on the MTL fault system. (2) Temporal transition of the active fault trace is inferred from the migration of depocenters of the sedimentary basins. (3) Low-angle detachment in the acoustic basement, which was regarded as a material boundary in the upper crust [26], is clearly reactivated in an extensional sense as shown by the dragging deformation of the adjacent Plio-Pleistocene sediments.

A previous study [27] attributed the along-arc difference in deformation style (east, contraction; west, extension) to counterclockwise rotation of the forearc sliver in response to the relative motion of the Philippine Sea and Pacific Plates, and the backarc spreading of the Okinawa Trough. Further quantitative investigation of the three-dimensional structure of the island arc crust is necessary for constructing a probable tectonic model.

2.2. Niigata-Kobe Tectonic Zone (NKTZ)

Based on geodetic analyses, Sagiya et al. [28] described a NE-SW zone of deformation in southwest Japan (Figure 1), and named it as the Niigata-Kobe Tectonic Zone (NKTZ). It is characterized by right-lateral shear deformation [29], and obliquely crosses over the Itoigawa-Shizuoka Tectonic Line (ISTL; Figure 1) with pure reverse motions. As shown by Nakajima and Hasegawa [30], the NKTZ is a deeply rooted crustal weakness accompanied by a P-wave velocity anomaly in the mid-crust. Iio et al. [31] argued that the high water content of the lower crust, linked to dehydration of the subducting slab, is responsible for the formation of such a weak zone.

Paleomagnetic studies have shown that the NKTZ is not a short-lived feature but contributes to cumulative deformation of the island arc. Itoh et al. [32] compiled reliable paleomagnetic data around the eastern part of southwest Japan, and confirmed significant clockwise rotation on the NKTZ during the Quaternary. They pointed out that similar rotational events were identified on both flanks of the ISTL, and stated that the two crossing tectonic zones with different deformation senses may be alternately activated in response to fluctuation of the regional tectonic stress, which is a theory to comprehend the paradox of a geophysically-assessed low activity level of the ISTL showing geologic significance.

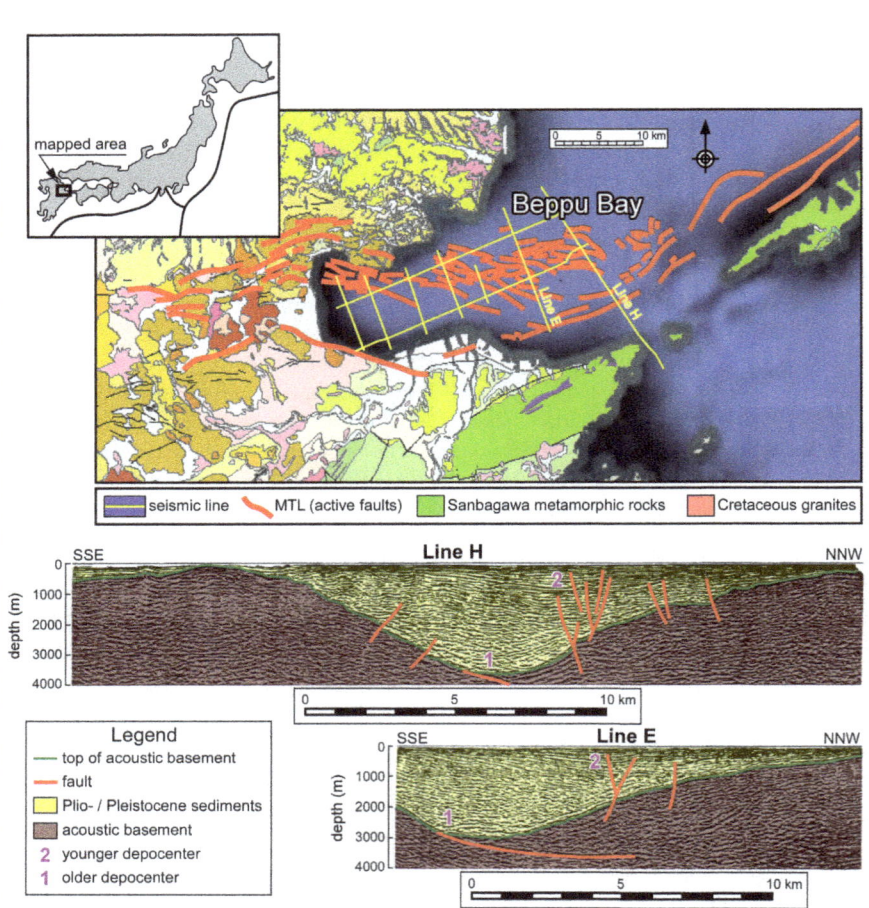

Figure 5. Upper: Westernmost part of active trace of the Median Tectonic Line after [9] and [27] with seismic line map [26]. Lower: Reinterpreted depth-converted seismic profiles without vertical exaggeration. Original seismic data is after Yusa et al. [26]

2.3. Echizen-Shima Tectonic Line (ESTL)

Huzita [33] stated that a triangular portion in the central Kinki district is characterized by intensive deformation and basin formation, and named it the 'Kinki Triangle'. The southern border of this tectonic area coincides with the MTL, and its western border roughly corresponds to the NKTZ. The tectonic context of the eastern border, however, has not been clearly discussed. Here, the authors attempt to redefine a tectonic line from the viewpoint of consistency in deformation trend including the forearc and backarc regions.

Itoh et al. [5] described the geologic structures of the backarc of southwest Japan. A sediment onlapping pattern depicts an inversion trend nearly normal to the elongation of the arc (Figure 1). They showed a seismic profile suggesting that the inversion developed from the Pliocene to Pleistocene. On the forearc side, the Shima Spur built up in early Quaternary [25]. These structural trends are linearly connected with onshore active faults, and constitute a regional zone of contraction. We regard it as a significant neotectonic boundary and name it the Echizen-Shima Tectonic Line (ESTL). At present, the origin of the ESTL is not fully understood. It probably has a close relation with the Miocene bending event in southwest Japan caused by the collision of the Izu-Bonin arc, the reason for this theory being that paleomagnetic studies [34,35] clarified that the hinge line of arc bending was located around the ESTL.

3. Neotectonic domains in southwest Japan

The authors have presented the characteristics of major neotectonic zones (lines) around the eastern part of southwest Japan. Next, we attempt to describe neotectonic domains bordered by these features. We identify the Chugoku, Kinki and Chubu domains from west to east (Figure 1).

The Chugoku domain is characterized by quite broad dextral wrenching and inactive basin formation. It is a crust sliver between the MTL and the Southern Japan Sea Fault Zone (SJSFZ). The SJSFZ is a reactivated right-lateral fault along the late Miocene backarc inversion zone [36]. Itoh and Takemura [7] pointed out that the recent absence of arc volcanism in southwest Japan has resulted in the homogeneous crustal strength and uniform strain rate of the fault-bounded sliver.

Among the geographically defined Kinki district, we take notice of the tectonic domain surrounded by the MTL, NKTZ and ESTL. It is a damage zone accompanied by countless faults and enormous intra-arc basins, which are delineated by low gravity anomalies (see Figure 1). The mechanism of paradoxical basin formation at a contraction step of the MTL is discussed in this book (Itoh, Y., Kusumoto, S. and Takemura, K.). After the incipient subsidence stage in the Pliocene, an accelerated strain rate during the Quaternary provoked rapid sedimentation. A geophysical view of the architecture of the crust and the general trend of subordinate structures within this domain are discussed in the next section.

The Chubu domain is bordered by the ISTL and ESTL, and subdivided by the NKTZ into northern and southern sectors. The northern Chubu sector seems to be under the influence of the backarc inversion zone of northeast Japan, and all the active faults show dominant reverse slip. In contrast, the southern sector is characterized by numerous conjugate faults suggestive of an E-W regional compression. Although large-scale intra-arc basins do not develop in this area, Itoh et al. [37] demonstrated that conspicuous small basins are formed around terminations and stepping parts of the strike-slip faults.

4. Discussion

4.1. Characteristics of gravity anomaly

We show a Bouguer gravity anomaly map for our study area in Figure 1. This Bouguer gravity anomaly map is based on gravity mesh data [8], and the Bouguer density is 2670 kg/m³.

In this region, positive gravity anomalies are dominant, and there are conspicuous positive anomalies over the Pacific Ocean and the Japan Sea. The Bouguer gravity anomaly of the Japan Sea side is relatively flat, while the Pacific Ocean side has a large gradient (Figure 1). The Bouguer gravity anomaly (Δg_B) in a marine area is generally positive in an area with a deep water, this is inferred form the Bouguer gravity anomaly given by the following (e.g., [38]).

$$\Delta g_B = \Delta g_F - 2\pi G (\rho_w - \rho) D \tag{1}$$

Here, Δg_F, D and G are the free-air gravity anomaly, the depth of water and the universal gravitational constant, respectively; ρ_w and ρ are water density and surface crust density, and generally $\rho_w < \rho$. Consequently, it is expected that these positive gravity anomaly areas have deep water. In fact, the areas correspond to the subduction zone along the Nankai Trough and the back-arc basins in the Japan Sea. There are negative gravity anomalies indicating the existence of subsidence structures between these positive gravity anomalies, and the subsidence structures forming the negative anomalies would be due to intra-arc basins.

These negative anomalies correspond to the active tectonic zone during the Quaternary called the 'Kinki Triangle' [33], and it is divided into the Osaka Bay and Lake Biwa areas. Negative gravity anomalies around Osaka Bay and the Lake Biwa reach -15 mGal and -60 mGal, respectively.

It is known that negative gravity anomalies in the Osaka Bay area can be explained by sediments accumulated in and around Osaka Bay (e.g., [39,40]), and these negative gravities are divided by some active faults (Figure 1). In contrast, it is known that negative gravity anomalies in the Lake Biwa area can not be explained by the distribution of soft sediments in the lake (e.g., [41]). Nishida et al. [41] have suggested that depression of the Conrad surface or the existence of very low-density materials due to faulting is necessary to explain the gravity low reaching -60 mGal.

Figure 6 depicts the first order horizontal derivative of the Bouguer gravity anomalies larger than 2 mGal/km that is shown by color gradation with an interval of 1 mGal/km. The first order horizontal derivative of the Bouguer gravity anomalies is defined by the following equation.

$$\sqrt{\left[\frac{\partial g(x,y)}{\partial x}\right]^2 + \left[\frac{\partial g(x,y)}{\partial y}\right]^2} \tag{2}$$

Here, $g(x, y)$ is the gravity anomaly on xy mesh data at a constant interval. Since the first order horizontal derivative of the Bouguer gravity anomaly emphasizes the shorter wavelength signals of subsurface structures, it is a good indication of a conspicuous density change and/or a large fault.

In Figure 6, high gradient anomalies greater than 2 mGal/km, except in the Lake Biwa area, have the same direction (roughly parallel to the Nankai Trough), and most of them in the land area correspond well with large faults or tectonic lines (Figure 1). The distribution of high gradient anomalies around Lake Biwa is very complex, and it could be considered that they reflect subsurface structures caused by extreme crustal activity including faulting during the Quaternary.

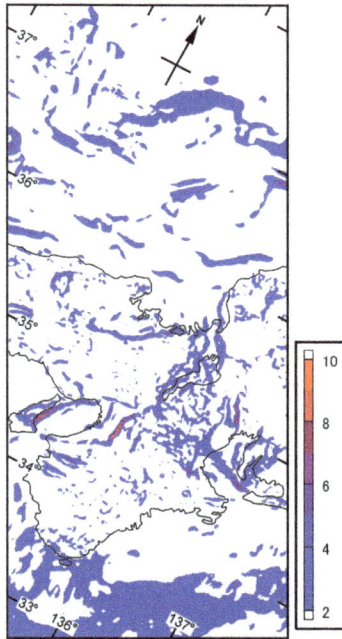

Figure 6. First order horizontal derivative of Bouguer gravity anomalies larger than 2 mGal/km, shown by color gradation with an interval of 1 mGal/km

4.2. Development of subordinate structure

As shown in the previous section, gravimetric analysis indicates that the crust of the Kinki domain is damaged under the influence of the complicated activity of surrounding tectonic zones. Numerous faults provoke the formation of intra-arc basins, among which Lake Biwa

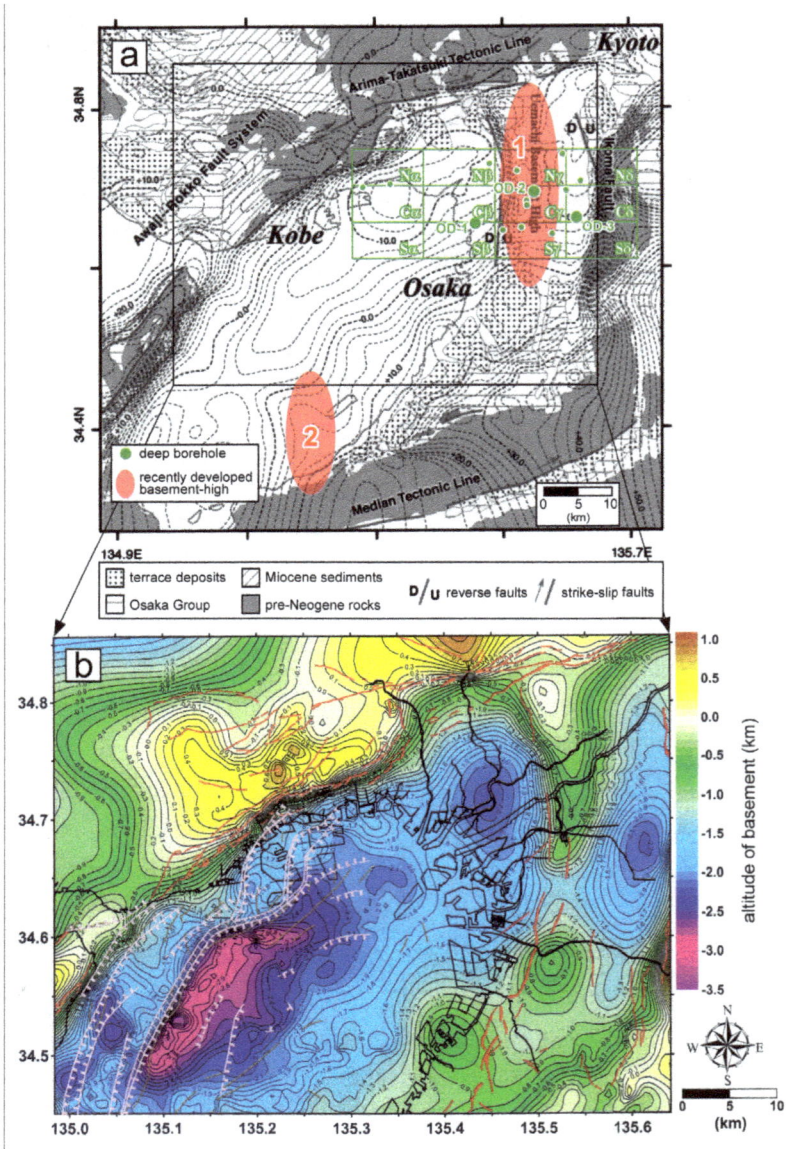

Figure 7. a) Bouguer gravity anomaly around the Osaka Bay at 2 mGal contour interval. The Bouguer density is 2670 kg/m³. Green grid shows domains for calculation of sediment thickness [42]. (b) Altitude of basement around the Osaka Bay inferred from gravity data [45]

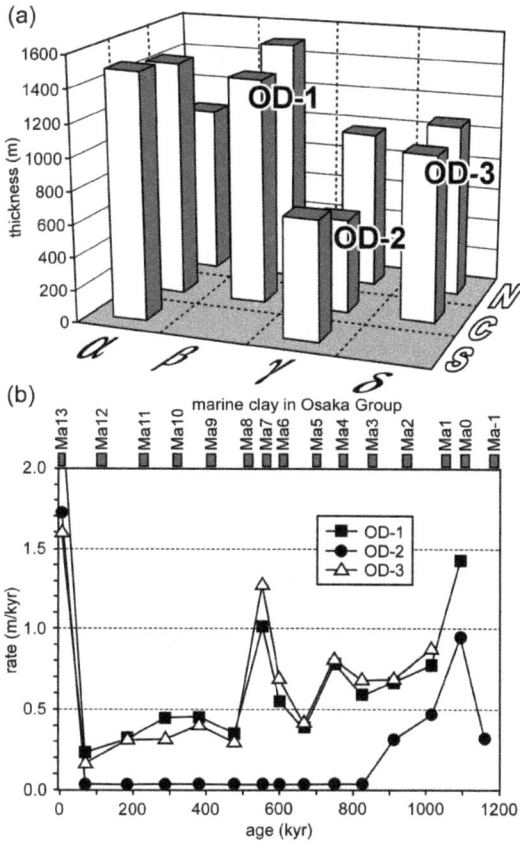

Figure 8. a) Sediment thickness diagram from the late Pliocene to early Pleistocene (from top of the basement to the Ma 2 marine clay intercalated in the Osaka Group) [42]. See Figure 7 for plan view and nomenclature of the analyzed domains. (b) Interval subsidence rates through the late Quaternary in the Osaka Plain [42]. See Figure 7 for locations of the selected boreholes

and Osaka Bay are the largest and most important for understanding the paleoenvironmental changes in southwest Japan. Takemura and others in this book present a comprehensive history of the Lake Biwa sedimentary basin.

It is noted that the majority of the subordinate faults have a N-S azimuth (Figure 1). Their activity results in the formation of N-S warping zones within the island arc as shown in Figure 7. Based on detailed well stratigraphy, Itoh et al. [42] showed that the largest warping in the Osaka sedimentary basin (Uemachi basement-high; 1 in Figure 7a) has been developing since the late Pliocene (Figure 8a), and episodically grew around 550 kyr, which is shown by synchronous acceleration of subsidence on the both flanks of the basement-high (Figure 8b). Similar events of crustal deformation are also confirmed in the Osaka Bay area. Itoh et al. [43] and Inoue et al. [44] indicated that a N-S warping (2 in Figure 7a) emerged around the mid-Quaternary and acted as a sedimentation divide in Osaka Bay. Basement altitude estimated from gravity (Figure 7b; [45]) implies that other subordinate structures emerged synchronous with the basin development. Thus, the differential motion of crustal blocks in a damage zone is closely related with complicated basin formation and conspicuous environmental changes.

5. Summary

A summary of basin-forming processes in an island arc was presented in connection with the development of a damaged area on an active plate margin. The Kinki district in southwest Japan has been a site of vigorous basin formation since the Pliocene. An accelerated Quaternary strain rate around the area is generally interpreted as a result of compressive stress linked to the westerly subduction of the Pacific Plate. Recent geodetic analyses demonstrated a NE-SW tectonic zone (Niigata-Kobe Tectonic Zone), which is an oblique trend of the geologically-detected active structure with a N-S azimuth (the Itoigawa-Shizuoka Tectonic Line). Based on the contrast in fault architecture and the subsurface structures depicted using geophysical methods, the authors define another cross-arc structural component, the Echizen-Shima Tectonic Line. Forearc deformation closely linked to activity of this tectonic line is discussed in a chapter of this book [46]. Westerly subduction of the Philippine Sea Plate has provoked the transcurrent motion of the forearc sliver and active faulting upon the along-arc Median Tectonic Line. Surrounded by these regional tectonic zones, the Kinki district is studded by countless subordinate faults and suffers from differential motion of crustal blocks, which results in great diversity of basin formation.

Acknowledgements

The authors are grateful to A. Noda for his constructive review. The early version of our manuscript was greatly improved based on his comments.

Author details

Yasuto Itoh[1], Keiji Takemura[2] and Shigekazu Kusumoto[3]

1 Graduate School of Science, Osaka Prefecture University, Osaka, Japan

2 Graduate School of Science, Kyoto University, Kyoto, Japan

3 Graduate School of Science and Engineering for Research, University of Toyama, Toyama, Japan

References

[1] Otofuji Y, Matsuda T. Paleomagnetic evidence for the clockwise rotation of Southwest Japan. Earth and Planetary Science Letters 1983; 62: 349-359.

[2] Huzita K. Role of the Median Tectonic Line in the Quaternary tectonics of the Japanese islands. Memoir of Geological Society of Japan 1980; 18: 129-153.

[3] Okamura Y. Relationships between geological structure and earthquake source faults along the eastern margin of the Japan Sea. Journal of Geological Society of Japan 2010; 116: 582-591.

[4] Itoh Y, Nagasaki Y. Crustal shortening of Southwest Japan in the Late Miocene. The Island Arc 1996; 5: 337-353.

[5] Itoh Y, Uno K, Arato H. Seismic evidence of divergent rifting and subsequent deformation in the southern Japan Sea, and a Cenozoic tectonic synthesis of the eastern Eurasian margin. Journal of Asian Earth Sciences 2006; 27: 933-942.

[6] Nakamura K, Renard V, Angelier J, Azema J, Bourgois J, Deplus C, Fujioka K, Hamano Y, Huchon P, Kinoshita H, Labaume P, Ogawa Y, Seno T, Takeuchi A, Tanahashi M, Uchiyama A, Vigneresse J-L. Oblique and near collision subduction, Sagami and Suruga Troughs - preliminary results of the French-Japanese 1984 Kaiko cruise, Leg 2. Earth and Planetary Science Letters 1987; 83: 229-242.

[7] Itoh Y, Takemura K. Quaternary geomorphic trends within Southwest Japan: extensive wrench deformation related to transcurrent motions of the Median Tectonic Line. Tectonophysics 1993; 227: 95-104.

[8] Gravity CD-ROM of Japan, Ver. 2, Digital Geoscience Map P-2. [CD-ROM] Tsukuba: Geological Survey of Japan; 2004.

[9] Geological Survey of Japan, AIST, editor. Seamless Digital Geological Map of Japan 1: 200,000 (July 3, 2012 Version), Research Information Database DB084. Tsukuba: National Institute of Advanced Industrial Science and Technology; 2012.

[10] Engebretson DC, Cox A, Gordon RC. Relative motions between oceanic and continental plates in the Pacific Basin. Geological Society of America Special Paper 1985; 206: 1-59.

[11] Noda A, Toshimitsu S. Backward stacking of submarine channel-fan successions controlled by strike-slip faulting: The Izumi Group (Cretaceous), southwest Japan. Lithosphere 2009; 1: 41-59.

[12]   Wang CL, Maekawa H. Albite-biotite zone of the Sanbagawa metamorphic belt in the northwestern part of the Kii Peninsula, Japan. Journal of Mineralogy, Petrology and Economic Geology 1997; 92: 43-54.

[13]   Wallis SR, Anczkiewicz R, Endo S, Aoya M, Platt JP, Thirlwall M, Hirata T. Plate movements, ductile deformation and geochronology of the Sanbagawa belt, SW Japan: tectonic significance of 89-88 Ma Lu-Hf eclogite ages. Journal of Metamorphic Geology 2009; 27: 93-105.

[14]   Hirota Y. Geology of the Sambagawa metamorphic belt in western Kii Peninsula, Japan. Memoirs of Faculty of Science, Shimane University 1991; 25: 131-142.

[15]   Takasu A, Dallmeyer RD, Hirota Y. $^{40}Ar/^{39}Ar$ muscovite ages of the Sambagawa schists in the Iimori district, Kii Peninsula, Japan: Implications for orogen-parallel diachronism. Journal of Geological Society of Japan 1996; 102: 406-418.

[16]   Nakatsuka T, Okuma S. Aeromagnetic Anomalies Database of Japan, Digital Geoscience Map P-6. Tsukuba: Geological Survey of Japan; 2005.

[17]   Research Group for Active Faults. The Active Faults in Japan: Sheet Maps and Inventories Rev. Ed. Tokyo: Univ. of Tokyo Press; 1991.

[18]   Itoh Y, Takemura K. Mode of Quaternary crustal deformation of Kyushu Island, Japan. In: Shichi R, Heki K, Kasahara M, Kawasaki I, Murakami M, Nakahori Y, Okada Y, Okubo S, Ota Y, Takemoto S. (eds.) Proceedings of the 8th International Symposium on Recent Crustal Movement, CRCM'93, 6-11 December 1993, Kobe, Japan. Kyoto: The Local Organizing Committee for the CRCM'93; 1994.

[19]   National Institute of Advanced Industrial Science and Technology. Active Fault Database of Japan, 23 June 2009 Version, Research Information Database DB095. http://riodb02.ibase.aist.go.jp/activefault/index_e.html (accessed 3 March 2013).

[20]   Yoshikawa S, Kadosawa H, Mitsuhashi A, Iwasaki Y. Geological structure of the Median Tectonic Line in Tomogashima Strait area. Journal of the Japan Society for Marine Surveys and Technology 1996; 8: 1-10.

[21]   Yoshikawa S, Iwasaki Y, Ikawa T, Yokota H. Geological structure of the MTL in west Wakayama by reflection seismic study. Memoir of Geological Society of Japan 1992; 40: 177-186.

[22]   Mizuno K. Preliminary report on the Plio-Pleistocene sediments distributed along the Median Tectonic Line in and around Shikoku, Japan. Bulletin of the Geological Survey of Japan 1987; 38: 171-190.

[23]   Ueki T, Mitusio T. Uplift history of the Asan Mountains, Shikoku, southwest Japan, based on stratigraphy and lithofacies of the Plio-Pleistocene Mitoyo Group. Journal of Geological Society of Japan 1998; 104: 247-267.

[24] Oka Y, Sangawa A. The formation of the sedimentary basin in the east of Inland Sea and the uplift of the Awaji Island, Japan. Journal of Geography 1981; 90: 393-409.

[25] Takano O, Nishimura M, Fujii T, Saeki T. Sequence stratigraphic distribution analysis of methane-hydrate-bearing submarine-fan turbidite sandstones in the eastern Nankai Trough area: relationship between turbidite facies distributions and BSR occurrence. Journal of Geography 2009; 118: 776-792.

[26] Yusa Y, Takemura K, Kitaoka K, Kamiyama K, Horie S, Nakagawa I, Kobayashi Y, Kubotera A, Sudo Y, Ikawa T, Asada M. Subsurface structure of Beppu Bay (Kyushu, Japan) by seismic reflection and gravity survey. Zisin (Bulletin of Seismological Society of Japan) 1992; 45: 199-212.

[27] Ikeda M, Toda S, Kobayashi S, Ohno Y, Nishizaka N, Ohno I. Tectonic model and fault segmentation of the Median Tectonic Line active fault system on Shikoku, Japan. Tectonics 2009; doi:10.1029/2008TC002349.

[28] Sagiya T, Miyazaki S, Tada T. Continuous GPS array and present-day crustal deformation of Japan. Pure and Applied Geophysics 2000; 157: 2303-2322.

[29] Toya Y, Kasahara M. Robust and exploratory analysis of active mesoscale tectonic zones in Japan utilizing the nationwide GPS array. Tectonophysics 2005; 400: 27-53.

[30] Nakajima J, Hasegawa A. Deep crustal structure along the Niigata-Kobe Tectonic Zone, Japan: Its origin and segmentation. Earth Planets Space 2007; 59: e5-e8.

[31] Iio Y, Sagiya T, Kobayashi Y, Shiozaki I. Water-weakened lower crust and its role in the concentrated deformation in the Japanese Islands. Earth and Planetary Science Letters 2002; 203: 245-253.

[32] Itoh Y, Kusumoto S, Miyamoto K, Inui Y. Short- / long-term deformation of upper crust: integrated and quantitative approach for neotectonics. In: Sharkov EV. (ed.) New Frontiers in Tectonic Research - General Problems, Sedimentary Basins and Island Arcs. Rijeka: InTech; 2011. p283-308.

[33] Huzita K. The Quaternary tectonic stress states of Southwest Japan. Journal of Geosciences, Osaka City University 1976; 20: 93-103.

[34] Itoh Y, Ito Y. Confined ductile deformation in the Japan arc inferred from paleomagnetic studies. Tectonophysics 1989; 167: 57-73.

[35] Otofuji Y, Enami R, Yokoyama M, Kamiya K, Kuma S, Saito H. Miocene clockwise rotation of southwest Japan and formation of curvature of the Median Tectonic Line: Paleomagnetic implications. Journal of Geophysical Research 1999; 104: 12895-12907.

[36] Itoh Y, Tsutsumi H, Yamamoto H, Arato H. Active right-lateral strike-slip fault zone along the southern margin of the Japan Sea. Tectonophysics 2002; 351: 301-314.

[37] Itoh Y, Kusumoto S, Furubayashi T. Quantitative evaluation of Quaternary crustal deformation around the Takayama Basin, central Japan: A paleomagnetic and numerical modeling approach. Earth and Planetary Science Letters 2008; 267: 517-532.

[38] Torge W. Gravimetry. Berlin: Walter de Gruyter; 1989.

[39] Nakagawa K, Ryoki K, Muto N, Nishimura S, Ito K. Gravity anomaly map and inferred basement structure in Osaka Plain, central Kinki, south-west Japan. Journal of Geosciences, Osaka City University 1991; 34: 103-117.

[40] Inoue N, Kitada N, Itoh Y, Takemura K, Nakagawa K. Integrated study of high resolution geophysical and geological information of Osaka Bay, Southwest Japan. Journal of Asian Earth Sciences 2003; 22: 1-11.

[41] Nishida J, Katsura I, Nishimura S. Gravity survey around Lake Biwa, Southwest Japan. Journal of Physics of the Earth 1990; 38: 1-17.

[42] Itoh Y, Takemura K, Ishiyama T, Tanaka Y, Iwaki H. Basin formation at a contractional bend of a large transcurrent fault: Plio-Pleistocene subsidence of the Kobe and northern Osaka Basins, Japan. Tectonophysics 2000; 321: 327-341.

[43] Itoh Y, Takemura K, Kawabata D, Tanaka Y, Nakaseko K. Quaternary tectonic warping and strata formation in the southern Osaka Basin inferred from reflection seismic interpretation and borehole sequences. Journal of Asian Earth Sciences 2001; 20: 45-58.

[44] Inoue N, Kitada N, Takemura K, Fukuda K, Emura T. Three-dimensional subsurface structure model of Kansai International Airport by integration of borehole data and seismic profiles. Geotechnical and Geological Engineering 2012; DOI 10.1007/s10706-012-9568-4.

[45] Research Committee on Ground in Osaka Bay. Ground and Construction of Bay Area. Osaka: Association of Research on Geothechnical Information in Osaka Bay; 2002.

[46] Takano O, Itoh Y, Kusumoto S. Variation in forearc basin configuration and basin-filling depositional systems as a function of trench slope break development and strike-slip movement: examples from the Cenozoic forearc basins along the Japan arcs. In: Itoh Y. (ed.) Mechanism of Sedimentary Basin Formation: Multidisciplinary Approach on Active Plate Margins. Rijeka: InTech; 2013. in press.

Geophysical Methods for Basin Research: From Micro-Fabric to Quantification of Evolutionary Process

Rock Magnetic Properties of Sedimentary Rocks in Central Hokkaido — Insights into Sedimentary and Tectonic Processes on an Active Margin

Yasuto Itoh, Machiko Tamaki and Osamu Takano

Additional information is available at the end of the chapter

1. Introduction

Reflecting a complicated subduction and collision history on the eastern Eurasian margin, central Hokkaido has been a site of various types of basin formation. Thick piles of the Cretaceous and Paleogene sediments (Figure 1; [1]) buried a regional forearc basin subducted by the Izanagi/Kula and Pacific Plates. Paleomagnetic studies of the Cretaceous Yezo Supergroup [2,3] showed that the present forearc is divided into some basins developed in different areas. Sedimentary system and forearc basin architecture in the Paleogene was studied in detail by Takano and Waseda [4] and Takano et al. [5].

Under the influence of arc-arc collision on the Pacific convergent margin, vigorous mountain building and formation of foreland basins became active since the late Cenozoic. The Ishikari-Teshio belt (see Figure 2) is underlain with thick middle Miocene clastic strata. These are the Kawabata and its correlative formations, derived from the longitudinal mountainous ranges that were uplifted and eroded during that time [6]. It is generally regarded as a typical foreland setting, and the burial history of turbidites and associated coarse clastics of the Kawabata Formation has previously been studied from a sedimentological viewpoint (e.g., [7]). The process through which the Miocene basin developed in central Hokkaido is not only governed by compressive stress in the collision zone, but also by coeval tectonic events like back-arc spreading in the Japan Sea (e.g., [8]) and dextral transcurrent faulting along the Eurasian margin (e.g., [9]).

In this paper, we present preliminary results of rock magnetic analyses of the Cretaceous Yezo Supergroup, the Eocene Ishikari Group and the Miocene Kawabata Formation in order to detect tectonic movements around the basin and to describe the microfabric of sedimentary

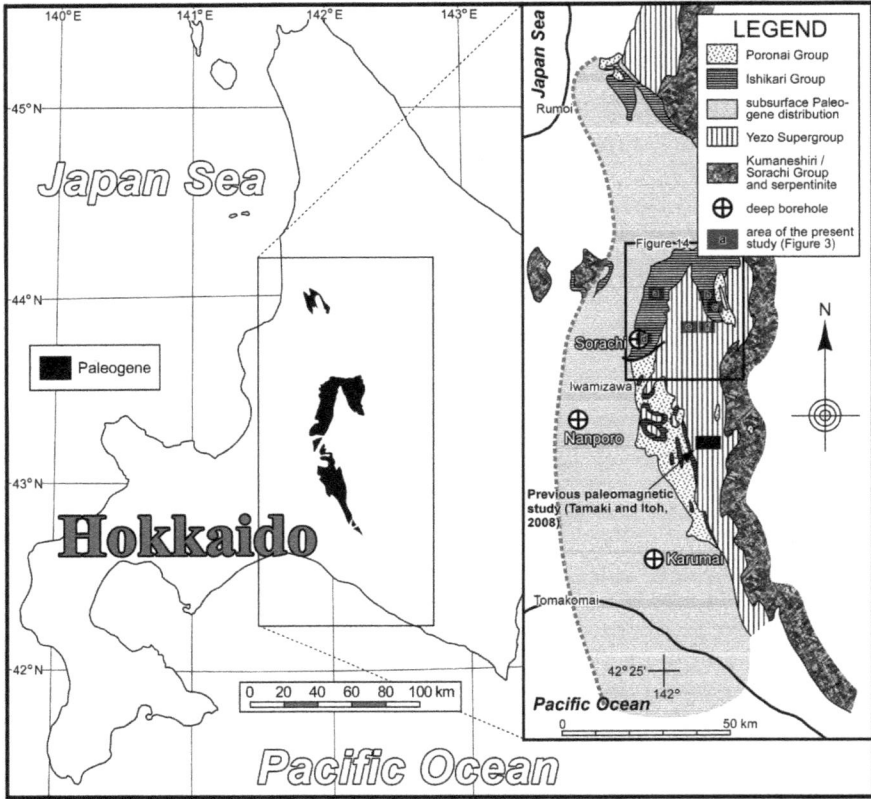

Figure 1. Index map of the study area of the Cretaceous and Paleogene strata. Geologic map is after Editorial Committee of Hokkaido, Regional Geology of Japan [1]

rocks related to the tectonic regime and sedimentation processes in the mobile zone. This study is an attempt to apply magnetic properties to tectono-sedimentology.

2. Geology

2.1. Background

The Yezo Supergroup deposited on the Cretaceous forearc and consists of monotonous mudstone intercalated by coarse clastics and ash layers. After a stagnant subsidence stage at the beginning of the Cenozoic, fluvial sediments of the Ishikari Group and its correlative units

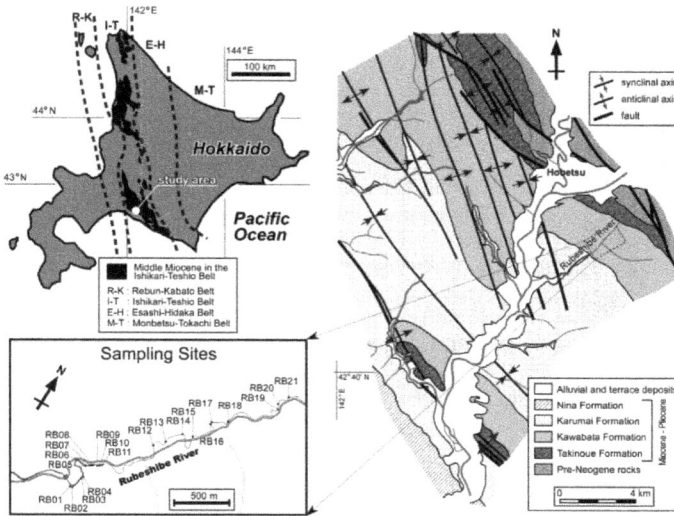

Figure 2. Cenozoic tectonic context of Hokkaido, geology of the study area of the Neogene strata (simplified from Kawakami et al. [7]), and locations of rock magnetic samples

began to bury depressions on the forearc. As a result of strong deformation and continued sedimentation on the active margin, surface distribution of the Eocene Ishikari Group is rather restricted. However, numerous exploration drilling clarified that voluminous Paleogene units are concealed under the alluvial plain (Figure 1). Paleogene depositional sequence and facies classification were described by Takano et al. [5]. They are shown in Figure 3 using abbreviations.

The study area of the Kawabata Formation is located in the southern part of the middle Miocene basins of the Ishikari-Teshio belt. Folded sedimentary units are distributed with a NNW-SSE trend, and are cut by numerous faults (Figure 2). The area is divided into the following formations in ascending order [7]: the Takinoue Formation, the Kawabata Formation, the Karumai Formation, and the Nina Formation (Figure 4). They represent the sequence by which an elongate N-S foreland basin was filled. The middle Miocene Kawabata Formation comprises mainly turbidites and associated coarse clastic rocks derived from the eastern hinterland [7].

2.2. Sedimentary facies of the Miocene unit

This study conducted sedimentary facies analysis for the Kawabata Formation along the Rubeshibe River (Figure 2). The analysis revealed that the turbidites of the Kawabata Formation mainly consisted of sheet-flow turbidite facies association and channel-levee facies association (Figure 5). The sheet-flow turbidite facies association comprises aggradational stacking of rhythmic alternating beds of turbidite sandstone and mudstone with rare upward

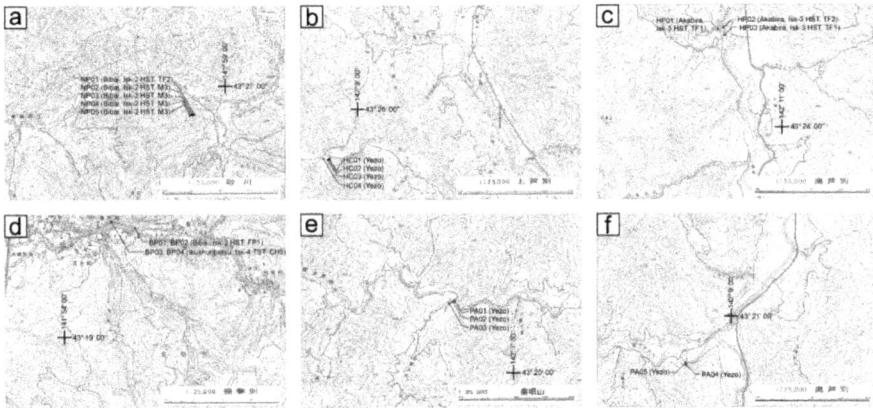

Figure 3. Sampling localities for rock magnetic analyses of the Cretaceous and Paleogene strata. The base maps are parts of the "Sunagawa", "Kamiashibetsu", "Okuashibetsu", "Ikushunbetsu" and "Bibaiyama" 1:25,000 topographic maps published by the Geographical Survey Institute. As for the Paleogene sites (a, c and d), geologic units (Yezo, Yezo Supergroup; Bibai, Bibai Formation; Akabira, Akabira Formation; Ikushunbetsu, Ikushunbetsu Formation), depositional sequence and facies classification are shown in parentheses after Takano et al. [5]

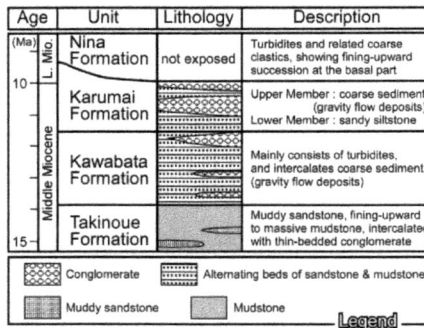

Figure 4. Neogene stratigraphy of the study area of the Kawabata Formation

thickening or thinning successions, and is interpreted to be sheet-like turbidites with minor occurrences of depositional lobes, which occupied major part of the trough-like foreland basin fill [7,10]. The channel-levee facies association is composed of thick amalgamated sandstone facies with slump blocks and thinly bedded alternating beds of sandstone and mudstone. These two facies appearing coupled is indicative of an elongated channel-levee system made of the main channel with levees on both sides. These two facies associations are believed to have been deposited in an elongated trough-like foredeep in the foreland basin [7]. The turbidites of the Kawabata Formation commonly contain sedimentary structures indicating paleocurrent directions; e.g., sole marks (mostly flute marks) at the bottom of individual turbidite bed, and current ripples in Bouma Tc division [11].

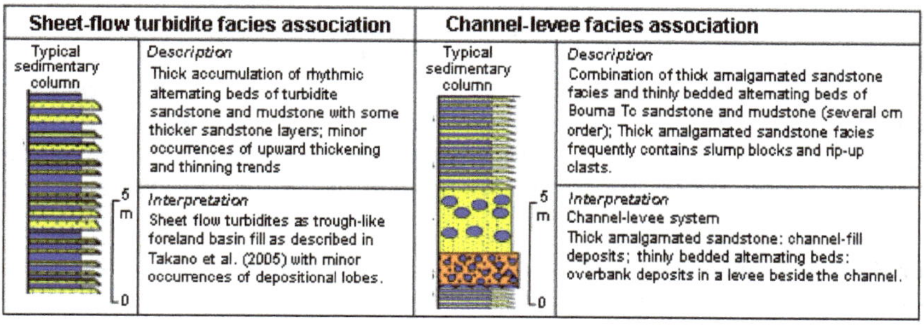

Sheet-flow turbidite facies association		Channel-levee facies association	
Typical sedimentary column	Description Thick accumulation of rhythmic alternating beds of turbidite sandstone and mudstone with some thicker sandstone layers; minor occurrences of upward thickening and thinning trends	Typical sedimentary column	Description Combination of thick amalgamated sandstone facies and thinly bedded alternating beds of Bouma Tc sandstone and mudstone (several cm order); Thick amalgamated sandstone facies frequently contains slump blocks and rip-up clasts.
	Interpretation Sheet flow turbidites as trough-like foreland basin fill as described in Takano et al. (2005) with minor occurrences of depositional lobes.		Interpretation Channel-levee system Thick amalgamated sandstone: channel-fill deposits; thinly bedded alternating beds: overbank deposits in a levee beside the channel.

Figure 5. Facies association classification of the Kawabata Formation along the Rubeshibe River

3. Rock magnetism

We obtained samples for rock magnetic analyses exclusively from fine-grained parts of the target sedimentary units, since fine sedimentary rocks generally preserve stable detrital remanent magnetization (DRM). Few visible markers of the sedimentation process accompany such sediments, so we attempted to measure their microscopic magnetic fabric, which may be related to paleocurrent directions (e.g., [12]).

3.1. Basic measurements

The Cretaceous and Eocene samples were taken from outcrops along the streambed in central Hokkaido (Figure 3) using an engine or electric drill at 21 sites. Samples of the Kawabata Formation were collected with a battery-powered electric drill at 21 sites along the Rubeshibe River (Figure 2). The bedding attitudes were measured on outcrops to allow us to compensate for tectonic tilting later. Between seven and sixteen independently oriented cores 25 mm in diameter were obtained at each site using a magnetic compass. Cylindrical specimens 22 mm in length were cut from each core and the natural remanent magnetization (NRM) of each specimen was measured using a cryogenic magnetometer (model 760-R SRM, 2-G Enterprises). Low-field magnetic susceptibility was measured on a Bartington MS2 susceptibility meter, and the anisotropy of magnetic susceptibility (AMS) was measured using an AGICO KappaBridge KLY-3 S magnetic susceptibility meter. After the basic measurements, pilot specimens with average NRM intensities, directions and susceptibility levels were selected from each site for subsequent demagnetization tests.

3.2. Demagnetization tests

In order to isolate stable components of the remanent magnetization, progressive alternating field demagnetization (PAFD) and progressive thermal demagnetization (PThD) tests were carried out on two pilot specimens per site that had average NRM directions. The PAFD test loading ranged from 0 to 80 mT using a three-axis tumbling system with specimens contained in a μ-metal envelope. The PThD test was performed using an electric furnace, with a residual magnetic field less than 10 nT, beginning at 100 °C and continuing until the specimen was either fully demagnetized and a characteristic remanent magnetization (ChRM) component was isolated, or until the thermal treatment provoked erratic behavior of the magnetic direction. Specimens' low-field bulk magnetic susceptibilities were measured using a susceptibility meter after each PThD step in order to monitor chemical changes in ferromagnetic minerals.

Figure 6 presents typical PThD and PAFD results for the Yezo Supergroup and Ishikari Group. It is obvious that the ChRM direction was not isolated because of unstable behavior in thermal treatment (Figure 6a), overlapping spectra of primary and secondary magnetization (Figure 6b) and partial remagnetization within a site (Figure 6c,d). Therefore further analyses for magnetic granulometry were not applied on the Cretaceous and Eocene samples. On the other hand, PThD treatment was effective for isolating stable ChRM in the sedimentary rocks of the Kawabata Formation. Figure 7 shows typical results of the progressive demagnetization tests.

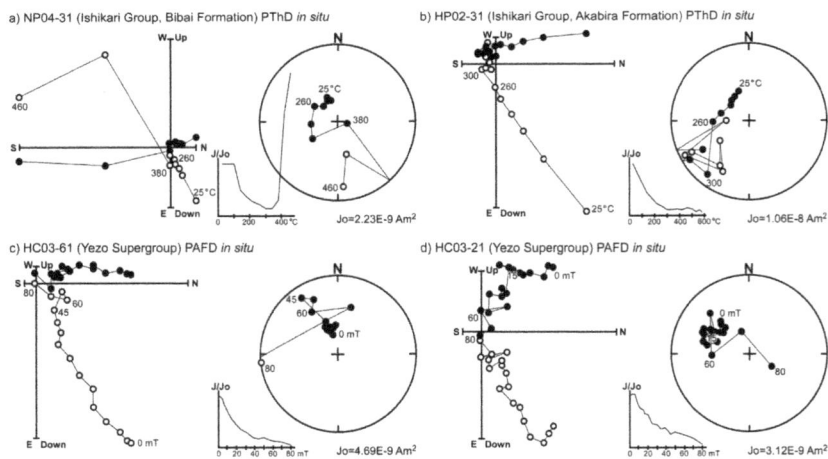

Figure 6. Typical results of progressive thermal demagnetization (PThD) and progressive alternating field demagnetization (PAFD) in geographic coordinates for the Paleogene Ishikari Group (a,b) and the Cretaceous Yezo Supergroup (c,d). On the vector-demagnetization diagrams, solid (open) circles are projection of vector end-points on horizontal (N-S vertical) plane. Equal-area projection and normalized intensity decay curve are shown on the right-side of each vector diagram. Solid (open) circles in equal-area nets are projections on the lower (upper) hemisphere. Numbers attached on data points are demagnetization levels in °C or mT

Rock Magnetic Properties of Sedimentary Rocks in Central Hokkaido — Insights into Sedimentary and
Tectonic Processes on an Active Margin

233

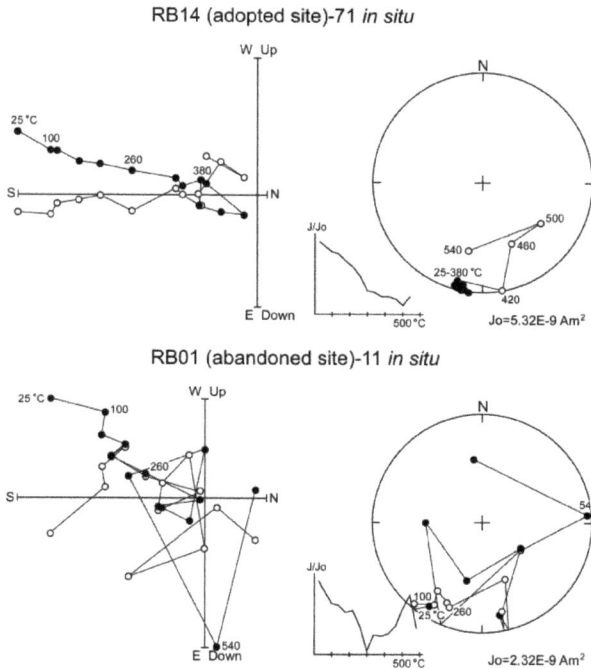

Figure 7. Results of progressive thermal demagnetization for samples of the Neogene Kawabata Formation with stable (upper) and unstable (lower) magnetization. All coordinates are geographic (*in situ*). Units are bulk remanent intensity. The solid and open circles in the vector-demagnetization diagrams (left) are projections of vector end-points on the horizontal and north-south vertical planes, respectively. The solid and open circles in the equal-area Schmidt nets (right) are projections on the lower and upper hemispheres, respectively

3.3. Hysteresis properties

Hysteresis parameters were determined for the Kawabata samples with an alternating gradient magnetometer (Princeton Measurements Corporation, MicroMag 2900). Ten sample chips up to 1 mm in size were randomly selected from site RB16, where stable ChRM has been successfully isolated. Figure 8 displays typical hysteresis of the Kawabata mudstones. The raw diagram seems to suggest the absence of ferromagnetic material. After correcting the linear gradient of paramagnetism, a weak ferromagnetic behavior signature can be recognized. Saturation magnetization (Js), saturation remanence (Jrs) and coercive force (Hc) values were determined for all samples from their hysteresis loops. Their relatively low Hc (~ 100 mT) implies that magnetite is the dominant remanence carrier. After acquiring coercivity of remanence (Hcr) values through backfield demagnetization experiments, we constructed a correlation plot of Jrs/Js versus Hcr/Hc [13] as shown in Figure 9. All the data are plotted in the pseudo-single domain (PSD) region of magnetite.

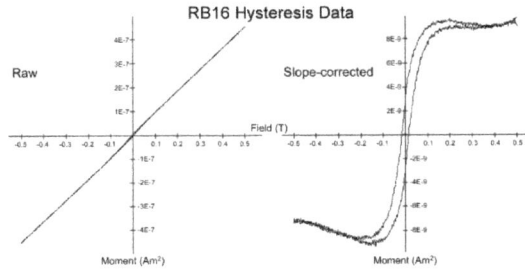

Figure 8. An example of hysteresis loop for a sample of the Kawabata Formation from site RB16 (Left: raw data, Right: data corrected for slope of paramagnetism)

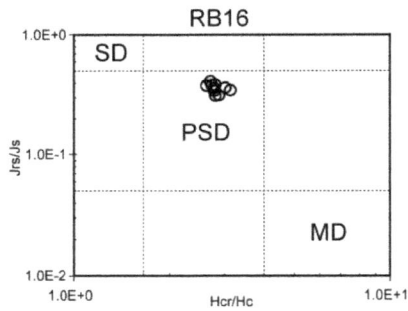

Figure 9. Logarithmic plot of hysteresis parameters [13] of ten samples of the Kawabata Formation from site RB16. Abbreviations: SD, single domain; PSD, pseudo-single domain; MD, multi-domain

4. Discussion

4.1. Rotational motions

We found stable magnetic components at three sites of the Kawabata Formation. Their directions were determined with a three-dimensional least squares analysis technique [14]. Figure 10 and Table 1 present site-mean ChRM directions obtained from the Kawabata Formation. They exhibit antipodal directions, and precision parameter (κ) improves after tilt correction. Although the number of data points is minimal for tectonic discussion, we can interpret the site-mean directions as a record of the Earth's dipole magnetic field, acquired before the strata tilted. The declination of the formation mean exhibits a significant westerly deflection, which suggests counterclockwise rotation of the study area.

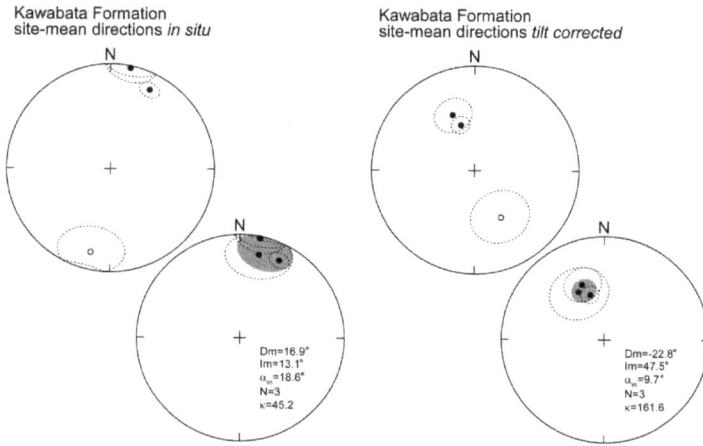

Figure 10. Site-mean ChRM directions of the Kawabata Formation in the study area. The solid and open circles in all the equal-area nets are projections on the lower and upper hemispheres, respectively. Dotted ovals show 95 % confidence limits. Lower diagrams are polarity-converted for calculating formation mean directions and Fisher's precision parameters as annotated in the diagrams (Shaded ovals depict 95 % confidence for the formation means)

Site	Latitude	Longitude	D	I	Dc	Ic	α₉₅	κ	N	φ	λ
RB14	42.7361	142.1771	-167.1	-18.7	151.2	-47.0	21.9	13.1	5	62.5	29.5
RB16	42.7379	142.1793	11.4	2.8	-21.8	42.5	14.0	14.4	9	64.5	13.9
RB17	42.7381	142.1793	26.8	17.4	-17.4	52.7	6.8	66.5	8	73.3	23.1

D and I, *in situ* site-mean declination and inclination before tilt correction in degrees, respectively; Dc and Ic, site-mean declination and inclination after tilt correction in degrees, respectively; α95, radius of 95% confidence circle in degrees; κ, precision parameter; N, number of specimens; φ and λ, latitude (N) and longitude (E) of north-seeking virtual geomagnetic pole for untilted site-mean direction in degrees, respectively.

Table 1. Paleomagnetic directions of the Kawabata Formation

A previous study [15] suggested a clockwise tectonic rotation around central Hokkaido based on a paleomagnetic study of the Kawabata Formation. Takeuchi et al. [16] proposed a coherent rotational model with 'domino-style' rigid crustal blocks. However, Tamaki et al. [17] criticized the block rotation scheme as being overly simplistic based on differential rotations inferred from Oligocene paleomagnetic data. They restored crustal deformation in central Hokkaido using dislocation modeling, and found complicated vertical-axis rotations around terminations of the faults that contributed to the formation of N-S elongate sedimentary basins. Figure 11 demonstrates differential rotation in central Hokkaido since the middle Miocene.

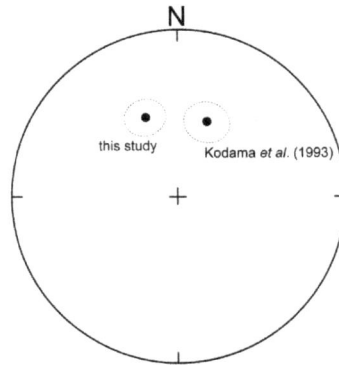

Figure 11. Comparison of the mean paleomagnetic directions of the Kawabata Formation in central Hokkaido between this study and [15]. Data are plotted on the lower hemisphere of the equal-area projection. Dotted ovals represent 95 % confidence limits

4.2. Sedimentation process inferred from AMS fabric

We found that the AMS fabric (orientation of principal axes) were precisely determined at all the sampled localities. Tables 2 and 3 show the AMS parameters for the Cretaceous/Eocene units and the Miocene unit, respectively. Figure 12 delineates typical AMS fabric obtained from the Ishikari (left) and Yezo (right) samples. After tilt-correction, the maximum (K_1) and intermediate (K_2) axes of AMS are bound to the horizontal plane with a subtle imbrication suggestive of hydrodynamic forcing.

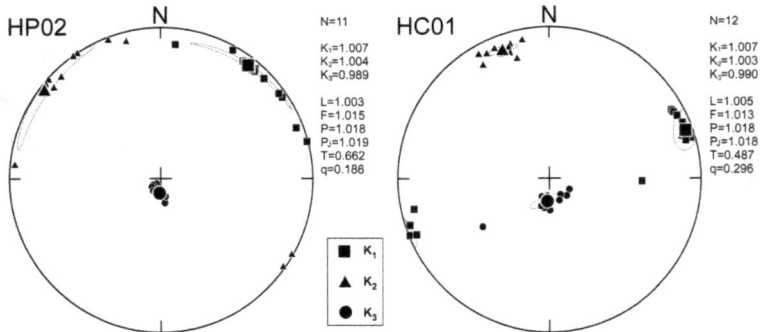

Figure 12. Anisotropy of magnetic susceptibility (AMS) fabric (principal susceptibility axes) for all specimens of typical sites of the Ishikari Group (HP02) and Yezo Supergroup (HC01) plotted on the lower hemisphere of equal-area projections. Data are shown in stratigraphic coordinates. Ovals surrounding mean directions of three axes (shown by larger symbols) are 95% confidence regions. See Table 2 for all the AMS parameters

Site	N	K_1 Str. (D, I)	K_2 Str. (D, I)	K_3 Str. (D, I)	L (K_1/K_2)	F (K_2/K_3)	P (K_1/K_3)	P_J	T	q	Unit / Sequence
Paleogene											
BP01	11	1.010	1.005	0.984	1.005	1.021	1.026	1.028	0.619	0.213	Bb / Isk-2HST
		(207, 4)	(297, 1)	(35, 86)							
BP02	11	1.009	1.008	0.982	1.001	1.026	1.027	1.031	0.917	0.043	Bb / Isk-2HST
		(226, 3)	(136, 4)	(0, 85)							
BP03	11	1.010	1.005	0.985	1.005	1.020	1.025	1.027	0.593	0.229	Ik / Isk-4TST
		(233, 9)	(141, 7)	(16, 78)							
BP04	13	1.010	1.002	0.988	1.008	1.014	1.022	1.022	0.259	0.458	Ik / Isk-4TST
		(225, 6)	(134, 3)	(15, 83)							
HP01	18	1.010	1.004	0.987	1.006	1.017	1.023	1.024	0.479	0.303	Ak / Isk-3HST
		(271, 4)	(180, 8)	(29, 81)							
HP02	11	1.007	1.004	0.989	1.003	1.015	1.018	1.019	0.662	0.186	Ak / Isk-3HST
		(38, 6)	(307, 4)	(187, 83)							
NP01	15	1.009	1.006	0.985	1.003	1.022	1.025	1.027	0.762	0.128	Bb / Isk-2HST
		(264, 2)	(174, 7)	(11, 83)							
NP02	9	1.013	1.006	0.981	1.006	1.026	1.032	1.034	0.597	0.227	Bb / Isk-2HST
		(252, 10)	(162, 2)	(59, 80)							
NP03	10	1.006	1.004	0.989	1.002	1.016	1.018	1.019	0.779	0.118	Bb / Isk-2HST
		(253, 1)	(163, 7)	(349, 83)							
NP04	7	1.009	1.006	0.985	1.002	1.022	1.024	1.027	0.791	0.111	Bb / Isk-2HST
		(80, 10)	(170, 4)	(282, 79)							
NP05	9	1.007	1.003	0.990	1.004	1.013	1.017	1.018	0.536	0.264	Bb / Isk-2HST
		(39, 28)	(141, 21)	(261, 54)							
Cretaceous											
HC01	12	1.007	1.003	0.990	1.005	1.013	1.018	1.018	0.487	0.296	
		(71, 5)	(340, 11)	(187, 78)							
HC02	10	1.011	1.005	0.984	1.006	1.021	1.027	1.028	0.541	0.262	
		(66, 2)	(156, 4)	(315, 86)							
HC03	11	1.010	1.002	0.988	1.008	1.014	1.022	1.022	0.274	0.447	
		(51, 4)	(321, 11)	(163, 78)							
PA01	13	1.006	1.003	0.991	1.003	1.012	1.015	1.016	0.650	0.193	
		(336, 11)	(70, 19)	(217, 68)							
PA04	14	1.017	1.014	0.969	1.004	1.046	1.049	1.055	0.852	0.078	
		(8, 9)	(99, 11)	(240, 76)							

N denotes the number of specimens. Directions of AMS principal axes are in stratigraphic coordinates. Abbreviations for the Paleogene geologic units: Ak, Akabira Formation; Bb, Bibai Formation; Ik, Ikushunbetsu Formation. Depositional sequence is after Takano & Waseda [4] and Takano et al. [5].

Table 2. Site-mean AMS parameters of the Paleogene and Cretaceous units in central Hokkaido

Figure 13 delineates typical AMS fabrics of the Kawabata Formation. Site RB08 typifies an elongate (prolate) fabric reflecting aligned detrital grains. Site RB14 has highly oblate fabric, as shown by a positive T parameter near unity. This fabric is essentially confined to the bedding plane under gravitational force. As the hysteresis study showed a negligible amount of ferromagnetic material in the Kawabata samples (Figure 8), we consider the AMS fabric as being governed simply by the shape anisotropy of paramagnetic minerals, i.e. alignments of elongate or platy grains such as amphibole or mica.

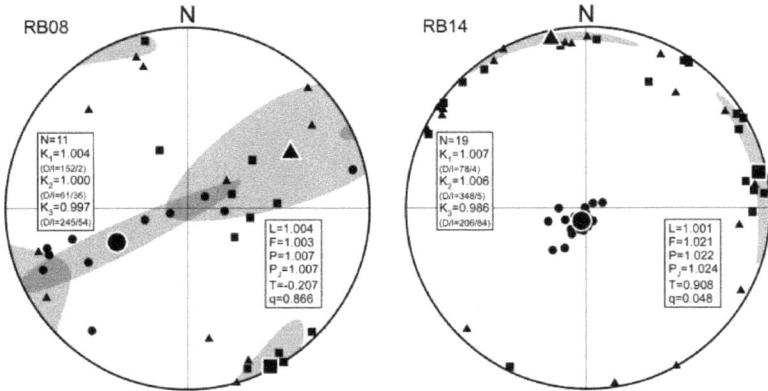

Figure 13. Typical tilt-corrected AMS fabric for the Kawabata Formation muddy samples. Prolate (left) and oblate (right) fabrics are numerically described by negative and positive T parameters, respectively, posted on the equal-area diagrams. All the data are plotted on the lower hemisphere. Square, triangular and circular symbols represent orthogonal maximum (K_1), intermediate (K_2), and minimum (K_3) AMS principal axes, respectively, and larger symbols show their mean directions. Shaded areas are 95 % confidence limits based upon Bingham statistics

Site	N	K_1 (D, I)	K_2 (D, I)	K_3 (D, I)	L (K_1/K_2)	F (K_2/K_3)	P (K_1/K_3)	P_J	T	q
RB01	11	1.013	1.006	0.981	1.007	1.025	1.033	1.034	0.546	0.260
		(16, 6)	(107, 7)	(242, 81)						
RB02	11	1.007	1.003	0.991	1.004	1.012	1.016	1.017	0.508	0.282
		(3, 20)	(98, 11)	(215, 66)						
RB03	12	1.006	1.002	0.992	1.004	1.010	1.014	1.015	0.475	0.304
		(326, 11)	(58, 11)	(190, 74)						
RB04	13	1.008	1.002	0.990	1.005	1.013	1.018	1.019	0.406	0.352
		(159, 2)	(249, 6)	(52, 83)						
RB05	15	1.005	1.001	0.993	1.004	1.008	1.012	1.012	0.330	0.404
		(4, 19)	(101, 19)	(232, 62)						

Rock Magnetic Properties of Sedimentary Rocks in Central Hokkaido — Insights into Sedimentary and Tectonic Processes on an Active Margin

239

Site	N	K_1 (D, I)	K_2 (D, I)	K_3 (D, I)	L (K_1/K_2)	F (K_2/K_3)	P (K_1/K_3)	P_J	T	q
RB06	14	1.007	1.001	0.992	1.005	1.009	1.014	1.015	0.256	0.459
		(336, 5)	(66, 7)	(208, 82)						
RB07	12	1.005	0.999	0.996	1.007	1.002	1.009	1.009	-0.459	1.151
		(135, 24)	(41, 9)	(291, 64)						
RB08	11	1.004	1.000	0.997	1.004	1.003	1.007	1.007	-0.207	0.866
		(152, 2)	(61, 36)	(245, 54)						
RB09	17	1.005	1.001	0.994	1.004	1.007	1.011	1.011	0.313	0.416
		(121, 1)	(31, 7)	(218, 83)						
RB10	10	1.010	1.007	0.983	1.003	1.024	1.027	1.030	0.777	0.120
		(343, 2)	(252, 12)	(84, 78)						
RB11	12	1.002	0.999	0.998	1.003	1.001	1.004	1.004	-0.410	1.090
		(313, 42)	(105, 45)	(210, 15)						
RB12	12	1.005	1.004	0.991	1.002	1.013	1.015	1.016	0.763	0.127
		(3, 6)	(93, 1)	(192, 84)						
RB13	9	1.004	1.002	0.994	1.001	1.009	1.010	1.011	0.732	0.144
		(119, 26)	(214, 11)	(325, 61)						
RB14	19	1.007	1.006	0.986	1.001	1.021	1.022	1.024	0.908	0.048
		(78, 4)	(348, 5)	(206, 84)						
RB15	12	1.009	1.004	0.986	1.005	1.018	1.023	1.024	0.558	0.251
		(188, 10)	(96, 10)	(324, 76)						
RB16	15	1.013	1.010	0.977	1.003	1.034	1.036	1.041	0.848	0.080
		(281, 4)	(11, 5)	(152, 83)						
RB17	14	1.009	1.003	0.987	1.006	1.016	1.022	1.023	0.455	0.318
		(292, 6)	(201, 6)	(66, 81)						
RB18	7	1.009	1.001	0.990	1.008	1.011	1.018	1.018	0.169	0.528
		(120, 1)	(30, 2)	(234, 87)						
RB19	13	1.012	1.003	0.985	1.009	1.018	1.028	1.028	0.324	0.411
		(26, 15)	(277, 50)	(127, 36)						
RB20	11	1.013	1.005	0.982	1.008	1.023	1.031	1.032	0.458	0.317
		(300, 19)	(205, 15)	(78, 66)						
RB21	17	1.015	1.005	0.980	1.010	1.026	1.036	1.037	0.433	0.335
		(215, 75)	(90, 9)	(358, 12)						

N is the number of specimens. Directions of AMS principal axes are in stratigraphic coordinates.

Table 3. Site-mean AMS parameters of the Kawabata Formation

Sedimentological context of the AMS fabric is demonstrated in Figures 14 and 15. Paleocurrent directions inferred from the Eocene AMS data tend to align in N-S azimuth (Figure 14), and accord with development process of the forearc basin [4]. Takano and Waseda [4] demonstrated that the Eocene paleo-Ishikari basin experienced differential subsidence during deposition. Such deformation may be related to longstanding strike-slip faulting around central Hokkaido [17], and tectono- / sedimentological context of the AMS fabric will be better evaluated in the light of quantitative study of basin-forming processes described in this book. For reliable interpretation of AMS data, it is necessary to assess properties of ferromagnetic minerals, such as composition, grain size and contribution to bulk magnetic susceptibility, as shown in this paper.

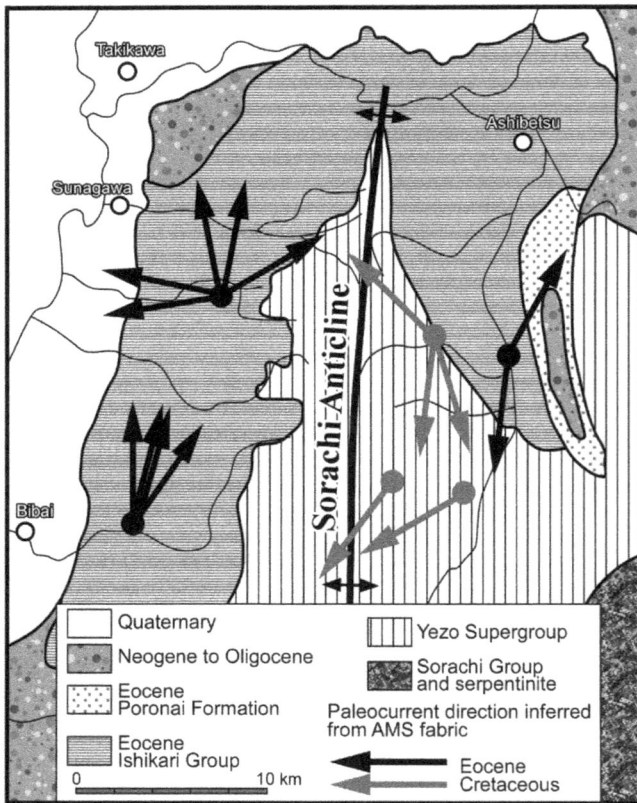

Figure 14. Paleocurrent directions inferred from AMS fabric of the Paleogene and Cretaceous samples. Geologic map is compiled from Editorial Committee of Hokkaido, Regional Geology of Japan [1] and Takano and Waseda [4]

Our field survey revealed indicators of paleocurrent directions in the Kawabata Formation along the Rubeshibe River as depicted in Figure 15. After correction for the counterclockwise rotation identified in our paleomagnetic study, most of the markers indicate a westward current direction with minor southward flow contributions. This is consistent with a tectono-sedimentary model of rapid burial of the Miocene N-S foreland basin by clastics derived from the eastern collision front presented in such research as Kawakami et al. [7]. Notably, the imbrication of the oblate AMS fabric matches visible sedimentary structures. Although the transport direction of muddy detrital material spilled out of a levee is not necessarily parallel to the turbidity current within a channel, AMS data can serve to indicate paleocurrents after the contributors to the magnetic fabric have been identified. Also note that K_1 of prolate samples (with negative T parameters) tend to align perpendicular to the paleocurrent direction, implying that elongate grains roll on the sediment surface.

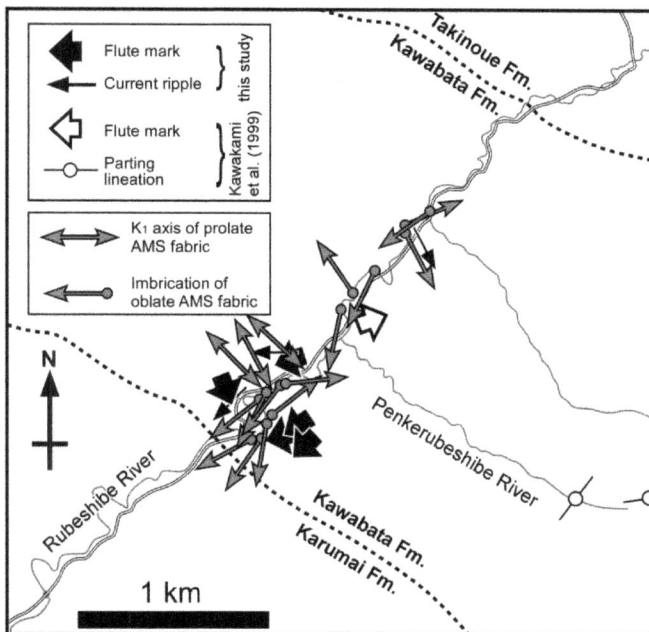

Figure 15. Paleocurrent map of the Kawabata Formation around the Rubeshibe River route. Formation boundaries are after Kawakami et al. [7]

Figure 16 delineates groups of microscopic fabrics identified in the Kawabata Formation as a function of the AMS shape parameter (T). The intensity of alignment forcing inferred from AMS data is closely related to sedimentary facies (shown on the right in the figure) determined by field observation. For example, weak hydrodynamic forcing corresponds to fine rhythmically alternating facies in channel-levee systems. Thus, the sedimentological context of muddy sediments' AMS fabric can be interpreted in the light of sandy sediments' facies analysis.

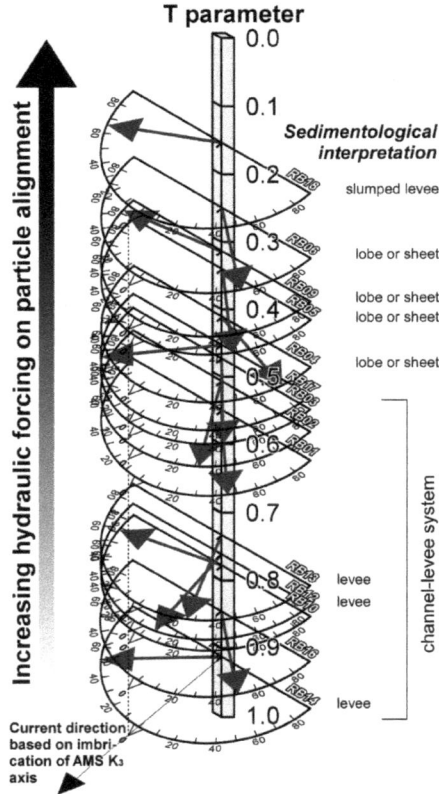

Figure 16. AMS paleocurrent indicators of the Kawabata Formation. Directions of K_1 (gray arrows) are shown as acute angles from the dotted baseline of K_3 axis imbrication. Vertical positions of the data are based on the T parameter. Samples with negative T values are excluded from the diagram because such cases have a large scatter in the K_3 directions

Azimuths of AMS maxima in natural sediments vary significantly, reflecting the size or shape of magnetic grains and changes in current velocities (e.g., [18]). Figure 16 presents the relationship between paleocurrent proxies estimated from the imbrication of the AMS minimum axis (K_3) and the K_1 trend. Tarling and Hrouda [19] stated that the angle between K_3 and K_1 changes as a function of current velocity and the slope of the sedimentary surface. Our result suggests that the orientation between those AMS sedimentary indicators can vary, regardless of the level of hydraulic forcing, based on the shape parameter (T), which implies development of a preferred orientation. Although the AMS fabric is a diagnostic tool for patterns of sediment transportation, laboratory-based experiments that analyze natural sediments under conditions where a few of the prevailing factors are controlled, are essential to allow firm sedimentological interpretation of formation processes.

4.3. Re-deposition experiment and the origin of AMS

In order to consider the origin of the AMS in the Kawabata samples, we organized a re-deposition experiment. A silty sandstone (SP1C-1) and a mudstone (SP2F-1) samples were crushed and sieved into coarse, medium and fine fractions. The fine fraction (< 63 μm) was then separated into magnetic and non-magnetic fractions with an isodynamic separator. The 'magnetic' fraction actually contained no ferromagnetic opaque minerals such as magnetite, but had abundant biotite and common hornblende. It also contained garnet, probably derived from metamorphic rocks exposed around the hinterlands during the rapid deposition of the Miocene turbidite.

A suspension of the fine fraction was poured into a vertically settled plastic tube 1 m in length and 2.5 cm in diameter, filled with water. This deposit of artificial sediment was dehydrated at room temperature. After being soaked in an adhesive resin, the samples were trimmed into standard-sized specimens for rock-magnetic measurements. The AMS was measured with an AGICO KappaBridge KLY-3 S magnetic susceptibility meter. The AMS parameters for the artificial samples are summarized in Table 4.

Figure 17 presents the magnitudes of magnetic fabrics in natural sedimentary rocks and the re-deposited sediments of the Kawabata Formation. Obviously, the magnetic separation results in remarkable decrease of both the bulk susceptibility and the degree of anisotropy (P_J). It is also noteworthy that the shape parameter (T) of the artificial sediments is almost null, suggesting a neutral magnetic fabric. The directions of the principal AMS axes (see Table 4) are not bound to the horizontal plane or to geomagnetic north. Thus, the detrital particles, free from paramagnetic minerals having shape anisotropy, like platy biotite, are deposited without any gravitational or geomagnetic forcing, creating an isotropic sediment.

Sample	N	K_1 (D, I)	K_2 (D, I)	K_3 (D, I)	L (K_1/K_2)	F (K_2/K_3)	P (K_1/K_3)	P_J	T	q
SP1C-1	1	1.0009	1.0000	0.9992	1.001	1.001	1.002	1.002	-0.080	0.740
		(167, 75)	(265, 2)	(356, 15)						
SP2F-1	1	1.0014	1.0002	0.9984	1.001	1.002	1.003	1.003	0.180	0.517
		(250, 27)	(343, 6)	(84, 62)						

N is the number of specimens. Directions of principal axes of AMS are shown in *in situ* coordinates.

Table 4. AMS parameters of re-deposited non-magnetic fine fraction of the Kawabata Formation

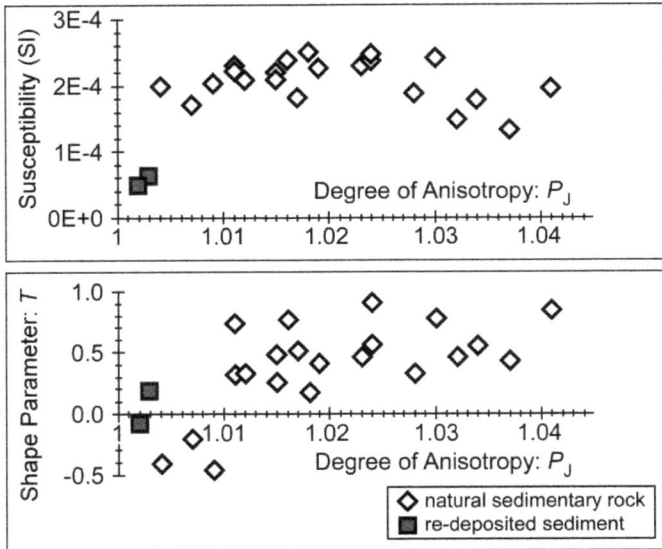

Figure 17. Magnitudes of magnetic fabrics in natural samples and re-deposited non-magnetic fine particles of the Kawabata Formation

5. Summary

Rock-magnetic investigation of sedimentary rocks provides insights into the basin's formation and sedimentation processes on an active margin. Cretaceous (Yezo Supergroup) ~ Eocene (Ishikari Group) strata and middle Miocene (Kawabata Formation) turbidites in central Hokkaido represent forearc and foreland settings, respectively. Progressive demagnetization successfully isolated characteristic remanent magnetization (ChRM) of the Kawabata Formation. Mean declination of the formation's ChRM exhibited significant westerly deflection, suggesting counterclockwise rotation of the study area since the middle Miocene. This differs from previous reports that indicated clockwise rotation. We attribute the difference to complicated deformation around the terminations of faults that form the N-S elongate Kawabata sedimentary basin. Anisotropy of magnetic susceptibility (AMS) principal axes were clearly determined for both the Cretaceous/Paleogene samples and Neogene samples, and regarded as a proxy of sediment influx directions. Paleocurrent directions inferred from the Eocene AMS data tend to align in N-S azimuth (Figure 14), and accord with the results of sedimentological paleoenvironment reconstruction, which suggest a northward downstream trend in fluvial to tidal estuarine systems [4]. As for the Cretaceous, further acquisition of AMS data is necessary to assess the effect of intensive syn-depositional deformation of the forearc [20]. After correcting for the tectonic rotation, most of the paleocurrent markers in the Kawabata Formation indicated a westward current direction with minor southward flow contributions, consistent with a sedimentary model that envisions burial of the Miocene N-S foreland basin by clastics derived from the eastern collision front. The intensity of alignment forcing of sedimentary particles inferred from the shape parameter (T) of the AMS data was closely related to sedimentary facies observed in the field. In investigating the origin of the AMS fabrics of turbidite deposits of the Kawabata Formation, we conducted a re-deposition experiment of fine detrital particles with no magnetic fraction including paramagnetic minerals with relatively high magnetic susceptibility, which demonstrated the significance of the alignment of paramagnetic minerals having shape anisotropy.

Acknowledgements

The authors are grateful to N. Ishikawa for the use of the rock-magnetic laboratory at Kyoto University and for thoughtful suggestions in the course of the magnetic analyses. We thank S. Oshimbe and S. Nishizaki for their help with field work. Thanks are also due to N. Yamashita and Y. Danhara for their support in mineral separation. Constructive review comments by G. Kawakami greatly helped to improve early version of the manuscript.

Author details

Yasuto Itoh[1*], Machiko Tamaki[2] and Osamu Takano[3]

*Address all correspondence to: itoh@p.s.osakafu-u.ac.jp

1 Graduate School of Science, Osaka Prefecture University, Osaka, Japan

2 Japan Oil Engineering Co. Ltd., Tokyo, Japan

3 JAPEX Research Center, Japan Petroleum Exploration Co. Ltd., Chiba, Japan

References

[1] Editorial Committee of Hokkaido, Regional Geology of Japan. Regional Geology of Japan, Part 1: Hokkaido. Tokyo: Kyoritsu Shuppan; 1990.

[2] Tamaki M, Itoh Y. Tectonic implications of paleomagnetic data from upper Cretaceous sediments in the Oyubari area, central Hokkaido, Japan. Island Arc 2008; 17: 270-284.

[3]   Tamaki M, Oshimbe S, Itoh Y. A large latitudinal displacement of a part of Creta-
      ceous forearc basin in Hokkaido, Japan: paleomagnetism of the Yezo Supergroup in
      the Urakawa area. Journal of Geological Society of Japan 2008; 114: 207-217.

[4]   Takano O, Waseda A. Sequence stratigraphic architecture of a differentially subsi-
      ding bay to fluvial basin: the Eocene Ishikari Group, Ishikari Coal Field, Hokkaido,
      Japan. Sedimentary Geology 2003; 160: 131-158.

[5]   Takano O, Waseda A, Nishita H, Ichinoseki T, Yokoi K. Fluvial to bay-estuarine sys-
      tem and depositional sequences of the Eocene Ishikari Group, central Hokkaido.
      Journal of Sedimentological Society of Japan 1998; 47: 33-53.

[6]   Miyasaka S, Hoyanagi K, Watanabe Y, Matsui M. Late Cenozoic mountain-building
      history in central Hokkaido deduced from the composition of conglomerate. Mono-
      graph of the Association for the Geological Collaboration in Japan 1986; 31: 285-294.

[7]   Kawakami G, Yoshida K, Usuki T. Preliminary study for the Middle Miocene Kawa-
      bata Formation, Hobetsu district, central Hokkaido, Japan: special reference to the
      sedimentary system and the provenance. Journal of the Geological Society of Japan
      1999; 105: 673-686.

[8]   Otofuji Y, Kambara A, Matsuda T, Nohda S. Counterclockwise rotation of northeast
      Japan: paleomagnetic evidence for regional extent and timing of rotation. Earth and
      Planetary Science Letters 1994; 121: 503-518.

[9]   Itoh Y, Tsuru T. Evolution history of the Hidaka-oki (offshore Hidaka) basin in the
      southern central Hokkaido, as revealed by seismic interpretation, and related tecton-
      ic events in an adjacent collision zone. Physics of the Earth and Planetary Interiors
      2005; 153: 220-226.

[10] Takano O, Tateishi M, Endo M. Tectonic controls of a backarc trough-fill turbidite
      system; the Pliocene Tamugigawa Formation in the Niigata-Shin'etsu inverted rift
      basin, Northern Fossa Magna, central Japan. Sedimentary Geology 2005; 176:
      247-279.

[11] Bouma AH. Sedimentology of Some Flysch Deposits; A Graphic Approach to Facies
      Interpretation. Amsterdam: Elsevier; 1962.

[12] Kawamura K, Ikehara K, Kanamatsu T, Fujioka K. Paleocurrent analysis of turbidites
      in Parece Vela Basin using anisotropy of magnetic susceptibility. Journal of the Geo-
      logical Society of Japan 2002; 108: 207-218.

[13] Day R, Fuller M, Schmidt VA. Hysteresis properties of titanomagnetites: grain-size
      and compositional dependence. Physics of Earth and Planetary Interiors 1977; 13:
      260-267.

[14] Kirschvink JL. The least-squares line and plane and the analysis of palaeomagnetic
      data. Geophysical Journal of the Royal Astronomical Society 1980; 62: 699-718.

[15] Kodama K, Takeuchi T, Ozawa T. Clockwise tectonic rotation of Tertiary sedimentary basins in central Hokkaido, northern Japan. Geology 1993; 21: 431-434.

[16] Takeuchi T, Kodama K, Ozawa T. Paleomagnetic evidence for block rotations in central Hokkaido-south Sakhalin, Northeast Asia. Earth and Planetary Science Letters 1999; 169: 7-21.

[17] Tamaki M, Kusumoto S, Itoh Y. Formation and deformation processes of late Paleogene sedimentary basins in southern central Hokkaido, Japan; paleomagnetic and numerical modeling approach. Island Arc 2010; 19: 243-258.

[18] Ledbetter MT, Ellwood BB. Spatial and temporal changes in bottom-water velocity and direction from analysis of particle size and alignment in deep-sea sediment. Marine Geology 1980; 38: 245-261.

[19] Tarling DH, Hrouda F. The Magnetic Anisotropy of Rocks. London: Chapman & Hall; 1993.

[20] Tamaki M, Tsuchida K, Itoh Y. Geochemical modeling of sedimentary rocks in the central Hokkaido, Japan: Episodic deformation and subsequent confined basin-formation along the eastern Eurasian margin since the Cretaceous. Journal of Asian Earth Sciences 2009; 34: 198-208.

Chapter 11

Numerical Modeling of Sedimentary Basin Formation at the Termination of Lateral Faults in a Tectonic Region where Fault Propagation has Occurred

Shigekazu Kusumoto, Yasuto Itoh,
Osamu Takano and Machiko Tamaki

Additional information is available at the end of the chapter

1. Introduction

1.1. Pull-apart basin forming at the termination of lateral faults

When a sedimentary basin forms at the termination of a lateral faults, it is known as a pull-apart basin. It is well known that tectonic basins, such as pull-apart basins, are generally formed at the termination of right-lateral, right-stepping and left-lateral, left-stepping fault systems (e.g., [1]). This is mainly caused by the formation of subsidence at the fault termination by the lateral motion of the faults. Subsidence is therefore likely to be found piled up at the termination of right lateral right-stepping and left lateral left-stepping fault systems. In contrast, uplift structures are formed at the termination of right-lateral left-stepping and left-lateral right-stepping fault systems, because the terminations are located in an area where uplift is piled up, due to the lateral motion of the fault (Figure 1). Such structures are found in many places globally, and their fundamental formation mechanisms have been numerically simulated by numerous researchers (e.g., [2, 3]).

Katzman et al. [3] attempted to restore the subsurface structures of the Dead Sea, estimated from gravity anomalies (e.g., [4]), by means of Boundary Element Modeling (BEM), and indicated that it is necessary to assume long overlapping faults and a very high Poisson's ratio in order to restore the Dead Sea pull-apart basin. Rodgers [2] attempted to simulate the formation of a pull-apart basin by means of dislocation modeling (e.g., [5]), and this was probably the first study which discussed the formation of a pull-apart basin using numerical modeling.

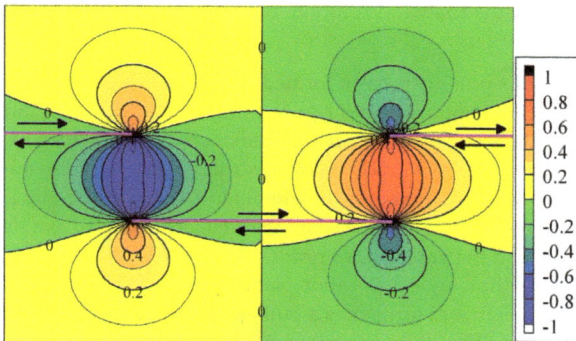

Figure 1. Schematic illustration of sedimentary basin and uplift at the terminations of lateral fault system. In this figure, right lateral motion was assumed to each vertical fault (pink line). Vertical displacements are normalized by the maximum value of absolute values of the total vertical displacement field, and they do not have a unit.

1.2. Dislocation modeling

In general, dislocation modeling is used for the quantitative interpretation of crustal deformation caused by earthquakes and/or volcanic activity (e.g., [6, 7]). Surface or interior displacements or strains can be calculated by considering dislocations on a plane embedded in an elastic isotropic half-space (e.g., [8, 9]).

Studies on dislocation theory and its applications began with Steketee [10] and were refined by Okada [9], who derived closed analytical solutions for surface and internal deformations or strains due to shear and tensile faults, with arbitrary dip angles in a half-space. During this period, many researchers applied dislocation theory to inhomogeneous media and considered the effects of viscoelasticity and poroelasticity (e.g., [11-18]). Such basic theories and their applications to earth science have been described in many textbooks (e.g., [19-21]). In addition, the fundamental idea of dislocation theory has been applied to theory for the interpreting the gravity and potential changes due to large earthquakes (e.g., [22-24]). In particular, Okubo's formula [23] was applied to an interpretation of gravity changes due to the 2011 Tohoku-Oki earthquake, as observed by the Gravity Recovery And Climate Experiment (GRACE) [25], and Wang et al. [26] were successful in discussing co- and post-seismic deformation independently of GPS data.

As mentioned above, many useful solutions have been derived, although all solutions cannot be referred to as being "closed analytical solutions." Closed analytical solutions (derived by [8, 9]) for a dislocation plane embedded in an elastic isotropic half-space are often employed for the quantitative interpretation of crustal movements, because they are very simple and are successful in explaining observational data. Consequently, some useful simulation software have been developed using Okada's formula [9], (for example Coulomb (e.g., [27, 28])) and/or stress evaluation methods (for example ΔCFF (e.g., [29])), and the method has been applied to the interpretation of crustal movements and/or the evaluation of stress changes due to earthquakes (e.g., [30-32]).

In this study, we employ the analytical solutions of Okada [8] for the numerical simulation of sedimentary basin formation.

For a typical example of a pull-apart basin formed by dislocation, it would be appropriate to use a restoration model of Beppu Bay and Figure 1 for a numerical simulation. Beppu Bay is located at the junction between the Honshu and Ryukyu arcs, and it is formed at the western end of the Median Tectonic Line (MTL), which is the largest right-lateral tectonic line in southwestern Japan (c.f., [33]). Another right-lateral fault, the Kurume-Hiji Line (KHL: [34]), is located 30 km north of the MTL. The MTL and the KHL are arranged as a right-lateral right-stepping fault system, and the potential for producing a pull-apart basin exists. Kusumoto et al. [35] therefore applied the simple dislocation plane concept of Okada [8] to this fault system and showed that Beppu Bay was formed as a pull-apart basin by the right-lateral faulting of the MTL and KHL. In addition, they suggested that two major tectonic events (namely, the formation of half-graben caused by a north–south extension, and the formation of the pull-apart basin caused by east–west compression) occurred in this region.

In Figure 1 and in [35], a single motion of the fault was assumed on the dislocation planes. However, it is well known that active faults undergo multiple movements over geological time scales. In order to reflect this effect in the simulation, as the Okada's solution [8] is based on linear elasticity, we attempted to introduce the history of fault activity into the numerical model by superimposing analytical solutions for different fault parameters on a single fault. For example, if the geological evidence suggests that the active zone of a lateral fault shifted along its strike direction over time, we express its tectonic history by superimposing analytical solutions for different fault lengths (Figure 2).

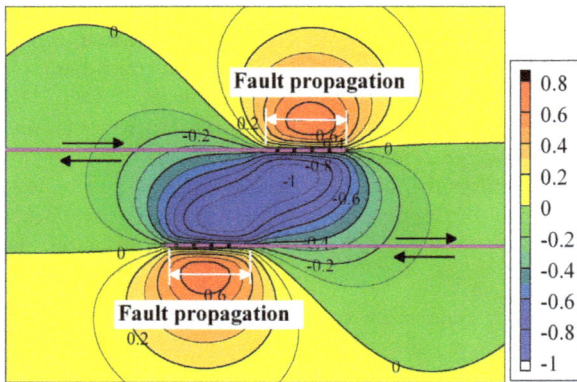

Figure 2. Illustration reflecting fault propagations in the dislocation modeling. Analytical solutions for different fault parameters are superimposed onto a single fault plane. Vertical displacements are normalized by the maximum value of absolute values of the total vertical displacement field, and they do not have a unit.

Itoh et al. [36] introduced this simulation technique while attempting to interpret topography and paleomagnetic data for the Takayama basin, central Japan. They showed that the basic structure of the Takayama basin, and changes in the declination of the thermoremanent magnetization, can be restored by considering the cumulative activity of two right-lateral faults (the Enako Fault and the Makigahora Fault) and a reverse fault (the Harayama Fault). In addition, they showed that the reverse fault can be divided into two segments that moved independently with a time lag between their active periods. Because this modeling technique can take into account the type and amount of fault motion, we can reflect the history of fault activity based on certain geological evidence, including paleomagnetic studies, in the numerical modeling, and can discuss the tectonics in detail.

This modeling technique has also been applied to the formation of pull-apart basins located in Hokkaido by Itoh et al. [37] and Tamaki et al. [38]. Tamaki et al. [38] found that a strike-slip fault motion reaching 30 km, is required to restore the distribution and volume of the Minami-Naganuma Basin located in southern central Hokkaido.

Here, we describe the disadvantages of the dislocation modeling defined using elasticity, and the solutions for the dislocation plane are given as a range of the linear elasticity. Since fracture, flow and other non-linear phenomena seen in the general solid material are not considered in the model, it is difficult to directly compare the amount of displacement between the modeled structure and the actual structure in long-time scale modeling. In general, dislocation modeling (including visco-elasticity effects) is often employed in discussions on crustal movements over a long time-scale such as on a geological time scale (e.g., [18, 39, 40]). In addition, Finite Element Modeling (FEM), Finite Difference Modeling (FDM) and Discrete Element Modeling (DEM) can simulate the formation processes of sedimentary basins or the building processes of mountains over a geological time-scale and can quantitatively discuss the mechanisms of their formation on a time axis (e.g., [41-43]).

As mentioned above, dislocation modeling defined using a range of the linear elasticity has disadvantages in the ability to directly compare the amount of displacement between the modeled structures and the actual structures. However, when we simply discuss the essential aspects of tectonics from the distribution pattern of structures caused by fault motions, dislocation modeling is a very useful tool because it provides the pattern of displacement using easy calculations.

1.3. Aims of this study

The aims of this study are to simplify the complex formation processes of sedimentary basins through numerical simulations and to show that the simplification enables us to estimate deductively which processes cause materialization. Central Hokkaido was selected as the field in which to achieve these aims, as many sedimentary basins are distributed in this area.

As will be described later in the paper, many sedimentary basins were formed from 48 to 12 Ma in central Hokkaido, and it is difficult to discuss their formation processes using only observational data because of their complex distribution both at and below the surface. In order to simplify the complex formation processes of sedimentary basins, we attempted to restore sedimentary basins using the advanced technique already mentioned, and we evaluated the fault type and the amount of movement required to form these sedimentary basins.

In the following sections, we describe the basic background and gravity anomaly in central Hokkaido, and we attempt the restoration of the sedimentary basins.

2. Basic background of central Hokkaido

2.1. Geophysical background

Hokkaido is located on the North American plate, at a junction of the Northeast Japan arc and the Kurile arc (Figure 3). Using recent GPS observations, an east-west compressive strain field has been observed in the northern part of Hokkaido, and this strain field is considered to be caused by the convergence of the Eurasia plate with the northern part of Hokkaido functioning as a part of the plate boundary (e.g., [44]).

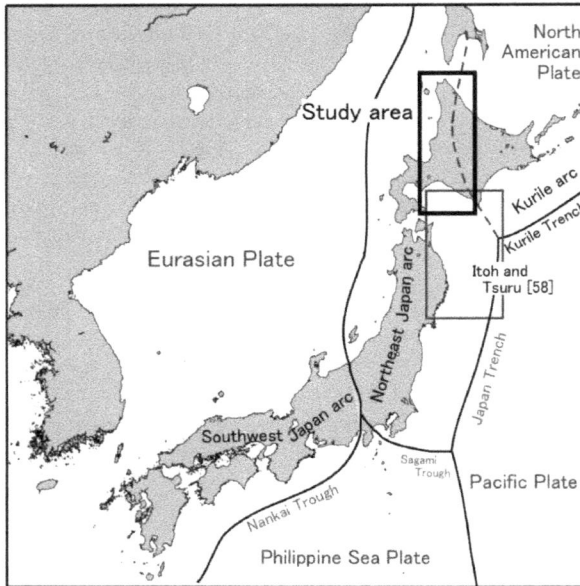

Figure 3. Location map of our study area. Hokkaido is located on the North American plate, at a junction of the Northeastern Japan arc and the Kurile arc. Gray dashed line indicates the old plate boundary between the Eurasian and North American Plates. Rectangular area by gray thin line indicates the study area of Itoh and Tsuru [58].

Although the present plate boundary between the North American plate and the Eurasian plate exists in the Sea of Japan, it is known that the plate boundary was located in central Hokkaido at around 13 Ma. This period of time corresponds to the stage when the uplifting of the Hidaka Mountains began (e.g., [45]). This tectonic framework is controlled by the dextral oblique collision between the Eurasian and North American Plates and the oblique subduction of the Pacific Plate beneath the Kurile Trench. It is considered that the Kurile arc migrated into the southwestward as a forearc sliver by the oblique subduction of the Pacific Plate and that the Hidaka Mountains would be formed by collision of the Kurile arc and the Northeast Japan arc (e.g., [46-48]).

Since, consequently, it is an important area for understanding characteristics and mechanism of collision zone, numerous geophysical surveys (e.g., seismic prospecting, gravity surveys, electromagnetic surveys) have been carried out around the Hidaka Mountains in order to obtain information regarding the subsurface structures and to apply such knowledge to a tectonic discussion of the Mountains and Hokkaido (e.g., [49-54]). Using these surveys, subsurface structures which indicate a collision between the Northeast Japan arc and the Kurile arc have been obtained, and have contributed to tectonic discussions. However, in contrast, geophysical studies in the sedimentary basin area in central Hokkaido are limited in number.

2.2. Geological and tectonic background

The geological characteristics in Hokkaido are that Cenozoic strata consist of island-arc-trench systems of the Northeast Japan arc and Kurile arc, and that each Cenozoic strata distributed in the Northeast Japan arc and the Kurile arc appear in the western half and eastern half area of Hokkado, respectively. This characteristic also appears in Neogene and Quaternary strata, volcanoes and their products, and topography (e.g., [41]).

Figure 4 shows the distribution of the Paleogene strata in a north-south direction in central Hokkaido. This distribution traces the old plate boundary. This N-S elongation area is included in the Ishikari-Teshio Belt that is underlain by the Cretaceous Yezo Group, and is regarded as a typical sequence in a forearc basin setting [55]. It is known that the Paleogene sedimentary strata were deposited during the early Eocene and Oligocene in almost the entire region (e.g., [56]).

In the study area, sedimentary basins and sedimentary layers were formed during 48–12 Ma and have been complexly distributed. They have been divided into 5 stages according to their formation: The Ishikari stage (48–40 Ma), The Horonai stage (40–32 Ma), The Minami Naganuma stage (34–20 Ma), and The Kawabata stage (15–12 Ma). The Ishikari stage is divided into early (48–45 Ma) and late (45–40 Ma) stages. The shape of the sedimentary basin and the distribution of sedimentary layers are shown in Figure 5.

It is known that the Ishikari Group differentially subsided and was then divided into several components [57]. Based on detailed sedimentological studies, Takano and Waseda [57] also points out that the rate of subsidence accelerated during deposition of the Ishikari Group.

The Ishikari stage is the sedimentation stage of the Ishikari Group, corresponding to the Eocene and is divided into early and later stages according to the sedimentation style. Sediments in the early Ishikari stage are distributed shallowly and widely (Figure 5A). In this stage, the sedimentary basins "A", "B" and "C" were formed (Figure 5A), and from well data their depths are estimated to be 600 m, 500 m and 1000 m, respectively. Sediments in the later Ishikari stage are distributed deeply and narrowly (Figure 5B). In this stage, the sedimentary basins named "A" and "C" were formed (Figure 5B), and from well data their depths are estimated to be

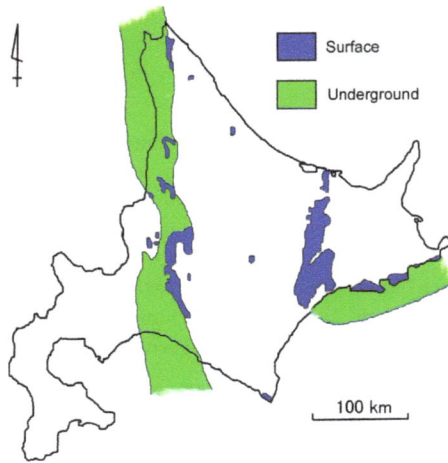

Figure 4. Distribution of the Paleogene strata. Green and blue areas indicate distribution areas of the Paleogene sedimentary layer under and on the surface, respectively. (After Kurita and Hoyanagi [56])

2800 m and 400 m. Itoh and Tsuru [58] identified a NNW-SSE trending deformation zone bounded by large transcurrent faults including T1 and T2 later describing from seismic reflecting data in the northern part of the Northeast Japan forearc (Figure 3) and their right lateral motions have been indicated by the clockwise rotation of Paleogene marine sediments and by paleogeographic reconstruction. Since, as already mentioned, the present study area (the western half of Hokkaido) has same Cenozoic strata distributed in the Northeast Japan arc, study area of Itoh and Tsuru [58] and our study area are geologically continuous in the Paleogene time. Consequently, it is expected that a right lateral motion of the crust was dominant.

The Horonai stage is the sedimentation stage of the Horonai Formation, the Tappu Group, the Sankebetsu Formation and the lower Magaribuchi Formation. This stage corresponds to the Eocene and the early Oligocene. In this stage, sedimentary basins "A", "B", "C", "D", "E" and "F" were formed (Figure 5C), and from well data their depths are estimated to be 3500 m, 1200 m, 1200 m, 600 m, 300 m, and 1500 m, respectively. It is expected that a right lateral motion was dominant in this stage, because right lateral motion was also dominant in the Eocene and the late Oligocene (see below).

The Minami-Naganuma stage is the sedimentation stage of the upper Magaribuchi Formation, the Minami-Naganuma Formation, the Horomui Formation, and the upper Sankebetsu Formation. This stage corresponds to the late Oligocene and the early Miocene. In this stage, sedimentary basins "B", "D", "E" and "F" were formed (Figure 5D), and from well data their depths (B, E and F) are estimated to be 2000 m, 300 m and 1500 m, respectively. The maximum depth of basin "D" is unknown because of lack of the well data and/or of outcrop section of the whole Minami-Naganuma Fromation. Itoh et al. [37], Tamaki et al. [38] and Itoh and Tsuru

[59] pointed out that all basins formed in the Ishikari-Teshio Belt in this stage are pull-apart basins. Tamaki et al. [38] showed that using dislocation modeling, a 30 km right-lateral strike-slip is required to restore the actual distribution and volume of the basin. Kurita and Yokoi [60] also stated that lateral faulting was dominant in forming some of the tectonic structures during the late Oligocene.

The Kawabata stage is the sedimentation stage of the Kawabata Formation, the Ukekoi Formation, the Fureoi Formation, the Kotanbetsu Formation and the Masuporo Formation. During the Neogene, Japan was affected by the opening event of the back-arc basin of the Sea of Japan. In this stage, sedimentary basins "A", "B", "D", "E", "F1" and "F2" were formed (Figure 5E), and from well data their depths are estimated to be 2000 m, 4000 m, 4000 m, 3500 m, 2000 m and 2000 m, respectively. As mentioned above, a lateral motion of the crust was dominant during the early Neogene [37, 38, 59]. Although a building of the Hidaka Mountains in around 13 Ma has been pointed out (e.g., [45]), details are unknown.

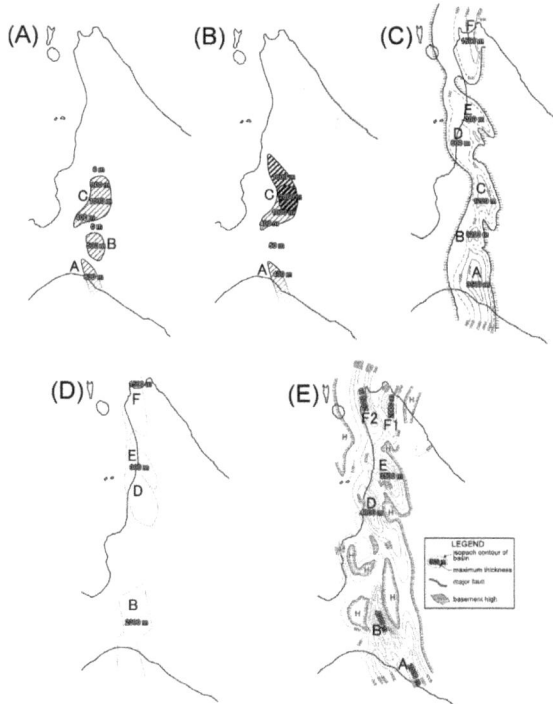

Figure 5. Shapes of sedimentary basins in (A) early Ishikari stage (48-45 Ma), (B) late Ishikari stage (45-40 Ma), (C) Horonai stage (40–32 Ma), (D) Minami-Naganuma stage (34–20 Ma) and (E) Kawabata stage (15–12 Ma). Isopach maps of the Horonai stage and the Kawabata stage are after Association of Natural Gas Mining and Association for Offshore Petroleum Exploration [75].

3. Bouguer gravity anomaly

Numerous geological and geophysical surveys have been carried out in the Hokkaido area, and each survey has played an important role in the understanding of crustal characteristics and tectonic events in the area. In particular, seismic prospecting has proved very useful in obtaining information relating to subsurface structures. However, seismic prospecting is almost two-dimensional, and it is difficult to intuitively understand the subsurface structures as three dimensional structures, even when provided with data from more than one profile. In contrast, the characteristics of gravity anomaly maps are easy to interpret and can be used to roughly estimate three dimensional subsurface structures from the data. Figure 6 shows the Bouguer gravity anomaly map of the study area. This map is based on the gravity mesh data by Komazawa [61]. The Bouguer density of 2670 kg/m^3 was employed.

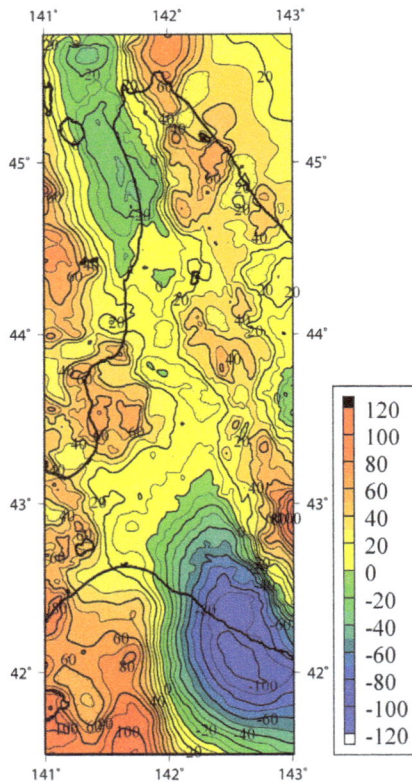

Figure 6. Bouguer gravity anomaly map. This map is based on the gravity mesh data by Komazawa [61], and the Bouguer density of 2670 kg/m^3 was assumed. Contour interval is 10 mGal.

There are negative gravity anomalies in the southern and northern parts of central Hokkaido. The negative gravity anomaly in the northern part reaches -20 mGal (Figure 6 and Area I in Figure 7) and the southern negative gravity anomaly is less than -100 mGal (Figure 6 and Area III in Figure 7). The southern negative gravity anomaly located at the curved subduction zone is the lowest in the country. From seismic prospecting, it is known that this negative gravity anomaly consists of a very thick sedimentary layer (5–8 km), with a velocity of 2.5–4.8 km/s (e.g., [50]). The sedimentary layer was formed by imbrications associated with the collision process of the Northeast Japan arc and the Kurile arc (e.g., [46-48]). In contrast, there is a positive gravity anomaly in the area of the mountains, and the mountain elevations are roughly less than 2000 m. It would not be necessary to consider isostasy for the mountains, because the mountain elevations are not very high and the gravity anomaly in this area is positive.

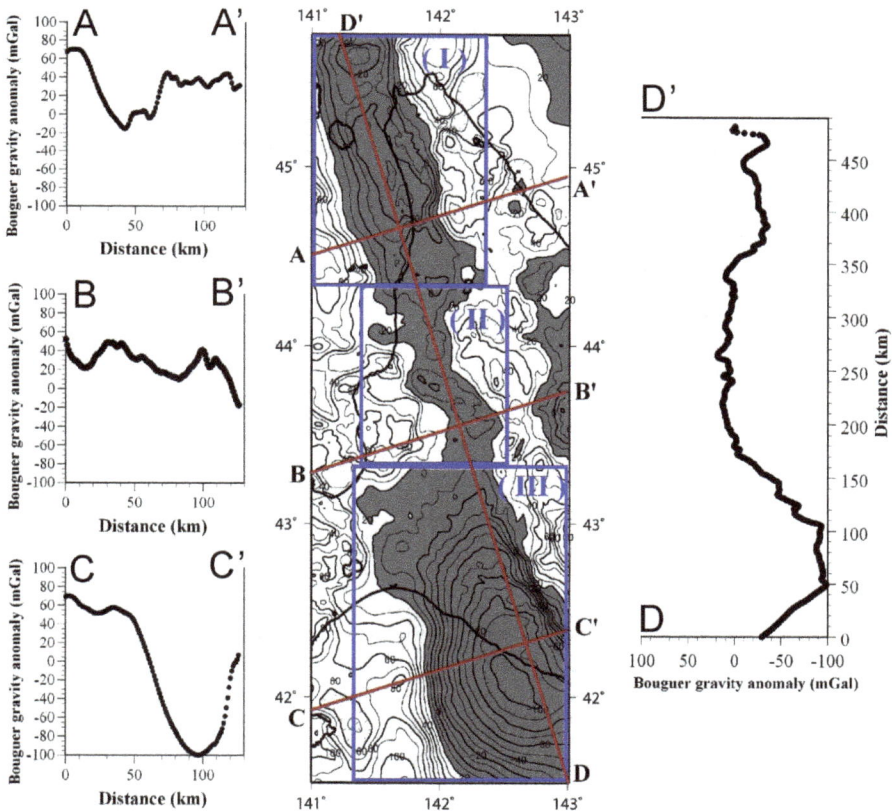

Figure 7. Bouguer gravity anomaly map. Gray indicates gravity low area less than 20mGal. A-A', B-B', C-C' and D-D' show gravity anomalies along each profile of four red lines shown in the Bouguer gravity anomaly map.

A flat gravity anomaly of less than 20 mGal is distributed like a belt between the northern and southern gravity anomalies (Figure 6 and Area II in Figure 7.). Figure 7 is the gravity anomaly map that the area less than 20 mGal was painted by gray. This painted area corresponds to the area where the Paleogene strata distribute under the surface (Figure 4). We show four cross section profiles (three E-W profiles, A-C, and one N-S profile, D) of the Bouguer gravity anomaly in Figure 7. From these profiles, the gravity anomalies in the region are shown to have the characteristics as follows:

1.   Gravity anomalies in the west-east direction have a regional trend which tilts toward the east. This could indicate the regional gravity field in Hokkaido.

2.   Gravity anomalies less than 20 mGal have a steep gradient on the east side, while those on the west side vary gently. These patterns of gravity anomalies indicate a depression structure called a "half-graben". Since there are many confirmed lateral faults and reverse faults in this region and no normal faults, it is considered that these patterns of gravity anomalies are caused by structures formed by the activities of lateral faults and/or reverse fault.

3.   Gravity anomalies in the north-south direction are relatively high and flat at the center of Hokkaido. It is possible that the high density of metamorphic belts near this region affect the observed gravity anomalies. Another cause to be considered could be the effect of subsurface structures such as a reduction of low density materials (e.g., a thin sedimentary layer) or an increase of high density material (e.g., uplift of the mantle).

In general, gravity anomalies are caused by spatial variations of subsurface structures, and indicate a deficiency or an excess of mass under the surface. In general, high gravity indicates the existence of a mass excess or of high density materials, and low gravity indicates the existence of a mass deficiency or of low density materials. These deficiencies or excesses, of mass can be evaluated quantitatively using Gauss's theorem (e.g., [62]).

$$\Delta M = \frac{1}{2\pi G} \iint \Delta g(x,y) dxdy \tag{1}$$

Here, $g(x, y)$ is the gravity anomaly data given on xy mesh with a constant interval. G, π and ΔM are the universal gravitational constant, circular constant and deficiency or excess of mass, respectively. Equation (1) is described by an infinite integration and it is difficult to perform an infinite integration with actual field data. Consequently, we understand this as being an approximate calculation and perform a numerical integration within a finite area (S) as follows:

$$\Delta M = \frac{1}{2\pi G} \iint_S \Delta g(x,y) dxdy \tag{2}$$

We applied equation (2) to three areas, I, II and III, and we attempted to estimate the magnitude of mass deficiency for the formation of a gravity anomaly less than 20 mGal in each area. In the calculations, we employed the Gauss-Legendre numerical integral formula (e.g., [63]).

As a result, mass deficiencies of 4.7×10^3 Gton, 8.6×10^2 Gton, and 1.5×10^4 Gton were estimated in areas I, II, and III, respectively. There are large mass deficiencies in areas I and III, where the negative gravity anomalies observed are very large and a small mass deficiency in area II. In central Hokkaido, the amount of mass deficiency is different by about two digits in both the maximum and the minimum values.

The amount of mass deficiency can be transformed into the volume (V) of sediment by the following equation, under a condition assuming an appropriate density contrast ($\Delta\rho$).

$$V = \frac{\Delta M}{\Delta\rho} \tag{3}$$

As an example, when a density contrast of 300 kg/m^3 is assumed, volumes of sediment of 1.6×10^4 km^3, 2.9×10^3 km^3 and 5×10^4 km^3 are estimated in areas I, II and III, respectively.

As mentioned above, gravity anomaly indicates also spatial variations of subsurface structures including the location of tectonic lines and/or faults. It is well known that if there is a tectonic line or a fault with a large gap in the vertical direction, the spatial distribution of the gravity anomaly varies steeply around these structures. The variation rate of the spatial distribution of the gravity anomaly is called the "horizontal gradient of gravity anomaly", and it is given by the first derivative (e.g., [64, 65]) or the second derivative (e.g., [4, 66]). In general, the first derivative of the gravity anomaly is more practical, because the calculation used is very simple and the geophysical and geological interpretations for the calculated results are straightforward.

We employed the first derivative of the gravity anomaly defined by the following equation (4), and calculated the horizontal gradient of the gravity anomaly (Figure 8):

$$\sqrt{\left[\frac{\partial g(x,y)}{\partial x}\right]^2 + \left[\frac{\partial g(x,y)}{\partial y}\right]^2} \tag{4}$$

Figure 8 shows the distribution of the horizontal gradient of the Bouguer gravity anomaly more than 2 mGal/km. The contour interval is 1 mGal/km. Although there are no continuous horizontal gradient anomalies within the area where the gravity anomaly is less than 20 mGal, the continuous horizontal gradient anomalies appear around this area. This may indicate that there are not tectonic lines including faults having large vertical deformation within this gravity low area less than 20 mGal and/or that gravity anomalies due to these tectonic lines are hidden by thick sediments, although faults with large vertical deformations actually exist.

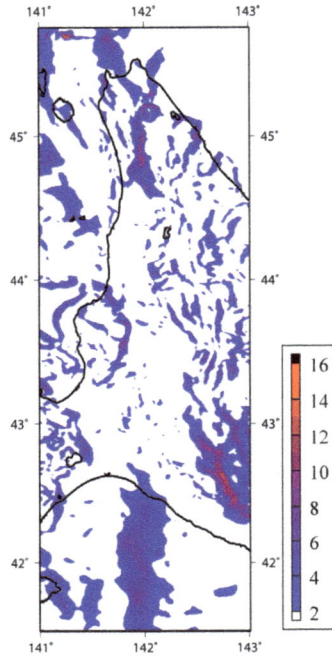

Figure 8. Distribution of the horizontal gradient of the Bouguer gravity anomalies more than 2 mGal/km. Contour interval is 1 mGal/km.

4. Restoration of sedimentary basins

In general dislocation modeling, the dislocation plane is assumed in the modeled crust by referring to the distribution of existing active faults and/or tectonic lines, and the surface deformations are calculated by assigning displacements on the plane. If the area for modeling is small, or if the tectonics and faults assumed for modeling are clear, such a modeling procedure is useful and practical (e.g., [35, 36, 38, 67]).

However, when details of the tectonics and/or the moved faults are not so clear (as in our study), the faults and their displacements for modeling are assumed experientially from characteristic distributions of target structures, by referring to more regional rough tectonics and fault distributions. The faults and their displacements (appropriately assumed) can then be considered as an initial model and can then be corrected by trial and error, so that the calculated results fit to the actual structures or their distribution pattern.

There are numerous small faults in central Hokkaido. As mentioned above, the details of tectonics and faulting in this area are unclear. It would, therefore, be impossible to attempt to model each fault for restoring the distribution of the sedimentary basins by trial and error. Consequently, in this study, we assumed that the dislocation plane used for the modeling was not a fault plane, but a typical or average plane of a fault zone. After trial and error, we defined the nine fault zones as shown in Table 1 and Figure 9, and employed them for numerical simulations. Each fault included in these fault zones is listed in Table 1 (with literature).

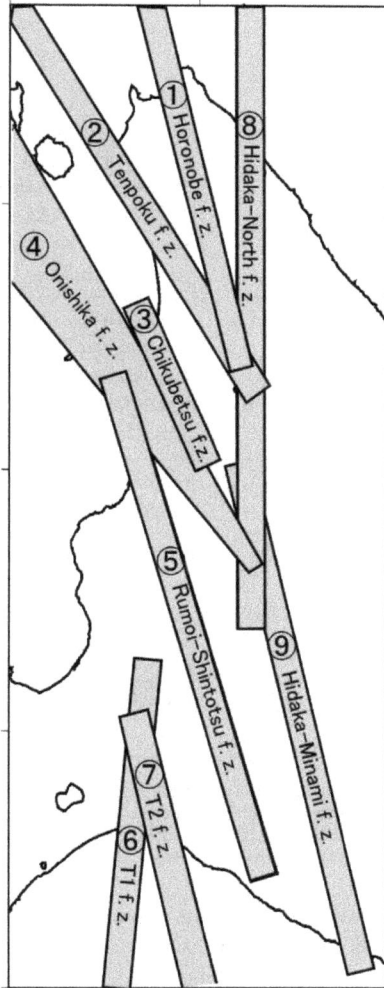

Figure 9. Fault zones defined in this study. Each fault included in these fault zones is listed in Table 1 with literature.

No.	Fault zone	Preference (recognition as tectonic zone)	Specific faults [reference]
1	Horonobe	Research Group for Active Faults [73]	Horonobe Fault [76] Higashino Fault [77, 78]
2	Tenpoku	Ikeda et al. [74]; Itoh et al. [37]	Enbetsugawa Tectonic Line [79]
3	Chikubetsu	this study	Shosanbetsu Fault [78] Chikubetsu Anticlinal Fault [78]
4	Onishika	this study	Onishika Fault [80] Tenguyama Thrust Fault [81] Okinaiyama Fault [82] Shimokinebetsuzawagawa Fault [82] Shisenzawa Fault [82] Poroshiriyama-Higashi Fault [82] Horoshin Fault [82]
5	Rumoi-Shintotsu	Association of Natural Gas Mining and Association for Offshore Petroleum Exploration [75]	Shozuzawa Fault [83] Horonuka Fault [83] Chashinai Fault [84] Sorachi Fault [85]
6	T1	Itoh and Tsuru [58, 59]	N/A (Most of the fault trace is in offshore area and buried with sediments.)
7	T2	Itoh and Tsuru [58, 59]	N/A (Most of the fault trace is in offshore area and buried with sediments.)
8	Hidaka-North [a]	this study	See footnote.
9	Hidaka-South [b]	this study	See footnote.

Table 1. Fault zones. Definition of each fault zone, and relationship fault zone and specific faults. (a) Hidaka-North Fault Zone is collectively defined as the eastern margin of N-S serpentinite zone along the longitudinal mountainous range in northern Hokkaido. (b) Hidaka-South Fault Zone is assigned to the western margin of the area of the highest recent uplift rate in Hokkaido (i.e. Hidaka Mountains), of which tectonic and structural context is still controversial.

In the following subsections, we give the results of numerical simulations, accompanied by simple explanations. The faults moved during each stage and their fault parameters are listed in Table 2. In calculations, the total movement of each fault plane was expressed by fault motions of 1000 times, and a high Poisson's ratio of 0.4 was assumed because of its application to modeling in the geological time scale (e.g., [3, 36-38, 68]).

No.	Fault zone (fault type)	Early Ishikari	Late Ishikari	Horonai	Minami Naganuma	Early Kawabata	Late Kawabata
1	Horonobe (Right lateral motion)	-	-	(88, 500)	(94, 500)	(36, 61)	-
2	Tenpoku (Right lateral motion)	-	-	(28, 24)	(14, 500)	(41, 32)	-
3	Chikubetsu (Right lateral motion)	-	-	-	-	(21, 36)	-
4	Onishika (Right lateral motion)	-	(48, 150)	(61, 63)	-	-	-
5	Rumoi-Shintotsu (Right lateral motion)	(22, 47)	-	(61, 94)	(48, 78)	(44, 73)	-
6	T1 (Right lateral motion)	(14, 500)	(14, 500)	(14, 500)	-	-	-
7	T2 (Right lateral motion)	(25, 500)	(25, 500)	(61, 500)	(44, 500)	(92, 500)	-
8	Hidaka-North (Right lateral motion)	(36, 500)	(36, 500)	(29, 500)	-	-	-
	Hidaka-North (Reverse motion)	-	-	-	-	-	(2.5, 500)
9	Hidaka-South (Reverse motion)	-	-	-	-	-	(2.5, 212)

Table 2. Fault zones moved in each stage. Values in (A, B) shown in Table indicate total slip (A: km) amount assumed on the fault plane and fault initial length (B: km). Width, depth and dip angle of fault moved as right lateral fault are assumed to be 15km, 15km and $\pi/2$, respectively. Width, depth and dip angle of fault moved as reverse fault are assumed to be 17.32km, 15km and $\pi/3$, respectively.

4.1. Early Ishikari stage (48–45 Ma)

We attempted to restore the sedimentary basins formed in the early Ishikari stage, named "A", "B" and "C" (Figure 5A). Results are shown in Figure 10A. Right lateral movements of fault zones were required, (including the T1 fault, T2 fault, the Rumoi-ShinTotsu tectonic line and the Hidaka-North fault), in order to restore these three sedimentary basins. The amount of movement of each fault was determined by trial and errors, and results are shown in Table 2 and are as follows:

1. T1 fault zone: 14 km

2. T2 fault zone: 25 km

3. Rumoi-ShinTotsu tectonic zone: 22 km

4. Hidaka-North fault zone: 36 km

From these amounts of displacement on each fault plane, it is estimated that during this tectonic stage, a horizontal movement reaching around 100 km occurred.

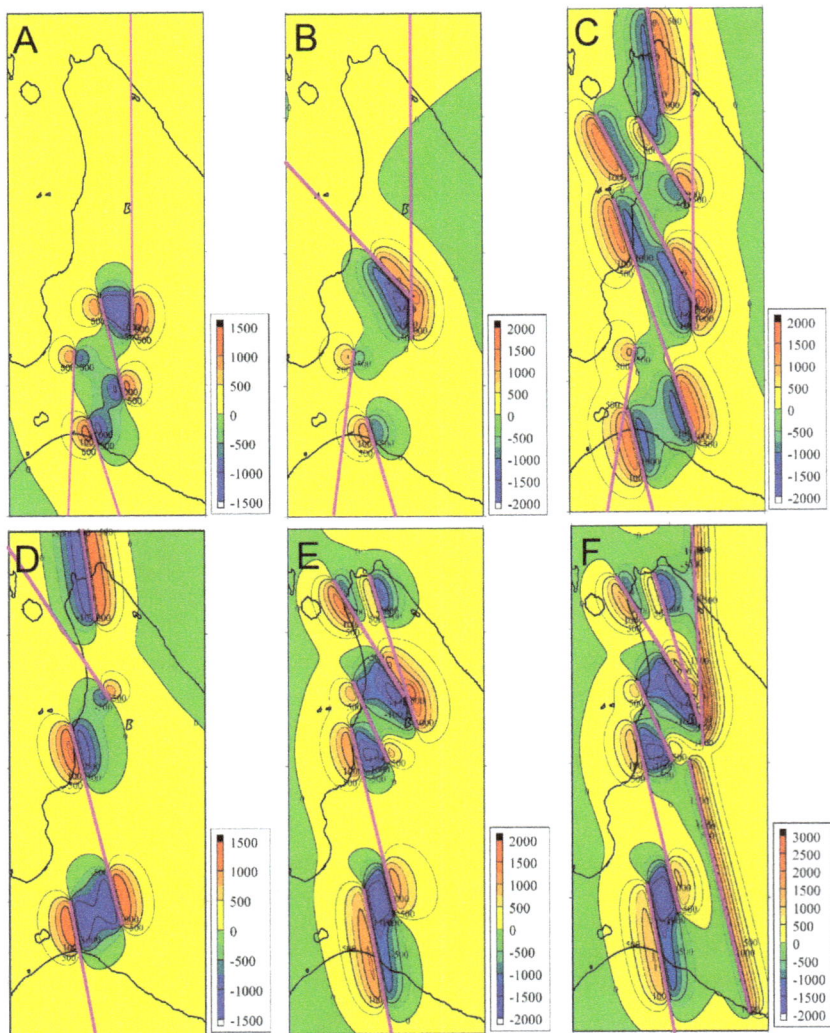

Figure 10. Distribution pattern of sedimentary basins in each stage restored by the dislocation modeling. A: early Ishikari stage. B: late Ishikari stage (45-40 Ma). C: Horonai stage (40–32 Ma). D: Minami-Naganuma stage (34–20 Ma). E: early Kawabata stage (15–13 Ma). F: late Kawabata stage (13-12Ma). Vertical displacement amounts are given in m.

From geological observations it is known that sedimentary basin "C" is the biggest basin and reaches to a depth of 1000 m. In our modeling, the depth of the modeled basin reached 1400 m. Sedimentary basins, "A" and "B" reach about 600 m and 500 m depths, respectively, and both depths of restored basins (modeled basins) corresponding to these basins are 1000m as a result of dislocation modeling. Differences between the actual basin depth and the modeled basin depth, namely "the modeled basin depth–actual basin depth," were +400 m, +400 m and +500 m for basins "C", "A" and "B", respectively. The amount of restored subsidence may be a little large.

Here, we assigned the right lateral motion to each fault as mentioned above, in order to restore the spatial patterns of basin distribution. If the amount of lateral motion of each fault is reduced to adjust to the depth component of each basin, it is not possible to restore the spatial distribution patterns of the basins.

4.2. Late Ishikari stage (45–40 Ma)

In the late Ishikari stage, sedimentary basins "A" and "C" (Figure 5B) were restored (Figure 10B). The right lateral movements of fault zones (including the T1 fault, T2 fault, Onishika fault and the Hidaka-North fault) were required in order to restore these two sedimentary basins. The amount of movement of each fault was determined by trial and error (see Table 2) and is as follows.

1. T1 fault zone: 14 km

2. T2 fault zone: 25 km

3. Onishika fault zone: 48 km

4. Hidaka-North fault zone: 36 km

From these amounts of displacement on each fault plane, it is estimated that during this tectonic stage, a horizontal movement reaching about 123 km occurred.

From geological observations, sedimentary basin "C" is known to be the largest basin and reaches a depth of 2800 m. The depth of the model basin in our modeling reached 1800 m. Sedimentary basin "A" reaches a depth of about 400 m, and the modeled basin corresponding to this had a depth of 500 m. The differences between the actual basin depth and the modeled basin depth were -1000 m and +100 m in basins "C" and "A", respectively. The restored subsidence amount of basin "C" was smaller than the actual basin depth. If the lateral motions of the Hidaka-North fault zone and Onishika fault zone were increased to adjust to the depth component of basin "C", it was not possible to restore the spatial distribution patterns of the basins.

4.3. Horonai stage (40 Ma–32 Ma)

In the Horonai stage, we attempted to restore six sedimentary basins, "A", "B", "C", "D", "E" and "F" (Figure 5C). Results are shown in Figure 10C. The right lateral movements of fault zones (including the T1 fault, T2 fault, the Rumoi-ShinTotsu tectonic line, Onishika fault, the Tenpoku fault, Horonobe fault and Hidaka-North fault) were required in order to restore these six sedimentary basins. The amount of movement of each fault was determined by trial and error and is shown in Table 2 and as follows:

1. T1 fault zone: 14 km

2. T2 fault zone: 61 km

3. Rumoi-ShinTotsu tectonic line zone: 61 km

4. Onishika fault zone: 61 km

5. Tenpoku fault zone: 28 km

6. Horonobe fault zone: 88 km

7. Hidaka-North fault zone: 29 km

From the amount of displacement of each fault plane, it is estimated that a horizontal movement reaching about 342 km occurred during this tectonic stage.

From geological observations, it is known that depths of the sedimentary basins "A", "B", "C", "D", "E" and "F" reach to 3500 m, 1200 m, 1200 m, 600 m, 300 m and 1500 m, respectively. In our model, depths of modeled sedimentary basins "A", "B," "C", "D", "E" and "F" reached 1000 m, 600 m, 1800 m, 1100 m, 900 m and 1000 m, respectively. The differences between the actual basin depth and the modeled basin depth were -2500 m, -600 m, +600 m, +500 m, +600 m, and -500 m in basin "A", "B", "C", "D", "E" and "F", respectively.

4.4. Minami-Naganuma stage (34–20 Ma)

We attempted to restore sedimentary basins "B", "D", "E" and "F" in the Minami-Naganuma stage (Figure 5D), and the results are shown in Figure 10D. The right lateral movements of fault zones, (including the T2 fault, the Rumoi-ShinTotsu tectonic line, Tenpoku fault and the Horonobe fault), were required in order to restore these four sedimentary basins.

In this stage, Tamaki et al. [38] have restored already the Minami-Naganuma basin (corresponding to basin "B": pull-apart basin) located in south central Hokkaido. We referred to their results and determined the amount of movement of each fault by trial and error. The amount of fault movement is shown in Table 2 and as follows:

1. T2 fault zone: 44 km

2. Rumoi-ShinTotsu tectonic zone: 48 km

3. Tenpoku fault zone: 14 km

4. Horonobe fault zone: 94 km

From the amounts of displacement on each fault plane, it is estimated that horizontal movement reaching about 200 km occurred during this tectonic stage.

From geological observations it is known that the depth of sedimentary basins "B", "E" and "F" reached 2000 m, 300 m and 1500 m, respectively. As already mentioned, the maximum depth of basin "D" is unknown. In our model, the depths of the modeled sedimentary basins "B", "D", "E" and "F" reached 1200 m, 1100 m, 700 m and 1200 m, respectively. The differences between the actual basin depth and the modeled basin depth were -800 m, +400 m, and -300 m for basins "B", "E" and "F", respectively.

4.5. Kawabata stage (15-12 Ma)

In the Kawabata stage, we attempted to restore six sedimentary basins, "A", "B", "D", "E", "F1" and "F2" (Figure 5E), and the result is shown in Figure 10E. The right lateral movements of fault zones, (including the T2 fault, Rumoi-ShinTotsu tectonic line, Onishika-chikubetsu fault, Tenpoku fault and the Horonobe fault), were required in order to restore these six sedimentary basins. The amount of movement of each fault is determined by trial and error and is shown in Table 2 and as follows:

1. T2 fault zone: 98 km

2. Rumoi-ShinTotsu tectonic line zone: 44 km

3. Onishika-chikubetsu fault zone: 21 km

4. Tenpoku fault zone: 41 km

5. Horonobe fault zone: 36 km

From the amount of displacement on each fault plane, it is estimated that a horizontal movement reaching about 240 km occurred during this tectonic stage.

As described above, these lateral motions could restore the basic distribution pattern of six sedimentary basins formed in the Kawabata stage. However, the whole distribution pattern of sedimentary basins in this stage could not be restored. After repeated trial and error, it was found that the reverse motion of two fault zones, (including the Hidaka-north fault and Hidaka-south fault), is necessary to restore the whole distribution pattern of the sedimentary basins in this stage (Figure 10F). In fact, such a reverse motion is required to successfully restore the whole basin distribution pattern. The amount of reverse motion required is shown in Table 2 and as follows:

1. Hidaka-north fault zone: 2.5 km

2. Hidaka-south fault zone: 2.5 km

Information on the depth of the sedimentary basins "A", "B", "D", "E", "F1" and "F2" is obtained from geological observations, and depths are found to reach 2000 m, 4000 m, 4000 m, 3500 m, 2000 m and 2000 m. In our model, the depths of modeled sedimentary basins "A", "B", "D", "E", "F1" and "F2" reached 1100 m, 1800 m, 1600 m, 1900 m, 1300 m and 800 m. The differences between the actual basin depth and the modeled basin depth were -900 m, -2200 m, -2400 m, -1600 m, -700 m, and -1200m in basins "A", "B", "D", "E", "F1" and "F2", respectively. Although the whole basin distribution pattern is good, the differences between the actual basin depth and the modeled basin depth are large in all the basins. If the lateral or reverse motion of each fault zone was increased to adjust the depth component of basins, it would not be possible to restore the spatial distribution patterns of the basins.

5. Discussion

As described in the previous sections within this paper, the distribution patterns of sedimentary basins restored by dislocation modeling are very similar to the actual distribution patterns of the sedimentary basins formed in each stage, although depth differences between the actual sedimentary basins and the restored sedimentary basins occurred because of the dislocation plane based on the linear elasticity. From these results, it is suggested that almost all the sedimentary basins in central Hokkaido can be explained as pull-apart basins, caused by right-lateral fault motions during the Paleogene. The results also show that sedimentary basins formed during the Kawabata stage were formed by a combination of right-lateral fault motions and reverse fault motions located at the western margin of the Hidaka Mountains.

Although we have simplified the distributions of each fault zone (Figure 9), it would be difficult for contiguous fault zones to move independently, simultaneously, as different faults, namely a reverse fault and a lateral fault, under their arrangement as shown in Figure 9. Consequently, we suggest that the Kawabata stage should be divided into two stages, namely the "early stage" and the "late stage", from the viewpoint of the stress field or fault motion. By considering the continuity of tectonics or the stress field, it is found that the lateral movements were made in the early Kawabata stage and that the reverse movements were made in the late Kawabata stage.

From a geological viewpoint, it has been illustrated that the building of the Hidaka Mountains was caused by reverse fault motions (e.g., [45]), and the timing of this event has been considered as being during the late Miocene or around 13 Ma (e.g., [45, 69]). This geological view supports our results and ideas, and our results also support the tectonics constructed based on geological data. Hence, we suggest that the boundary of the late Kawabata stage is around 13 Ma.

In Figures, we show the total vertical displacement field calculated from the vertical displacement field in each stage (Figures 10). The vertical displacement fields shown in Figures 11 are normalized by the maximum value of absolute values of the total vertical displacement field restored by dislocation modeling. Thus, the displacement fields shown in Figures 11 do not have a unit.

Figure 11A illustrates the normalized displacement field map that the negative displacement areas are shown in gray. This shows the distribution of the subsurface sedimentary basins restored in this study, and the distribution is seen to be similar to the distribution of actual buried sedimentary basins formed during the Paleogene (Figure 4). Figure 11B is the normalized total vertical displacement pattern restored in this study. From this figure, it is found the deepest sedimentary basin restored is located at the center of central Hokkaido. In actual depth distribution, the basin "C" is not the deepest basin. However, this sedimentary basin has a depth reaching 6000 m and is large and deep basin.

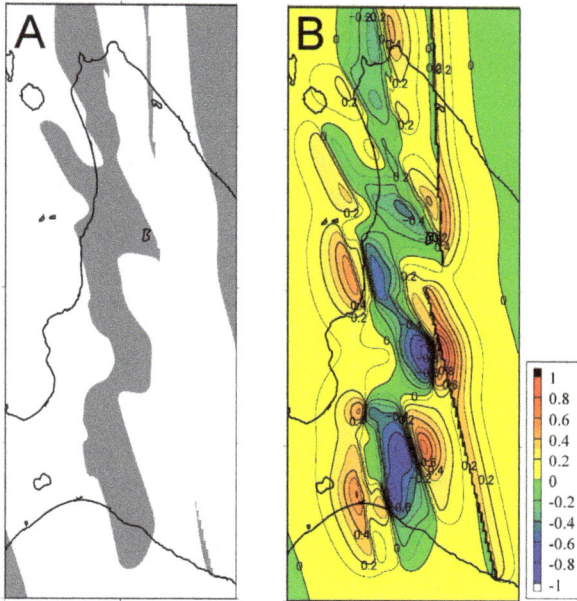

Figure 11. Total vertical displacement pattern. The displacement field was calculated by adding up the vertical displacements field restored in each stage by the dislocation modeling. A: Gray shows negative displacement areas, and indicates the distribution of the subsurface sedimentary basins restored in this study. B: Total vertical displacement pattern normalized by the maximum value of absolute values of the total vertical displacement field.

From results shown in Figures 10-11 and the discussion above, we conclude that the sedimentary basins formed from 48 to 12 Ma in central Hokkaido can be explained by the formation of pull-apart basins, due to right lateral motions (before 13 Ma), and by reverse motions of the Hidaka-North fault zone and the Hidaka-South fault zone after 13 Ma. Namely, although the right-lateral motions were predominant from 48 Ma to 13 Ma, the reverse motions were dominant in 13 Ma. This leads us to expect a significant change in the regional tectonic stress field during this stage. Although a change of the collision direction of the Northeast Japan arc and the Kurile arc could be cited as a possibility for this, a more accurate and quantitative future investigation is required.

We here reconsider the Bouguer gravity anomaly (Figure 6 and 7) and subsurface structures in viewpoint of tectonics mentioned above. As described in the section on the Bouguer gravity anomaly, gravity low area less than 20 mGal corresponds to the area where the Paleogene strata distribute under the surface. It is known that the southern negative gravity anomaly less than -100 mGal (area III in Figure 7) consists of a very thick sedimentary layer of 8 km, and the depth of the Moho discontinuity in this area is more than 30 km (e.g., [70, 71]), and that the negative gravity anomaly which reaches -20 mGal (area I in Figure 7) consists of a sedimentary layer of several kilometers thickness, and a Moho discontinuity of around 30 km in depth (e.g., [70, 71]). Consequently, we understand that the conspicuous gravity lows in areas I and III are caused by a thick sedimentary layer and a deep Moho. In contrast, the Bouguer gravity anomaly in area II is not negative, in spite of an area of subsidence of several kilometers depth. In actual fact, using receiver function analysis [71], it is reported that the depth of the Moho discontinuity in this area ranges from a depth between 26–31 km. The Moho in area II is about 6 km shallower than in areas I and III.

We made a simple subsurface structure model consisting of sedimentary layer, crust and mantle, based on information of subsurface structures mentioned above. We then calculated the gravity anomalies along a D-D' profile with assumption density contrasts of sedimentary layer-crust and crust-mantle being -300 kg/m^3 and -500 kg/m^3. We employed the two dimensional Talwani's method [72] in our calculations.

Figure 12 shows the simple subsurface structure model and the estimated gravity anomaly along the D-D' profile. The assumed subsurface structure model explains the Bouguer gravity anomalies very well, in spite of the model being very simple, and leads us to assume that the conspicuous subsidence at the surface would induce the crustal thinning-mantle uplift, via isostasy. Subsidence at the surface, caused by frequent lateral motions during the Paleogene, as shown and discussed in this paper, has reached to several km. In this study, right lateral motions of the crust reaching about 1000 km were required, in order to restore the sedimentary basins distributed in central Hokkaido, as shown in the previous section. Although the correct depths of sedimentary basins were not shown in our dislocation modeling because the model was based on linear elasticity, it is expected that deeper basins would be formed in the actual crust by lateral motions. Consequently, it is considered possible that conspicuous subsidence caused by a large lateral motion would induce the crustal thinning-mantle uplift. Or, since this area experienced a tension stress field over a period of about 35 million years in spite of the local stress caused by the pull-apart basin formation, this tension stress field might induce the crustal thinning-mantle uplift viscoelastically over a very long time scale. In either case, it is necessary to reconsider the estimated mass deficiency and volume of the sediment in area II from the Bouguer gravity anomalies, by correcting the effect of mantle uplift.

In this study, we suggested one tectonic model to explain the distribution of sedimentary basins located in central Hokkaido and formed between 48 Ma and 12 Ma. We then discussed the characteristics of the Bouguer gravity anomalies based on the tectonic model. In future studies, we will attempt to estimate the subsurface structure more accurately, and discuss the tectonics through simulation procedures considering a more realistic behavior of the crust and mantle (e.g., FEM, FDM, DEM and others), based on the model shown in this paper.

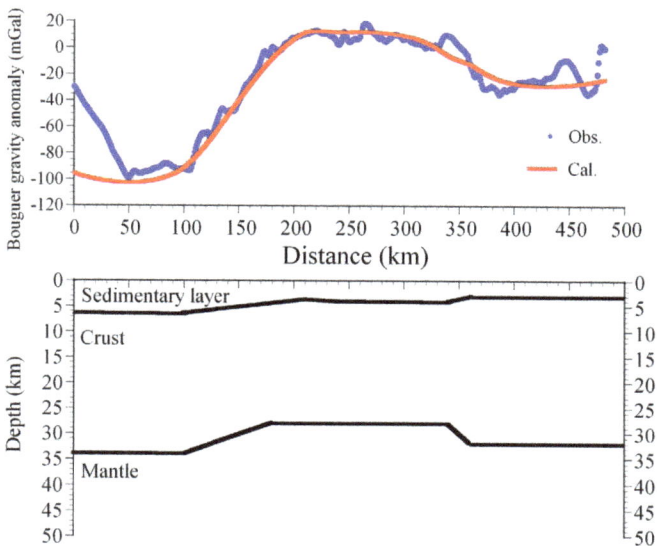

Figure 12. Two-dimensional gravity modelling along D-D' profile in Figure 7. Blue circle shows measured Bouguer gravity anomalies and red solid line shows calculated values. Bottom figure shows the density structure. Density contrasts of sedimentary layer-crust and crust-mantle were assumed to be -300 kg/m³ and -500 kg/m³, respectively.

6. Conclusion

We simply reviewed on dislocation modeling and on its applications to geological problems (including the disadvantages of the model), and make the following two points.

1. When discussing the essential aspects of tectonics, simply from a distribution pattern of structures caused by fault motions, the dislocation modeling was a very useful tool and should be used with an assumption of a high Poisson's ratio. However, it is difficult to directly compare the displacement amounts between the modeled structures and the actual structures.

2. It is useful to superimpose analytical solutions for different fault parameters on a single fault, in order to introduce the history of fault activity into the numerical model.

We then employed the suggested dislocation modeling technique for the restoration modeling of sedimentary basins formed from 48 Ma to 12 Ma (located in central Hokkaido, Japan), and evaluated the fault types (lateral faulting or reverse faulting). As a result it was found that:

3. Sedimentary basins that were formed from 48 to 12 Ma in central Hokkaido can be explained by the formation of pull-apart basins, due to right lateral motions before 13 Ma, and by reverse motions of the Hidaka-North fault zone and the Hidaka-South fault zone after 13 Ma. Although this makes us expect a significant change in the regional tectonic

stress field during this stage, its source should be investigated accurately and quantitatively in the future.

4. The distribution pattern of sedimentary basins restored in this study is similar to the actual distribution of buried sedimentary basins formed during the Paleogene in this area.

Finally, we discovered the following two points relating to the Bouguer gravity anomaly in central Hokkaido:

5. A gravity low area less than 20 mGal corresponds to the area where the Paleogene strata are distributed under the surface.

6. The Bouguer gravity anomalies in the gravity low belt less than 20 mGal can be roughly explained by the subsurface structure model that the mantle around the center of the belt was lifted upwards. Although it is considered that the conspicuous subsidence caused by large lateral motion would induce the crustal thinning-mantle uplift, this possibility should be discussed more accurately and quantitatively in the future.

Acknowledgements

This study was partially supported by the First Bank of Toyama Scholarship foundation and a Grants-in-Aid for Scientific Research (No. 21671003). We are most grateful to Ana Pantar and Book Editors for their editorial advices and cooperation.

Author details

Shigekazu Kusumoto[1], Yasuto Itoh[2], Osamu Takano[3] and Machiko Tamaki[4]

1 Graduate School of Science and Technology for Research, University of Toyama, Japan

2 Department of Physical Science, Graduate School of Science, Osaka Prefecture University, Japan

3 JAPEX Research Center, Japan Petroleum Exploration Co. Ltd., Japan

4 Japan Oil Engineering Co. Ltd., Japan

References

[1] Aydin A, Nur A. Evolution of Pull-apart Basins and Their Scale Independence. Tectonics 1982; 1 91-105.

[2] Rodgers D A. Analysis of pull-apart basin development produced by en echelon strike-slip faults. Special Publications, International Association of Sedimentologists 1980; 4 27–41.

[3] Katzman R, ten Brink U S, Lin J. Three-dimensional modeling of pull-apart basins: Implications for the tectonics of the Dead Sea Basin. Journal of Geophysical Research 1995; 100 6295–6312.

[4] ten Brink U S, Ben-Avarham Z, Bell R E, Hassouneh M, Coleman D F, Andreasen F, Tibor G, Coakley B. Structure of the Dead sea pull-apart basin from gravity analyses. Journal of Geophysical Research 1993; 98 21877-21894.

[5] Chinnery M A. The deformation of the ground around surface faults. Bulletin of Seismological Society of America 1961; 51, 355–372.

[6] Lasserre C, Peltzer G, Crampe F, Klinger Y, Van der Woerd J, Tapponnier P. Coseismic deformation of the 2001 Mw=7.8 Kokoxili earthquake in Tibet, measured by synthetic aperture radar interferometry. Journal of Geophysical Research 2005; 110. doi: 10.1029/2004JB003500.

[7] Miura S, Ueki S, Sato T, Tachibana K, Hamaguchi H. Crustal deformation associated with the 1998 seismo-volcanic crisis of Iwate volcano, northeastern Japan, as observed by a dense GPS network. Earth Planets Space 2000; 52 1003–1008.

[8] Okada Y. Surface deformation due to shear and tensile faults in a half-space. Bulletin of Seismological Society of America 1985; 75 1135–1154.

[9] Okada Y. Internal deformation due to shear and tensile faults in a half-space. Bulletin of Seismological Society of America 1992; 82 1018–1040.

[10] Steketee J A. Some geophysical applications of the elasticity theory of dislocations. Canadian Journal of Physics 1958; 36 1168-1198.

[11] Rybicki K, Kasahara K. A strike-slip fault in a laterally inhomogeneous medium. Tectonophysics 1977; 42 127-138.

[12] Savege J C. Effect of crustal layering on dislocation modeling. Journal of Geophysical Research 1987; 92 10595-10600.

[13] Du Y, Segall P, Gao H. Dislocations in inhomogeneous media via a moduli-perturbation approach; general formulation and 2D solutions. Journal of Geophysical Rearch 1994; 99 13767-13779.

[14] Matu'ura M, Tanimoto T, Iwasaki T. Quasi-static displacements due to faulting in a layered half-space with an intervenient viscoelastic layer. Journal of Physics of the Earth 1981; 29 23-54.

[15] Pollitz F. Postseismic relaxation theory on the spherical earth. Bulletin of Seismological Society of America 1992; 82 422-453.

[16] Rudnicki J W. Plane strain dislocations in linear elastic diffusive solids. Journal of Applied Mechanics 1987; 109 545-552.

[17] Rudnicki J W, Roeloffs E. Plane strain shear dislocations moving steadily in linear elastic diffusive solids. Journal of Applied Mechanics 1990; 112 32-39.

[18] Takada Y, Matsu'ura M. A unified interpretation of vertical movement in Himalaya and horizontal deformation in Tibet on the basis of elastic and viscoelastic dislocation theory. Tectonophysics 2004; 383 105 – 131. doi:10.1016/j.tecto.2003.11.012.

[19] Weertman J, Weertman J. Elementary dislocation theory. Oxford: Oxford University Press; 1992.

[20] Weertman J. Dislocation based fracture mechanics. Singapore: World Science Publication; 1996.

[21] Segall P. Earthquake and volcano deformation. Princeton: Princeton University Press; 2010.

[22] Okubo S. Potential and gravity changes raised by a point dislocation. Geophysical Journal International 1991; 105 573-586.

[23] Okubo S. Potential and gravity changes due to shear and tensile fault in a half-space. Journal of Geophysical Research 1992; 97 7137-7144.

[24] Sun W, Okubo S. Surface potential and gravity changes due to internal dislocations in a spherical earth – II, Application to a finite fault. Geophysical Journal International 1998; 132 79-88.

[25] Tapley B D, Bettadpur S, Ries J C, Thompson P F, Watkins M. GRACE measurements of mass variability in the Earth system science. Science 2004; 305 503–505, doi: 10.1126/science.1099192.

[26] Wang L, Shum C K, Simons F J, Tapley B, Dai C. Coseismic and postseismic deformation of the 2011 Tohoku-Oki earthquake constrained by GRACE gravimetry. Geophysical Research Letters 2012; 39 L07301, doi:10.1029/2012GL051104.

[27] Lin J, Stein R S. Stress triggering in thrust and subduction earthquakes, and stress interaction between the southern San Andreas and nearby thrust and strike-slip faults. Journal of Geophysical Research 2004; 109. doi:10.1029/2003JB002607.

[28] Toda S, Stein R S, Richards-Dinger K, Bozkurt S. Forecasting the evolution of seismicity in southern California: Animations built on earthquake stress transfer. Journal of Geophysical Research 2005; B05S16. doi:10.1029/2004JB003415.

[29] King G C P, Stein R S, Lin J. Static stress changes and the triggering of earthquakes. Bulletin of Seismological Society of America 1994; 84 935-953.

[30] Nostro C, Stein R S, Cocco M, Belardinelli M E, Marzocchi W. Two-way coupling be-
      tween Vesuvius eruptions and southern Apennine earthquakes (Italy) by elastic
      stress transfer. Journal of Geophysical Research 1998; 103 24487-24504.

[31] Toda S, Stein R S, Sagiya T. Evidence from the 2000 Izu Islands swarm that seismicity
      is governed by stressing rate. Nature 2002; 419 58-61.

[32] Ganas A, Sokos E, Agalos A, Leontakianakos G, Pavlides S., Coulomb stress trigger-
      ing of earthquakes along the Atalanti Fault, central Greece: Two April 1894 M6+
      events and stress change patterns. Tectonophysics 2006; 420 357–369. doi:10.1016/
      j.tecto.2006.03.028.

[33] Itoh Y, Kusumoto S, Takemura K. Characteristic basin formation at terminations of a
      large transcurrent fault: basin configuration based on gravity and geomagnetic data.
      In: Itoh, Y. (ed.), Mechanism of sedimentary basin formation: multidisciplinary ap-
      proach on active plate margins. Rijeka: InTech; 2013. (in press)

[34] Kido M. Tectonic history of the north-western area of 'Kuju–Beppu Graben'. Memoirs
      Geological Society of Japan 1993; 41 107–127. (in Japanese with English abstract)

[35] Kusumoto S, Takemura K, Fukuda Y, Takemoto S. Restoration of the depression
      structure at the eastern part of central Kyushu, Japan, by means of the dislocation
      modeling. Tectonophysics 1999; 302 287-296.

[36] Itoh Y, Kusumoto S, Furubayashi T. 2008. Quantitative evaluation of the Quaternary
      crustal deformation around the Takayama basin, central Japan: A paleomagnetic and
      numerical modeling approach. Earth and Planetary Science Letters 2008; 267 517-532.
      doi: 10.1016/j.epsl.2007.11.062.

[37] Itoh Y, Kusumoto S, Maeda J. Cenozoic basin-forming processes along the northeast-
      ern margin of Eurasia: Constraints determined from geophysical studies offshore of
      Hokkaido, Japan. Journal of Asian Earth Sciences 2009; 35 27-33.

[38] Tamaki M, Kusumoto S, Itoh Y. Formation and deformation process of the late Paleo-
      gene sedimentary basins in the southern central Hokkaido, Japan: paleomagnetic and
      numerical modeling approach. Islands Arc 2010; 19 243-258.

[39] Sato T, Matsu'ura M. A kinematic model for deformation of the lithosphere at sub-
      duction zones. Journal of Geophysical Research 1988; 93 6410–6418.

[40] Sato T, Matsu'ura M. A kinematic model for evolution of island arc–trench systems.
      Geophysical Journal International 1993; 114 512– 530.

[41] Panien M, Buiter S J H, Schreurs G, Pfiffner O A. Inversion of a symmetric basin: in-
      sights from a comparison between analogue and numerical experiments. In Buiter S J
      H, Schreurs G. (eds) Analogue and Numerical Modelling of Crustal-Scale Processes.
      London: Geological Society, London, Special Publications 2006; 253 p253-270.

[42] Le Pourhiet L, Mattioni L, Moretti I. 3D modeling of rifting through a pre-existing
      stack of nappes in the Gulf of Corinth (Greece): a mixed analogue/numerical ap-

proach. In Buiter S J H, Schreurs G. (eds) Analogue and Numerical Modelling of Crustal-Scale Processes. London: Geological Society, London, Special Publications 2006; 253 p233-252.

[43] Vietor T, Oncken O. Controls on the shape and kinematics of the Central Andean plateau flanks: Insights from numerical modeling. Earth and Planetary Science Letters 2005; 236 814-827. doi:10.1016/j.epsl.2005.06.004.

[44] Takahashi H, Kasahara M, Kimata F, Miura S, Heki K, Seno T, Kato T, Vasilenko N, Ivashchenko A, Bahtiarov V, Levin V, Gordeev E, Korchagin F, Gerasimenko M. Velocity field of around the Sea of Okhotsk and Sea of Japan regions determined from a new continuous GPS network data. Geophysical Research Letters 1999; 26 2533-2536.

[45] Niida K. Outline of Hokkaido. In: Niida K, Arita K, Kato M. (Eds.) Regional Geology of Japan, 1. Hokkaido. Tokyo: Asakura; 2010 p1-15. (in Japanese)

[46] Kimura G. Oblique subduction and collision: forearc tectonics of the Kuril arc. Geology 1986; 14 404–407.

[47] Kimura G, Tamaki K. Collision, rotation and back-arc spreading in the region of the Okhotsk and Japan Seas. Tectonics 1986; 5 389– 401.

[48] Kimura G. Collision orogeny at arc-arc junctions in the Japanese Islands. The Island Arc 1996; 5 262-275.

[49] Arita K, Ikawa T, Ito T, Yamamoto A, Saito M, Nishida Y, Satoh H, Kimura, G, Watanabe T, Ikawa T, Kuroda T. Crustal structure and tectonics of the Hidaka Collision Zone, Hokkaido (Japan), revealed by vibroseis seismic reflection and gravity surveys. Tectonophysics 1998; 290 197–210.

[50] Iwasaki T, Adachi K, Moriya T, Miyamachi H, Matsushima T, Miyashita K, Takeda T, Taira T, Yamada T, Ohtake K. Upper and middle crustal deformation of an arc–arc collision across Hokkaido, Japan, inferred from seismic refraction/wide-angle reflection experiments. Tectonophysics 2004; 388 59–73.

[51] Iwasaki T, Levin V, Nikulin A, Iidaka T. Constraints on the Moho in Japan and Kamchatka. Tectonophysics 2013; in press.

[52] Yamamoto A. 2004, Dense clustering of latest Cenozoic caldera-like basins of central Hokkaido, Japan, evidenced by gravimetric study. Journal of the Faculty of Science, Hokkaido University. Series 7, Geophysics 2004; 12 75-95.

[53] Kamiyama H, Yamamoto A, Hasegawa T, Kajiwara T, Mogi T. Gravity and density variations of the tilted Tottabetsu plutonic complex, Hokkaido, northern Japan: implications for subsurface intrusive structure and pluton development. Earth Planets Space 2005; 57 e21-e24.

[54] Ogawa Y, Nishida Y, Makino M. A collision boundary imaged by magnetotellurics, Hidaka Mountains, central Hokkaido, Japan. Journal of Geophysical Research 1994; 99 22373-22388.

[55] Ando H. Stratigraphic correlation of Upper Cretaceous to Paleocene forearc basin sediments in Northeast Japan: cyclic sedimentation and basin evolution. Journal of Asian Earth Sciences 2003; 21 921-935.

[56] Kurita H, Hoyanagi K. Outline (Paleogene) In: Niida K, Arita K, Kato M. (eds.) Regional Geology of Japan, 1. Hokkaido. Tokyo: Asakura; 2010. p97-99. (in Japanese)

[57] Takano O, Waseda A. Sequence stratigraphic architecture of a differentially subsiding bay to fluvial basin: the Eocene Ishikari Group, Ishikari Coal Field, Hokkaido, Japan. Sedimentary Geology 2003; 160 131-158.

[58] Itoh Y, Tsuru T. A model of late Cenozoic transcurrent motion and deformation in the fore-arc of northeast Japan: Constraints from geophysical studies. Physics of the Earth and Planetary Interiors 2006; 156 117-129.

[59] Itoh Y, Tsuru T. Evolution history of the Hidaka-oki (offshore Hidaka) basin in the southern central Hokkaido, as revealed by seismic interpretation, and related tectonic events in an adjacent collision zone. Physics of the Earth and Planetary Interiors 2005; 153 220-226.

[60] Kurita H, Yokoi S. Cenozoic tectonic settings and a current exploration concept in southern central Hokkaido, northern Japan. Journal of the Japanese Association for Petroleum Technology 2000; 65 58-70.

[61] Komazawa M. Gravity grid database of Japan, Gravity CD-ROM of Japan, ver. 2, Digital Geoscience Map P-2. [CD-ROM] Tsukuba: Geological Survey of Japan; 2004.

[62] Wangen M. Physical principles of sedimentary basin analysis. London: Cambridge University Press; 2010.

[63] Davis P J, Rabinowitz P. Methods of numerical integration, second edition. New York: Dover; 2007.

[64] Kusumoto S, Fukuda Y, Takemoto S, Yusa Y. Three-dimensional subsurface structure in the eastern part of the Beppu-Shimabara graben, Kyushu, Japan, as revealed by gravimetric data. Journal of the Geodetic Society of Japan 1996; 42 167-181.

[65] Cooper G R J. The removal of unwanted edge contours from gravity datasets. Exploration Geophysics. in press.

[66] Burger H R, Sheehan A F, Jones C H. Introduction to Applied Geophysics – Exploring the shallow subsurface. New York: Norton; 2006.

[67] Itoh Y, Kusumoto S, Inoue T. Magnetic properties of siliceous marine sediments in Northern Hokkaido, Japan: a quantitative tectono-sedimentological study of basins along an active margin. Basin Research 2013; doi:10.1111/bre.12020.

[68] ten Brink U S, Katzman R, Lin J. Three-dimensional models of deformation near strike-slip faults. Journal of Geophysical Research 1996; 101 16205-16220.

[69] Kawakami G. Foreland of the Hidaka Mountains In: Niida K, Arita K, Kato M. (eds.) Regional Geology of Japan, 1. Hokkaido. Tokyo: Asakura; 2010. P530-532. (in Japanese)

[70] Katsumata A. Depth of the Moho discontinuity beneath the Japanese islands estimated by travel time analysis. Journal of Geophysical Research 2010; 115. doi: 10.1029/2008JB005864.

[71] Igarashi T, Iidaka T, Miyabayashi S. Crustal structure in the Japanese Islands inferred from receiver function analysis. Zisin 2011; 63 139–151. (in Japanese with English abstract)

[72] Talwani M, Worzel J, Landisman L. Rapid gravity computations for two-dimensional bodies with application to the Mendocino submarine fracture zone. Journal of Geophysical Research 1959; 64 49-59.

[73] Research Group for Active Faults. The Active Faults in Japan: Sheet Maps and Inventories (Revised Edition). Tokyo: University of Tokyo Press 1991. (in Japanese with English abstract)

[74] Ikeda Y, Imaizumi T, Togo M, Hirakawa K, Miyauchi T, Sato H. Atlas of Quaternary Thrust Faults in Japan. Tokyo: University of Tokyo Press; 2002. (in Japanese)

[75] Association of Natural Gas Mining and Association for Offshore Petroleum Exploration. Petroleum and Natural Gas Resources of Japan, Tokyo; 1982. (in Japanese)

[76] Nagao S. Geology of the Toyotomi District, Scale 1:50,000. Sapporo: Geological Survey of Hokkaido; 1960. (in Japanese with English summary)

[77] Hata M, Tsushima S. Geology of the Enbetsu District, Scale 1:50,000. Tsukuba: Geological Survey of Japan; 1968. (in Japanese with English summary)

[78] Matsuno H, Kino Y. Geology of the Chikubetsu-Tanko District, Scale 1:50,000. Sapporo: Hokkaido Development Agency; 1960. (in Japanese with English summary)

[79] Ogura N, Kamon M. The subsurface structures and hydrocarbon potentials in the Tenpoku and Haboro area, the northern Hokkaido, Japan. Journal of the Japanese Association for Petroleum Technology 1992; 57 32-44. (in Japanese with English abstract)

[80] Tsushima S, Matsuno H, Yamaguchi S. Geology of the Onishika District, Scale 1:50,000. Tsukuba: Geological Survey of Japan; 1956. (in Japanese with English summary)

[81] Tsushima S, Tanaka K, Matsuno H, Yamaguchi S. Geology of the Tappu District, Scale 1:50,000. Tsukuba: Geological Survey of Japan; 1958. (in Japanese with English summary)

[82] Watanabe M, Yoshida F. Geology of the Ebishima District, Scale 1:50,000. Tsukuba: Geological Survey of Japan; 1995. (in Japanese with English summary)

[83] Kobayashi I, Hata M, Yamaguchi S, Kakimi T. Geology of the Moseushi District, Scale 1:50,000. Tsukuba: Geological Survey of Japan; 1969. (in Japanese with English summary)

[84] Matsuno H, Tanaka K, Mizuno A, Ishida M. Geology of the Iwamizawa District, Scale 1:50,000. Sapporo: Hokkaido Development Agency; 1964. (in Japanese with English summary)

[85] Matsui H, Kakimi T, Nemoto T. Geology of the Sunagawa District, Scale 1:50,000. Tsukuba: Geological Survey of Japan; 1965. (in Japanese with English summary)

Chapter 12

Characteristic Basin Formation at Terminations of a Large Transcurrent Fault — Basin Configuration Based on Gravity and Geomagnetic Data

Yasuto Itoh, Shigekazu Kusumoto and
Keiji Takemura

Additional information is available at the end of the chapter

1. Introduction

The formation of a pull-apart basin is a ubiquitous phenomenon associated with transcurrent faulting in the Earth's upper crust (e.g., [1]). Because an oblique mode of subduction provokes the detachment and transcurrent motion of a forearc crustal sliver [2], numerous pull-apart basins develop on active plate margins.

Southwest Japan is an island arc that has long been under the influence of the oblique subduction of oceanic plates. Figure 1 delineates its bisecting fault, the Median Tectonic Line (hereafter referred to as MTL). Regional pull-apart basins related to sinistral motion of the MTL during the late Cretaceous were described by Noda and Toshimitsu [3]. Vigorous sinistral faulting also caused regional wrench deformation of the adjoining terranes [4]. The northwestward motion of the Philippine Sea Plate, depicted as a positive gravity anomaly (red-colored) portion within the Pacific Ocean in Figure 1, resulted in dextral slips on the MTL since the Pliocene. Although the recent activity of the MTL is key to understanding the neotectonic regime of southwest Japan [5], basin-forming processes along the regional fault system have not been fully described.

The authors attempt to present a quantitative description of sedimentary basins at the western and eastern terminations of the MTL, namely, the Beppu-Iyo Basin and Osaka Basin, respectively (Figure 1). As shown by negative gravity anomalies, they are enormous depressions filled by recent clastics. Seismic surveys in the study areas became popular in recent decades, but interpretation has not reached a mature stage. Thus, we utilize gravimetric methods for estimating their morphology and volume. Geomagnetic anomaly modeling helps us to identify subsurface constituents of the sedimentary basins. Additionally, the evolutionary processes of the basins are argued based on geologic information. This is a multidisciplinary case study of basins ahead of a comprehensive visualization of the basin interior with seismic information.

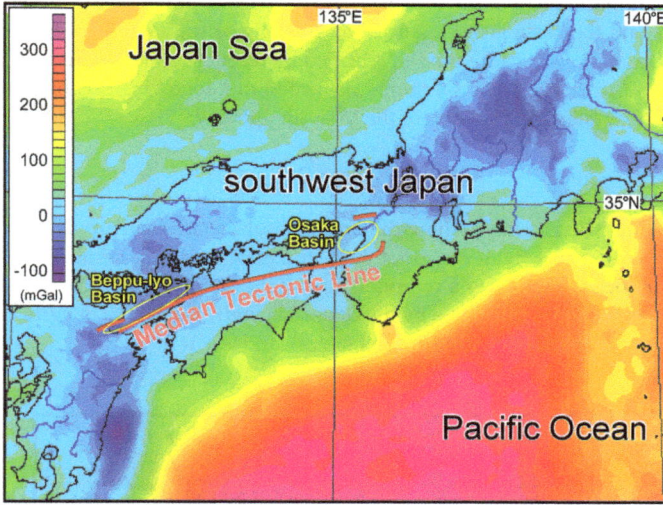

Figure 1. Index map of the studied basins. Base map shows Bouguer gravity anomaly [6]. Bouguer density is 2670 kg/m³. The northern part of the Philippine Sea Plate is depicted as a positive gravity anomaly (red-colored) portion within the Pacific Ocean. Shape of an oceanic plate is generally delineated by positive Bouguer anomaly in ocean, because an area of deep water is accompanied by positive gravity values as a result of correction procedures (cf. Y. Itoh, K. Takemura and S. Kusumoto in this book)

2. Volumetric analysis

Volume estimation of the basins is necessary in order to consider the mass balance of the upper crust around the island arc. Utilizing a gravity database [6], we present the analytical results for the Beppu-Iyo and Osaka Basins in the following sections.

2.1. Beppu-Iyo Basin

The large depression of the Beppu-Iyo Basin is divided into the Beppu Bay (west) and Iyonada Sea (east) areas by a relative high of the Bouguer anomaly. Beppu Bay is a younger part of an extensive tectonic depression, the Hohi Volcanic Zone, which has been developed since the Pliocene [7]. As for the Beppu Bay basin, active trace of the MTL is terminated around southwestern coast of the bay (e.g., [7]). Right-stepping lateral faults are aligned on the northern corner of the bay, and the parallel two fault strands are connected by normal listric faults constituting a rhomboidal depression surrounded by the faults [7]. These structural patterns are characteristic for releasing bend of strike-slip fault.

The Iyonada Sea is characterized by a remarkable negative gravity anomaly [8,9], but its origin has not been discussed so far. In contrast with the Beppu Bay basin, ambiguous points remain in the formation mechanism of the extensive Iyonada Sea depression. Paired dextral fault is not identified, and the basin is not regarded as an elongate sag in an area of propagation of lateral fault terminations because the sag is buried by recent sediments (Age assignment of a sedimentary unit upon the basin margin is presented in a following section.) simultaneous with those in the western termination of the MTL (Beppu Bay; [7]).

2.1.1. Beppu Bay

Kusumoto et al. [10] determined a three-dimensional subsurface structure around Beppu Bay on the basis of gravimetric data. We adopt their structural model and estimate the volume of the sedimentary basin. Figure 2 is a compiled map of the Hohi Volcanic Zone. The volume of the basement depression is calculated by the Gauss-Legendre numerical integration [11], from depth data given on the mesh with a 10 km interval, and is 4.1×10^3 km^3.

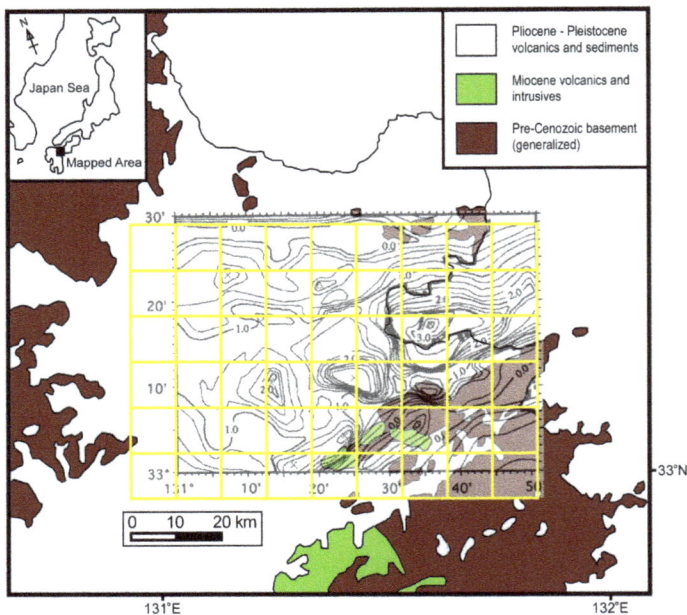

Figure 2. Simplified geology around the Hohi Volcanic Zone including Beppu Bay. Overlain basement structural contours are based on gravity anomaly modeling after Kusumoto et al. [10]. Grid shows data points for volumetric analysis

2.1.2. Iyonada Sea

Figure 3 is a Bouguer anomaly map around the Iyonada Sea with two-dimensional gravity modeling lines (1–10). Although a previous study [9] assumed there to be three units with different densities in the basement rock, its geologic context still remains ambiguous. Therefore we adopted a simple two-layered (sediment and basement) model. We applied Talwani's method [12] in order to estimate two-dimensional subsurface structures and assumed that the each structure at both ends of each profile was infinity to the depth. As shown in Figure 4, the remarkable elongate depression has a profile common with a half-graben, implying a tensile stress state. An isopach map indicating the top basement structure (Figure 5) demonstrates that the deepest part of the basin reaches 4 km from mean sea level. The volume of the basement depression is calculated by the Gauss-Legendre numerical integration [11], from depth data given on the mesh with a 10 km interval, and is 7.2×10^3 km^3.

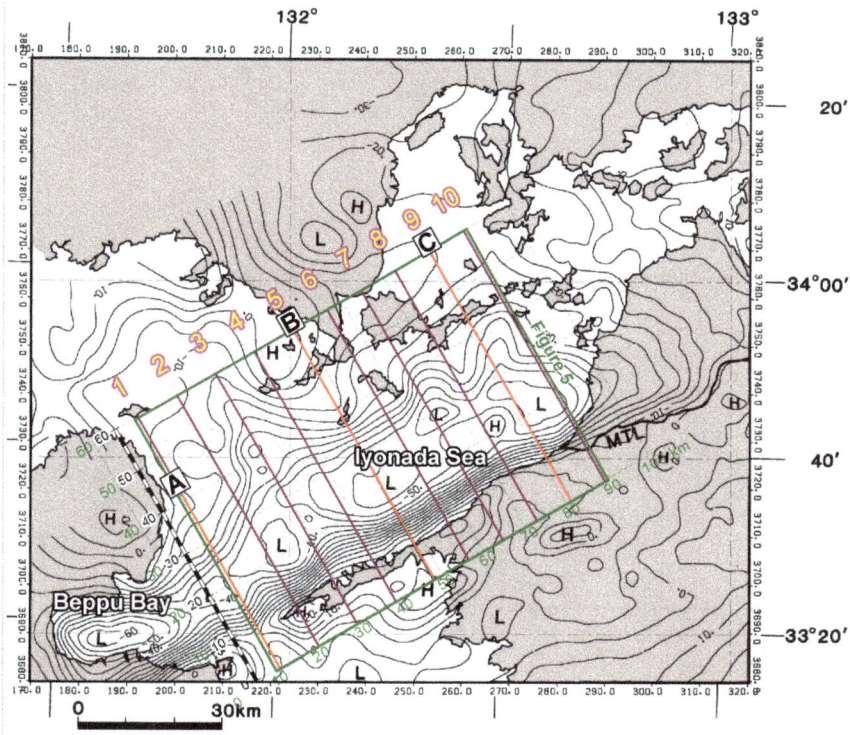

Figure 3. Bouguer gravity anomaly map around the Beppu-Iyo Basin [9]. Bouguer density is 2670 kg/m^3. Contour interval is 5 mGal. Lines 1 to 10 are for 2-D gravity anomaly modeling in the Iyonada Sea. A, B and C show previous modeling lines [9]

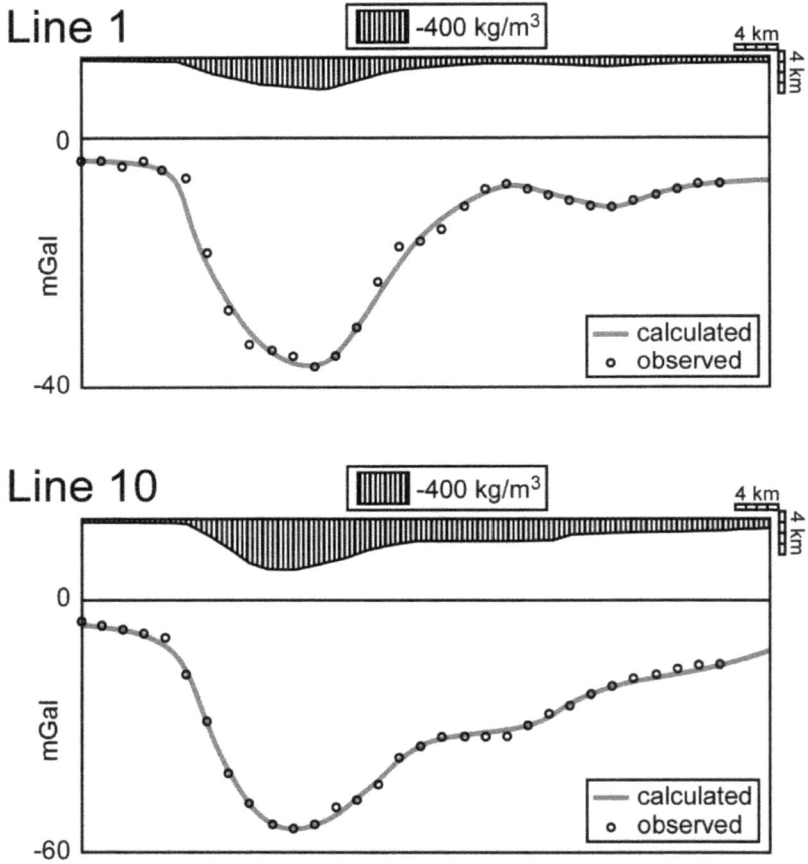

Figure 4. Examples of Bouguer anomaly profiles calculated for the two-layer models. See Figure 3 for line location

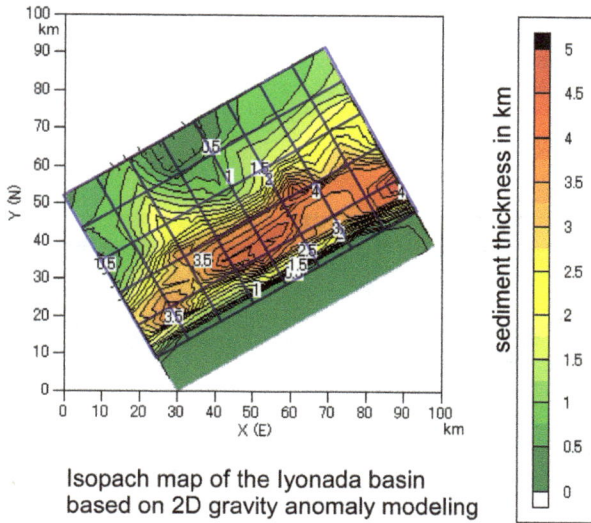

Figure 5. Isopach map of the Iyonada Basin based on 2D gravity anomaly modeling, indicating basement structure of the Iyonada Sea. See Figure 3 for mapped area. Grid shows data points for volumetric analysis

2.2. Osaka Basin

The active trace of the MTL tends to shrink compared to the older phase of faulting. Its eastern termination is now around 136ºE with a length of active segment of 400 km, whereas the Cretaceous MTL as a significant geological break reached ca. 140ºE with a total length of 800 km [13]. The Arima-Takatsuki Tectonic Line (E-W trending fault on the northern flank of the Osaka Basin; Figure 1) is the unique parallel fault around the shrunken eastern termination having comparable dextral slip rate during the Quaternary [14]. Although this fault alignment should act as a confining bend, the area is characterized by recent vigorous basin formation around the Osaka Bay. We, therefore, attempt to simulate the deformation pattern introducing effect of reverse slips on secondary faults which show complex arrangement as a result of longstanding differential motion of crustal blocks (cf. Y. Itoh, K. Takemura and S. Kusumoto in this book) in a following section.

A regional structural model around the eastern termination of the MTL has not been shown because complicated basin morphology does not allow two-dimensional modeling. Thus, we estimated mass deficiency from the gravity anomaly data of the Osaka Basin given on the mesh with a 5 km interval (Figure 6) by Gauss's theorem [15]:

$$\Delta M = \frac{1}{2\pi G} \int\int_{-\infty}^{\infty} \Delta g(x,y) dx dy \qquad (1)$$

Here, $\Delta g(x, y)$ is the gravity anomaly data given on an xy mesh with a constant interval. G, π and ΔM are the universal gravitational constant, circular constant and deficiency/excess of mass, respectively. Next, a first approximation of the volume of the sedimentary basin was calculated, from the estimated deficiency of mass by assuming a density contrast of 400 kg/m³ between sediment and basement, to be 9.1×10^2 km³.

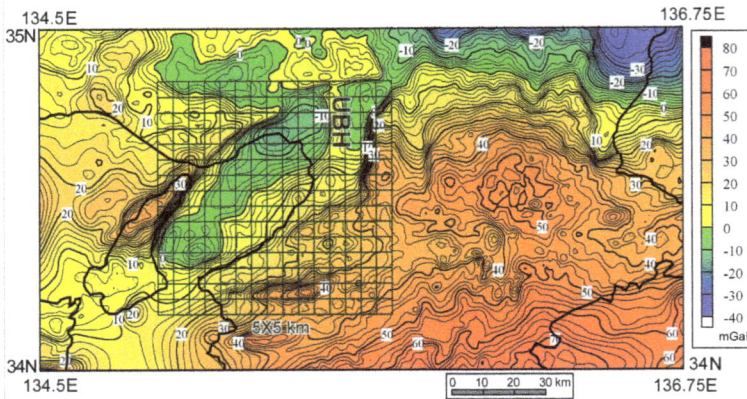

Figure 6. Bouguer gravity anomaly map [6] around the Osaka Basin. Bouguer density is 2670 kg/m³. Grid shows data points for volumetric analysis. UBH is the Uemachi Basement High in the Osaka Basin

The present result indicates that the total volume of the sedimentary basin at the western end of the MTL is 10 times larger than that of the eastern counterpart. It may be attributed to a difference in the mechanism of basin formation between releasing and confining steps of strike-slip faults, which will be discussed later.

3. Sedimentation process of the Beppu-Iyo Basin

Itoh et al. [7] demonstrated that the Hohi Volcanic Zone, including the Beppu Bay, has shifted its depocenter according to the transition of active segments of major faults, among which the MTL acted the most significant role for the development of tectonic sedimentary basin. As for the Iyonada Sea, which lacks seismic or drilling survey subsurface information, we aim at finding a temporal change in sediment supply pattern deduced from lithologic observation of an adjoining onshore sedimentary unit.

Figure 7 shows a geologic map around the eastern part of the Iyonada Sea [16]. It is known that a conspicuous sedimentary unit known as the Gunchu Formation is distributed along the northwestern coast of the Shikoku Island [17], which is a non-marine deposit containing abundant plant remains and has a steep homoclinal structure affected by the Quaternary

activity of the MTL running along the southern margin of the Iyonada Sea [18]. The Middle
Member of the Gunchu Formation contains a considerable amount of crystalline schist gravels
that were derived from the Sanbagawa metamorphic belt [19]. Its sedimentological description, however, has not been reported. Therefore the authors present the results of our preliminary geologic survey and chronological analysis in the following sections.

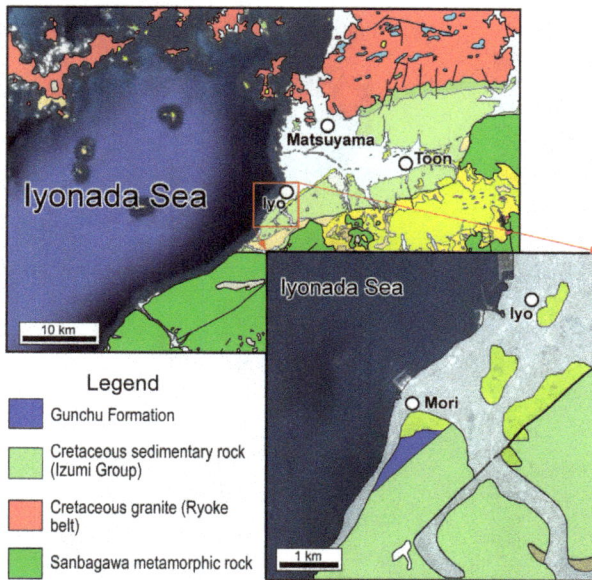

Figure 7. Geological index around the eastern part of the Iyonada Sea [16]

sample	mineral	No. of crystals	Spontaneous track		Total U count		r	P (χ^2) (%)	U (ppm)	Age±1σ (Ma)	Method
			ρ_s (cm^{-2})	N_s	ρ_u (cm^{-2})	N_u					
granite	zircon	30	1.11×10^7	4938	3.48×10^8	154845	0.856	0	280	77.9±6.1	IS

Table 1. Fission-track age of granite pebbles contained in basal part of the Gunchu Formation r is correlation
coefficient between ρ_s and ρ_u. P (χ^2) is the probability of obtaining χ_2-value for v degrees of freedom (where v = No. of
crystals - 1). IS means internal surface

3.1. Provenance and sedimentary structure

Figure 8 shows results of the geologic survey of the Middle Member of the Gunchu Formation
along the coastal section. It is clear that the composition of gravels fluctuates along the section

in response to changes in lithology. The content of schist gravel is related to sediment supply from the Outer Zone, a geologic zone on the southern side of the MTL. The paleocurrent direction, based on the imbrication structure of gravels, also shows considerable variation. It seems that the E-W elongate trough of the Iyonada Sea was alternately buried by the westward and northward influx of clastics. Further facies analysis would help provide an understanding of the sedimentary system of the pull-apart basin.

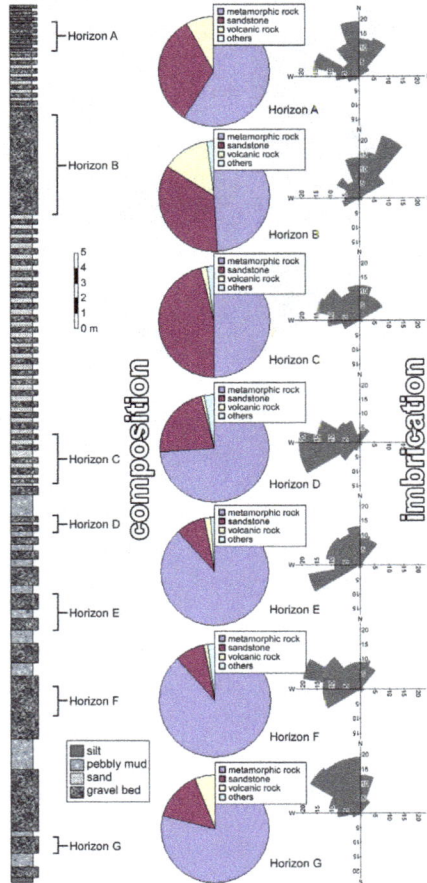

Figure 8. Sedimentological description of the Pleistocene Gunchu Formation. Left: Lithologic column of the Middle Member of the Gunchu Formation. Center: Gravel composition of selected horizons. Right: Rose diagrams showing paleocurrent directions (downcurrent directions) of the selected horizons measured using imbricate structure of gravels. Number of data is 100 for each measured horizon

3.2. Chronology

Kitabayashi et al. [20] executed fission-track dating of an ash layer intercalated in the Lower Member of the Gunchu Formation, and obtained an age of 2.2 Ma. Thus the subsidence and burial of the huge depression of the Beppu-Iyo Basin seems to be a recent event, maybe in response to an accelerated slip rate on the MTL. From the viewpoint of sediment provenance, we executed fission-track dating for pebble samples. It is noted that the Lower Member of the Gunchu Formation is lacking in schist gravels, and is characterized by sporadic granite pebbles. Table 1 and Figure 9 suggest that the granite pebbles were derived from the Cretaceous Ryoke intrusive rocks, which are distributed on the northern side of the MTL (Figure 7). Cessation of sediment supply from the northern terrane is thought to reflect entrapment of clastics in a deepening tectonic basin, and the succeeding emergence of voluminous schist gravels may correspond to an episodic uplift of the forearc sliver. Thus the drastic change in the pattern of sediment supply is a key to describe development processes of the tectonic basin.

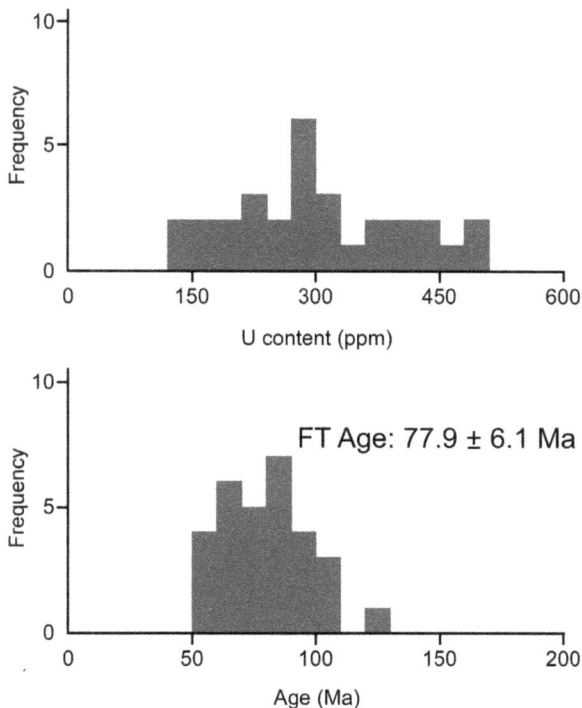

Figure 9. Result of fission-track dating of granite pebbles contained in the lowermost part of the Gunchu Formation. See Table 1 for detailed analytical data

4. Development of the Osaka Basin

In contrast to the simple half-graben on the western end of the MTL, the Osaka Basin at the eastern end is characterized by a complicated subsurface morphology reflected in the pattern of the gravity anomaly [21,22] (Figure 10). We argue a mechanism of basin formation based on numerical modeling, describe the seismic reflection profile showing the deformation pattern during the development of the basin, and interpret the origin of a concealed geologic unit on the basis of gravity and geomagnetic anomaly modeling in the following sections.

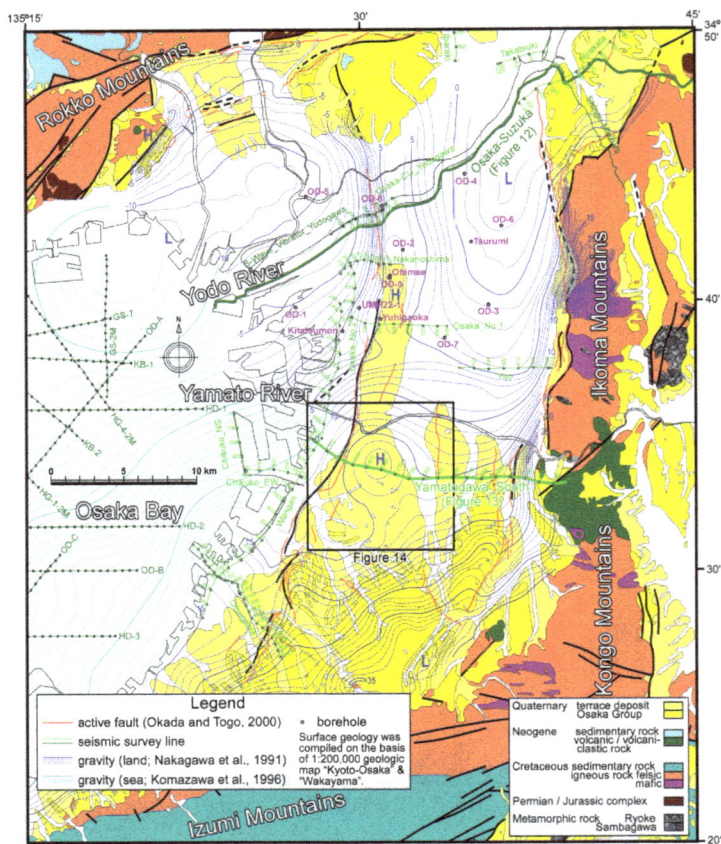

Figure 10. Geologic and geophysical database of the Osaka Basin (mainly for land area). Gravity contour is compiled after Nakagawa et al. [21] and Komazawa et al. [22]

4.1. Paradox of basin formation at a confining bend of a fault

Figure 11 delineates the deformation scheme at stepping parts of strike-slip faults. Generally speaking, a depression is formed at a right-stepping part of a dextral fault (Figure 11a), whereas an upheaval is formed at a left-stepping part of a dextral fault (Figure 11b). The MTL and the Arima-Takatsuki Tectonic Line (Figure 11c) are considered to act as a confining left-step of the regional dextral fault. However, geologic information shows that the Osaka Basin is a site of Quaternary basin formation. In order to solve the paradox, Kusumoto et al. [23] executed dislocation modeling for assessment of the vertical displacement at a complex termination of a strike-slip fault. They found that actual basin morphology could be restored by introducing reverse motions to secondary faults as shown in Figure 11c. The simulated deformation field predicts that a relative basement high, which corresponds to the Uemachi Basement High (UBH) in Figure 6, emerges within a depression surrounded by the modeled active faults.

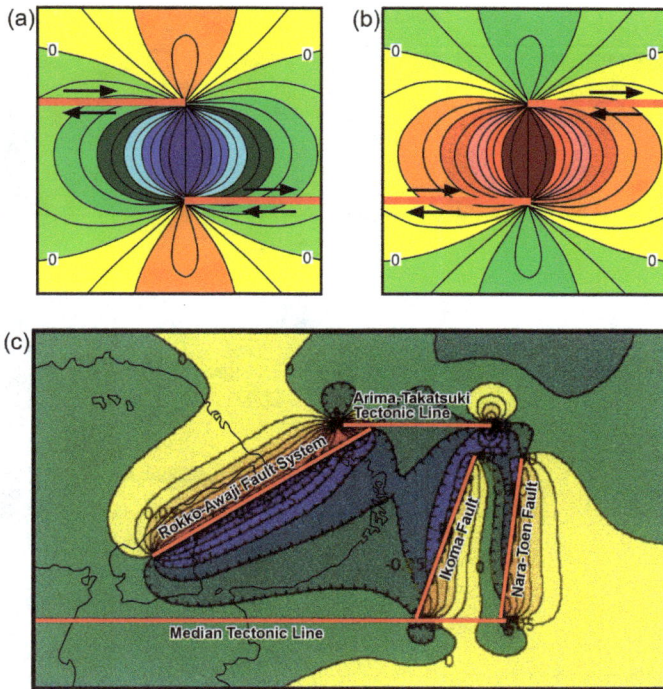

Figure 11. Architecture of numerical modeling of the formation of the Osaka Basin [23]. (a) Normalized vertical displacement at a releasing bend of a strike-slip fault. (b) Normalized vertical displacement at a confining bend of a strike-slip fault. (c) Dislocation model of the Osaka Basin. Red lines are major faults adopted for the modeling. Warm and cold color gradations indicate upheaval and subsiding areas, respectively

4.2. Seismic interpretation

Figure 12 is an E-W seismic reflection profile across the northern part of the Osaka Plain (land area of the Osaka Basin) [24]. Although the internal structure of the sedimentary basin should be discussed with detailed stratigraphic control utilizing borehole information in the future, it is noted that faults within the profile show normal and reversed displacements in the western and the eastern portions, respectively. A close-up of the western portion is characterized by a master normal fault accompanied by a collapsed anticline on its downthrown side. It coincides with the area of relative upheaval in the numerical modeling (shown by warm-colored portions within the fault-bounded basin in Figure 11c), and accords with the actual extensional structure, which is generally expected around an area of upheaval in the modeling of crustal deformation.

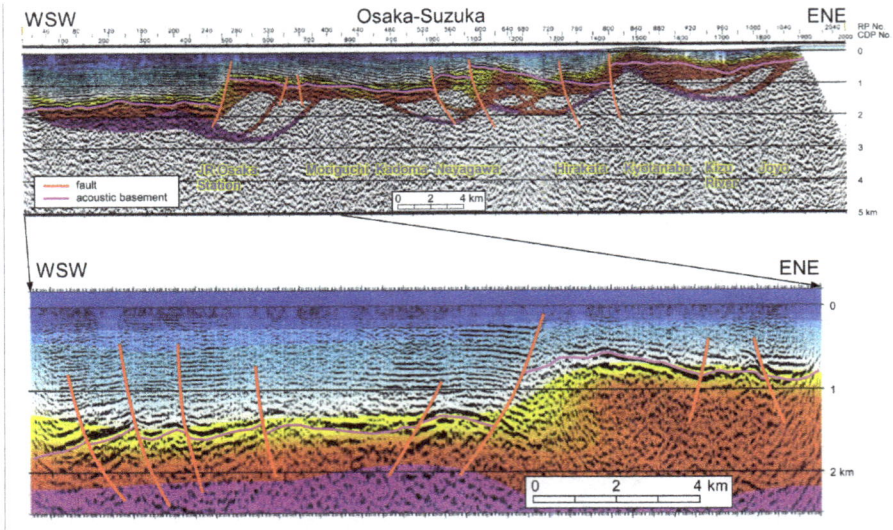

Figure 12. A depth-converted seismic profile (Osaka-Suzuka) crossing the northern part of the Osaka Plain [24]. See Figure 10 for line location

Seismic data indicate a large diversity in structural attitudes in the Osaka Basin. Figure 13 is an E-W seismic profile across the southern part of the Osaka Plain [25]. Remarkable vertical displacement with a steep gradient of the Bouguer gravity anomaly is observed at the easternmost part of the profile, and interpreted as the southern part of the Ikoma fault system (Figure 11c). A strong reflection in the eastern part of the basin is correlated with the Miocene volcanic surface based on surface geology along the survey line. It is noteworthy that a unit showing similarity in reflection pattern with the volcanic rocks is buried on the upthrown side of a reverse fault around the western part of the section. As the unit is accompanied by a positive gravity anomaly [21] and geomagnetic anomaly [26], we will construct three-dimensional models for the subsurface seismic unit.

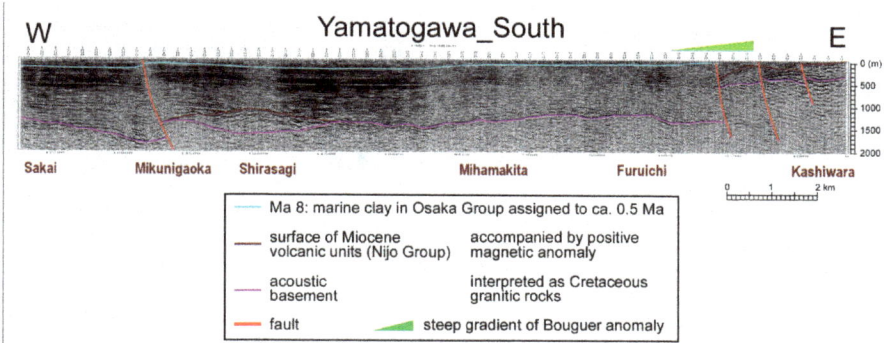

Figure 13. A depth-converted seismic profile (Yamatogawa South) crossing the southern part of the Osaka Plain [25] without vertical exaggeration. See Figure 10 for line location

4.3. Anomaly modeling

Middle Miocene volcanic rocks, collectively named the Nijo Group, are exposed in a hilly province upon the eastern margin of the Osaka Basin. It has been studied from stratigraphic (e.g., [27]) and paleomagnetic (e.g., [28]) points of view, and regarded as a volcanic product formed just after the clockwise rotation of southwest Japan related to the opening of the Japan Sea.

We assumed that conspicuous gravity and geomagnetic anomalies in the southern Osaka Plain are caused by a buried middle Miocene volcano, and estimated its three-dimensional structure based on Talwani's methods [29,30]. Figure 14 summarizes the modeling parameters. We assumed that the body of the volcano is magnetized parallel to the Earth's axial dipole field because the majority of the Nijo Group acquired its thermoremanent magnetization during the normal polarity Chron after the Miocene clockwise rotation of southwest Japan. As shown in Figure 14, gravity and geomagnetic anomalies are successfully simulated on the assumption of a conical volcano on the base of the sedimentary basin. In the Osaka Basin, there are isolated positive Bouguer anomalies associated with magnetic signatures from place to place. Combined anomaly modeling is a useful tool to estimate the origin of subsurface units in advance of a comprehensive interpretation of seismic data.

3D Talwani Modeling
Parameters

[shape of polygon]
Number of layer=40
Number of corner=20
Diameter of bottom=8km
Depth of bottom=1.3km
Height=800m
[Physical property]
Density=+300kg/m^3
Magnetization=0.1A/m
(Dec.=0°, Inc.=54°)

3D model of subsurface volcano

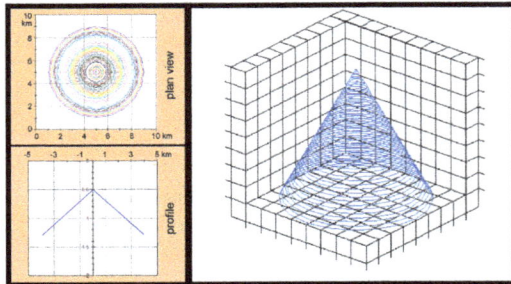

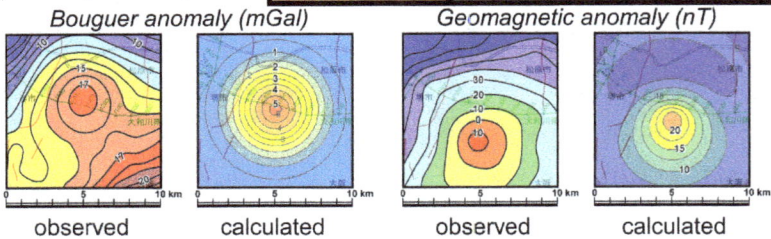

Bouguer anomaly (mGal)   *Geomagnetic anomaly (nT)*

observed   calculated   observed   calculated

Figure 14. Three-dimensional gravity and geomagnetic modeling of subsurface structure in the southern Osaka Plain. See Figure 10 for mapped area

5. Summary

An integrated geophysical study of sedimentary basins was executed on an active plate margin. Volumes of conspicuous depressions upon both terminations of the Median Tectonic Line (MTL), a dextral bisecting fault of the southwest Japan arc, were estimated by means of gravimetric methods. The western end of the MTL and its secondary faults constitute a releasing step and form a gigantic composite depression of the Beppu-Iyo Basin and has been developed since the Pliocene. Sedimentological and chronological investigation revealed that the major constituent, the Iyonada Sea depression, was rapidly buried during the Quaternary by clastics derived from different geologic terrains. On the other hand, the eastern end of the MTL is a site of basin formation (Osaka Basin), even though the fault architecture is regarded as a confining step. Numerical modeling showed that a combination of major strike-slip and secondary reverse faults can provoke complicated ups and downs within an area surrounded by faults. The stress regime predicted through the modeling of the vertical displacement was concordant with the deep structure of the basin visualized by seismic interpretation. Although the present study area is not accompanied by sufficient geologic evidence of a deep basin interior provided by drilling survey, geomagnetic anomaly modeling successfully delineated a buried volcanic unit, which was anticipated from the viewpoint of regional geology.

Acknowledgements

The authors are grateful to O. Takano for his constructive review; an early version of the manuscript was substantially improved by his comments.

Author details

Yasuto Itoh[1], Shigekazu Kusumoto[2] and Keiji Takemura[3]

1 Graduate School of Science, Osaka Prefecture University, Osaka, Japan

2 Graduate School of Science and Engineering for Research, University of Toyama, Toyama, Japan

3 Graduate School of Science, Kyoto University, Kyoto, Japan

References

[1] Aydin A, Nur A. Evolution of pull-apart basins and their scale independence. Tectonics 1982; 1: 91-105.

[2] Fitch TJ. Plate convergence, transcurrent faults, and internal deformation adjacent to southeast Asia and the western Pacific. Journal of Geophysical Research 1972; 77: 4432-4460.

[3] Noda A, Toshimitsu S. Backward stacking of submarine channel-fan successions controlled by strike-slip faulting: The Izumi Group (Cretaceous), southwest Japan. Lithosphere 2009; 1: 41-59.

[4] Wang CL. Metamorphism and tectonics of the Sanbagawa Belt in northwestern Kii Peninsula, southwest Japan. PhD thesis. Kobe University; 1997.

[5] Huzita K. Role of the Median Tectonic Line in the Quaternary tectonics of the Japanese islands. Memoir of Geological Society of Japan 1980; 18: 129-153.

[6] Gravity CD-ROM of Japan, Ver. 2, Digital Geoscience Map P-2. [CD-ROM] Tsukuba: Geological Survey of Japan; 2004.

[7] Itoh Y, Takemura K, Kamata H. History of basin formation and tectonic evolution at the termination of a large transcurrent fault system: deformation mode of central Kyushu, Japan. Tectonophysics 1998; 284: 135-150.

[8] Koizumi K, Fujimoto H, Inokuchi H, Uchitsu M, Kono Y. Marine gravity measurements over the Seto Inland Sea, western Japan. Journal of the Geodetic Society of Japan 1994; 40: 333-345.

[9] Ohno I, Kono Y, Fujimoto H, Koizumi K. Gravity anomaly in and around the western Seto Inland Sea and subsurface structure of negative anomaly belt. Zisin (Bulletin of the Seismological Society of Japan) 1994; 47: 395-401.

[10] Kusumoto S, Fukuda Y, Takemoto S, Yusa Y. Three-dimensional subsurface structure in the eastern part of the Beppu-Shimabara Graben Kyushu, Japan, as revealed by gravimetric data. Journal of the Geodetic Society of Japan 1996; 42: 167-181.

[11] Davis PJ, Rabinowitz P. Methods of Numerical Integration, Second Edition. New York: Dover; 2007.

[12] Talwani M, Lamar WJ, Landisman M. Rapid gravity computations for two-dimensional bodies with application to the Mendocino Submarine Fracture Zone. Journal of Geophysical Research 1959; 64: 49-59.

[13] Yamada N, Teraoka Y, Hata M. Geological Atlas of Japan. Tsukuba: Geological Survey of Japan; 1982.

[14] Research Group for Active Faults. The Active Faults in Japan: Sheet Maps and Inventories Rev. Ed. Tokyo: Univ. of Tokyo Press; 1991.

[15] Wangen M. Physical Principles of Sedimentary Basin Analysis. London: Cambridge University Press; 2010.

[16] Geological Survey of Japan, AIST, editor. Seamless Digital Geological Map of Japan 1: 200,000 (July 3, 2012 Version), Research Information Database DB084. Tsukuba: National Institute of Advanced Industrial Science and Technology; 2012.

[17] Nagai K. Geology of the Ehime Prefecture. Matsuyama: Tomoeya; 1957.

[18] Ogawa M, Okamura M, Shimazaki K, Nakata T, Chida N, Nakamura T, Miyatake T, Maemoku H, Tsutsumi H. Holocene activity on a submarine active fault system of the Median Tectonic Line beneath the northeastern part of Iyonada, the Inland Sea, Southwest Japan. Memoir of the Geological Society of Japan 1992; 40: 75-97.

[19] Mizuno K. Preliminary report on the Plio-Pleistocene sediments distributed along the Median Tectonic Line in and around Shikoku, Japan. Bulletin of the Geological Survey of Japan 1987; 38: 171-190.

[20] Kitabayashi E, Danhara T, Iwano H. Fission-track age of zircon from volcanic ash layer of the Gunchu Formation in Iyo City, Ehime Prefecture in Shikoku, Japan. Journal of the Geological Society of Oita 2012; 18: 61-64.

[21] Nakagawa K, Ryoki K, Muto N, Nishimura S, Ito K. Gravity anomaly map and inferred basement structure in Osaka Plain, central Kinki, south-west Japan. Journal of Geosciences, Osaka City University 1991; 34: 103-117.

[22] Komazawa M, Ohta Y, Shibuya S, Kumai M, Murakami M. Gravity survey on the sea bottom of Osaka Bay and its subsurface structure. Butsuri-Tansa (Geophysical Exploration) 1996; 49: 459-473.

[23] Kusumoto S, Fukuda Y, Takemura K, Takemoto S. Forming mechanism of the sedimentary basin at the termination of the right-lateral left-stepping faults and tectonics around Osaka Bay. Journal of Geography 2001; 110: 32-43.

[24] Disaster Prevention Research Institute. Integrated Research Project for the Uemachi Active Fault System by METI. Uji: DPRI, Kyoto University; 2011.

[25] Osaka Prefecture. Subsurface Structural Survey of the Osaka Plain in the Heisei 15 Fiscal Year. http://www.hp1039.jishin.go.jp/kozo/osaka8frm.htm (accessed 22 March 2013).

[26] Nakatsuka T, Okuma S. Aeromagnetic Anomalies Database of Japan, Digital Geoscience Map P-6. Tsukuba: Geological Survey of Japan; 2005.

[27] Miyachi Y, Tainosho Y, Yoshikawa T, Sangawa A. Geology of the Osaka-Tonanbu District, with Geological Sheet Map at 1:50,000. Tsukuba: Geological Survey of Japan; 1998.

[28] Hoshi H, Tanaka D, Takahashi M, Yoshikawa T. Paleomagnetism of the Nijo Group and its implication for the timing of clockwise rotation of southwest Japan. Journal of Mineralogical and Petrological Sciences 2000; 95: 203-215.

[29] Talwani M, Ewing WM. Rapid computation of gravitational attraction of three-dimensional bodies of arbitrary shape. Geophysics 1960; 25: 203-225, doi: 10.1190/1.1438687.

[30] Talwani M. Computation with the help of a digital computer of magnetic anomalies caused by bodies of arbitrary shape. Geophysics 1965; 5: 797-817.

Permissions

The contributors of this book come from diverse backgrounds, making this book a truly international effort. This book will bring forth new frontiers with its revolutionizing research information and detailed analysis of the nascent developments around the world.

We would like to thank Yasuto Itoh, for lending his expertise to make the book truly unique. He has played a crucial role in the development of this book. Without his invaluable contribution this book wouldn't have been possible. He has made vital efforts to compile up to date information on the varied aspects of this subject to make this book a valuable addition to the collection of many professionals and students.

This book was conceptualized with the vision of imparting up-to-date information and advanced data in this field. To ensure the same, a matchless editorial board was set up. Every individual on the board went through rigorous rounds of assessment to prove their worth. After which they invested a large part of their time researching and compiling the most relevant data for our readers. Conferences and sessions were held from time to time between the editorial board and the contributing authors to present the data in the most comprehensible form. The editorial team has worked tirelessly to provide valuable and valid information to help people across the globe.

Every chapter published in this book has been scrutinized by our experts. Their significance has been extensively debated. The topics covered herein carry significant findings which will fuel the growth of the discipline. They may even be implemented as practical applications or may be referred to as a beginning point for another development. Chapters in this book were first published by InTech; hereby published with permission under the Creative Commons Attribution License or equivalent.

The editorial board has been involved in producing this book since its inception. They have spent rigorous hours researching and exploring the diverse topics which have resulted in the successful publishing of this book. They have passed on their knowledge of decades through this book. To expedite this challenging task, the publisher supported the team at every step. A small team of assistant editors was also appointed to further simplify the editing procedure and attain best results for the readers.

Our editorial team has been hand-picked from every corner of the world. Their multi-ethnicity adds dynamic inputs to the discussions which result in innovative outcomes. These outcomes are then further discussed with the researchers and contributors who give their valuable feedback and opinion regarding the same. The feedback is then collaborated with the researches and they are edited in a comprehensive manner to aid the understanding of the subject.

Apart from the editorial board, the designing team has also invested a significant amount of their time in understanding the subject and creating the most relevant covers. They scrutinized every image to scout for the most suitable representation of the subject and create an appropriate cover for the book.

The publishing team has been involved in this book since its early stages. They were actively engaged in every process, be it collecting the data, connecting with the contributors or procuring relevant information. The team has been an ardent support to the editorial, designing and production team. Their endless efforts to recruit the best for this project, has resulted in the accomplishment of this book. They are a veteran in the field of academics and their pool of knowledge is as vast as their experience in printing. Their expertise and guidance has proved useful at every step. Their uncompromising quality standards have made this book an exceptional effort. Their encouragement from time to time has been an inspiration for everyone.

The publisher and the editorial board hope that this book will prove to be a valuable piece of knowledge for researchers, students, practitioners and scholars across the globe.

List of Contributors

Atsushi Noda
Geological Survey of Japan, National Institute of Advanced Industrial Science and Technology,
Tsukuba, Ibaraki, Japan

Osamu Takano
JAPEX Research Center, Japan Petroleum Exploration, Japan

Yasuto Itoh
Graduate School of Science, Osaka Prefecture University, Japan

Shigekazu Kusumoto
Graduate School of Science and Technology for Research, University of Toyama, Japan

Tetsuya Sakai
Department of Geoscience, Shimane University, Shimane, Japan

Mototaka Saneyoshi
Hayashibara Museum of Natural Sciences, Okayama, Japan

Yoshihiro Sawada
Department of Geoscience, Shimane University, Shimane, Japan

Yutaka Kunimtatsu and Masato Nakatsukasa
Department of Zoology, Graduate School of Science, Kyoto University, Kyoto, Japan

Emma Mbua
Department of Earth Sciences, National Museums of Kenya, Nairobi,

Kosuke Egawa
School of Earth and Environmental Sciences, Seoul National University, Seoul, Republic of Korea
Institute for Geo-Resources and Environment, National Institute of Advanced Industrial Science and Technology, Tsukuba, Japan
Methane Hydrate Research Center, National Institute of Advanced Industrial Science and Technology, Sapporo, Japan

Akira Takeuchi
Graduate school of Science and Engineering for Research, University of Toyama, Japan

Takeshi Nakajima
Institute for Geo-resources and Environment, Geological Survey of Japan, AIST, Japan

Gentaro Kawakami
Geological Survey of Hokkaido, Hokkaido Research Organization, Japan

Keiji Takemura
Beppu Geothermal Research Laboratory, Institute for Geothermal Sciences, Graduate School of Science, Kyoto University, Noguchibaru, Beppu, Japan

Tsuyoshi Haraguchi
Graduate School of Science, Osaka City University, Osaka, Japan

Shigekazu Kusumoto
Graduate School of Science and Engineering for Research, University of Toyama, Toyama, Japan

Yasuto Itoh
Graduate School of Science, Osaka Prefecture University, Osaka, Japan

Yasuto Itoh and Keiji Takemura
Graduate School of Science, Osaka Prefecture University, Osaka, Japan

Shigekazu Kusumoto
Graduate School of Science and Engineering for Research, University of Toyama, Toyama, Japan

Yasuto Itoh
Graduate School of Science, Osaka Prefecture University, Osaka, Japan
Department of Physical Science, Graduate School of Science, Osaka Prefecture University, Japan

Machiko Tamaki
Japan Oil Engineering Co. Ltd., Tokyo, Japan

Osamu Takano
JAPEX Research Center, Japan Petroleum Exploration Co. Ltd., Chiba, Japan

Shigekazu Kusumoto
Graduate School of Science and Technology for Research, University of Toyama, Japan

Osamu Takano
JAPEX Research Center, Japan Petroleum Exploration Co. Ltd., Japan

Machiko Tamaki
Japan Oil Engineering Co. Ltd., Japan

Yasuto Itoh
Graduate School of Science, Osaka Prefecture University, Osaka, Japan

Shigekazu Kusumoto
Graduate School of Science and Engineering for Research, University of Toyama, Toyama, Japan

Keiji Takemura
Graduate School of Science, Kyoto University, Kyoto, Japan